T0257645

Lipid Peroxidation: Current Topics

Lipid Peroxidation: Current Topics

Edited by **Donna Thompson**

New York

Published by Callisto Reference,
106 Park Avenue, Suite 200,
New York, NY 10016, USA
www.callistoreference.com

Lipid Peroxidation: Current Topics
Edited by Donna Thompson

International Standard Book Number: 978-1-63239-450-7 (Hardback)

Contents

Preface

This book was inspired by the evolution of our times; to answer the curiosity of inquisitive minds. Many developments have occurred across the globe in the recent past which has transformed the progress in the field.

Lipid peroxidation is essentially the oxidative degradation of lipids. This book focuses on current advances made in lipid peroxidation. The data compiled in this book has been contributed by researchers with extensive experience in various fields of study. We hope that the matter provided here is comprehensible to a wide audience, not only experts but also students interested in the above stated topics. This book presents modern professional views on the subject of lipid peroxidation. It discusses various topics related to lipid peroxidation in health and diseases.

This book was developed from a mere concept to drafts to chapters and finally compiled together as a complete text to benefit the readers across all nations. To ensure the quality of the content we instilled two significant steps in our procedure. The first was to appoint an editorial team that would verify the data and statistics provided in the book and also select the most appropriate and valuable contributions from the plentiful contributions we received from authors worldwide. The next step was to appoint an expert of the topic as the Editor-in-Chief, who would head the project and finally make the necessary amendments and modifications to make the text reader-friendly. I was then commissioned to examine all the material to present the topics in the most comprehensible and productive format.

I would like to take this opportunity to thank all the contributing authors who were supportive enough to contribute their time and knowledge to this project. I also wish to convey my regards to my family who have been extremely supportive during the entire project.

Editor

Lipid Peroxidation in Health and Disease

Lipid Peroxidation After Ionizing Irradiation Leads to Apoptosis and Autophagy

Juliann G. Kiang, Risaku Fukumoto and Nikolai V. Gorbunov

Additional information is available at the end of the chapter

1. Introduction

A living cell is a dynamic biological system composed primarily of nucleic acids, carbohydrates, lipids, and proteins that structurally and functionally interact with many other molecules--organic and inorganic--to carry out normal cell metabolism. Exposure of a cell to radiation can both directly and indirectly alter molecules within the cell to affect cell viability. Radiation energy absorbed by tissues and fluids is dissipated by the radiolysis of water molecules and biomolecules [1-3]. These reactions result in redox-reactive products such as hydroxyl radical (HO*), hydrogen peroxide (H_2O_2), hydrated electron (e-aq), and an array of biomolecule-derived carbon-, oxygen-, sulfur-, and nitrogen-centered radicals (i.e., RC*, RO*, RS*, and RN*) that can in turn lead to the formation of organic peroxides and superoxide anion radicals (O_2*-) in the presence of molecular oxygen [3, 4].

While the strongly electrophilic HO* has the capacity to damage molecules like polypeptides, amino acids, and polyunsaturated fatty acids (PUFAs) directly, the alterations caused by peroxide and superoxide radicals are usually produced indirectly via Fenton-type reactions [1-3, 5]. It is the interaction of these radiation-induced free radicals with important biomolecules within the cell that is the basis of the cellular sensitivity to radiation.

Free radical reactions generated after short-term radiation exposure are often quickly terminated by antiradical/antioxidant redox cycles. However, in certain cases and certain cellular environments, free radicals can initiate self-propagating chain reactions that can magnify the effects of the initial oxidations induced by radiation, leading to major disruptions that affect basic cell function [4, 6-9]. This especially true in cellular structures rich in poly-unsaturated fatty acids (PUFAs), such as cellular membranes, where radiation exposure can induce lipid peroxidation chain reactions that trigger reactions both within and beyond of the membrane [see 10-13]. Although the chemical mechanisms by which radiation induces the formation of redox-reactive products in cells are fairly well-understood, many of the mechanisms by which they impact specific cellular processes to produce radiation injury are only beginning to be elucidated.

Lipid peroxidation is a process in which free radicals remove electrons from lipids, producing reactive intermediates that can undergo further reaction. Cellular membranes, because of their high lipid content, are especially susceptible to damage. Because lipid peroxidation reactions can alter the structure and function of critical membrane lipids, they can lead to cell injury and cell death.

Lipid peroxidation reactions take place in three steps. The first step is initiation, which produces a fatty acid radical. In polyunsaturated fatty acids, methylene groups next to carbon-carbon double bonds possess especially reactive hydrogen atoms. Lipid peroxidation is most commonly initiated when reactive oxygen species (ROS) such as OH· and HO_2 interact with a reactive methylene hydrogen atom to produce water and a fatty acid radical:

$$PUFA + OH· \rightarrow PUFA^* + H_2O \tag{1}$$

The second step is propagation. Molecular oxygen reacts with the unstable lipid radical to produce a lipid peroxyl radical. This radical is also unstable; it reacts readily with an unsaturated fatty acid to regenerate a new fatty acid radical as well as a lipid peroxide. The generation of the new fatty acid radical in this step reinitiates the cycle. For this reason, this series of reactions is referred to as a lipid peroxidation chain reaction.

$$PUFA^* + O_2 \rightarrow PUFAOO^* \tag{2}$$

$$PUFAOO^* \rightarrow Fenton \rightarrow HO\text{-}PUFA^* + OH^- \tag{3}$$

The third step is termination. As the chain reaction continues, an increasing concentration of lipid radicals are produced, thereby increasing the probability two lipid radicals will react with each other, which can produce a non-radical species. This constitutes a chain-breaking step, which in combination with the activity of natural radical scavenging molecules in cellular systems ultimately quells the chain reaction.

There are four types of radiation capable of ionizing, and thus damaging, target molecules via mechanisms such as lipid peroxidation: alpha particles, beta particles, gamma rays, and neutrons. Gamma radiation has been shown to increase lipid peroxidation in a variety of biological systems. After gamma radiation exposure, levels of the lipid peroxidation indicator MDA have been shown to increase in brain [14], liver [15-19], lens [20], serum [21], and skeletal muscle [22] of rats as well as in bacteria [23].

The possibility of exposure to radiation doses significant enough to cause lipid peroxidation leading to tissue injury is more than a hypothetical hazard. It is estimated that more than 50% of cancer patients receive radiotherapy at some point during the course of their disease, and these exposures can injure normal tissues [24, 25]. Potentially harmful radiation exposures after a nuclear power plant accident are also possible, either as a plant worker or a citizen who lives in or moves through fallout areas. Such exposures are unlikely; however, as the Fukushima, Japan, reactor incident showed, the threat is still very real. The threat of general exposure to radiation via a nuclear or radionuclide-based terrorist device is unfortunately also a real-world scenario. In order to provide public health protection in such cases, it is important to understand more about how radiation affects cells and tissues and learn how to ameliorate radiation injury.

Living organisms have evolved a variety of free radical-scavenging molecules to help protect the cell membrane from damage. Endogenous antioxidants include the enzymes superoxide dismutase (SOD), catalase, and peroxidase. Other antioxidants such as exogenously derived vitamin E can also play a role.

Free radical-mediated lipid peroxidation is harmful not only because damaged lipids disrupt membrane structure and function, but also because the process produces potentially mutagenic and carcinogenic byproducts [26]. One such product is the highly reactive carbonyl compound, malondialdehyde (MDA), which can react with deoxyadenosine and deoxyguanosine in DNA to form DNA adducts, primarily pyrimido[1,2-a]purin-10(3H)-one (M_1G) [23, 26]. M_1G toxicity has been demonstrated in experiments with glutathione peroxidase 4 knockout mice, which have a diminished capacity to protect themselves from lipid peroxidation and thus M_1G toxicity; mice with this lethal phenotype do not survive past embryonic day 8 [27].

Lipid peroxidation reactions can occur at the both the cell membrane and mitochondria membranes, and either can subsequently trigger cell death through apoptosis and/or autophagy [28]. Apoptosis is typically executed by caspases, which are cysteine aspartic acid-specific proteases that cleave an amino acid sequence-motif located N-terminal to a specific aspartic acid residue. Caspases can be broadly divided into two functional subgroups: (1) those activated during apoptosis (caspases -2, -3, -6, -7, -8, -9, and -10) ; and (2) those implicated in the processing of proinflammatory cytokines during responses (caspases -1, -4, and -5) [29-31].

The apoptosis process can follow two pathways, extrinsic and intrinsic. The extrinsic pathway involves the activation of pro-caspase-8 by external, typically molecular signals such as FAS ligand binding to FAS or TNF binding to TNF receptors on the cell membrane. Through a series of steps pro-caspase-8 becomes activated caspase-8, which then acts on the mitochondrial membrane either directly or via the activation of caspase-3. Subsequent steps then follow those described for the intrinsic pathway. The intrinsic pathway involves mitochondria directly. When mitochondria are under stress, cytochrome c is released from the mitochondria. The released cytochrome c conjugates with Apf-1and caspase-9 tin the cytoplasm to form apoptosomes, which in turn activate caspase-3 and -7. Activated caspase-3 then activates caspase-2, -6, -8, and -10. It should be noted that caspase-independent apoptosis pathways also exist, such as the intrinsic, apoptosis inducible factor (AIF) pathway and the intrinsic, mitochondria-derived endonuclease G-related pathway [29-31].

Autophagy is a catabolic process involving the bulk degradation of cellular constituents in lysosomes [32]. Autophagy under normal conditions is a cytoprotective process involved in tissue remodeling, recovery, and rejuvenation. Autophagy dynamics in mammalian cells are well-described in several recent reviews [33-37]. The autophagic pathway is complex. To date there are over thirty genes identified in mammalian cells as regulators of various steps of autophagy. The misregulation of the autolysosomal pathway during autophagy can eventually cause cell death either by triggering apoptosis in apoptosis-sensitive cells or as a result of destructive self-digestion [38]. Light chain 3 (LC3) is a protein involved in the formation of autophagosomes in mammalian cells that serves as a biomarker for occurrence of autophagy [13].

Given the widespread potential for cellular damage from lipid peroxidation after exposure to ionizing radiation, we hypothesize that ionizing radiation-induced lipid peroxidation leads to caspase-mediated apoptotic cell death and LC-3-mediated autophagic cell death. The objective of this current chapter is to provide evidence of this hypothesis. We used human Jurkat T cultured cells and mouse ileum to investigate the relationship between lipid peroxidation and cell death both *in vitro* and *in vivo*.

2. Experimental procedures and technical approach

2.1. Cell culture

Human Jurkat T cells (American Type Cell Collection, Rockville, MD, USA) were grown in 75 cm^2 tissue culture flasks (Costar, Cambridge, MA, USA) containing RPMI 1640 medium supplemented with 0.03% glutamine, 4.5 g/L glucose, 25 mM HEPES, 10% fetal bovine serum, penicillin (50 µg/mL), and streptomycin (50 U/mL) (Gibco/BRL, Gaithersburg, MD, USA). Cells were incubated in a 5% CO2 atmosphere at 37 °C and fed every 3-4 d.

2.2. Animal

CD2F1 male mice (25-30 g) were purchased from Harlan Laboratories (Indianapolis, IN). All mice were randomly assigned to experimental groups. Eight mice were housed per filter-topped polycarbonate cage (MicroIsolator) in conventional holding rooms. Rooms were provided 20 changes per hour of 100% fresh air, conditioned to 72 ± 2 °F and a relative humidity of 50 ± 20%. Mice were maintained on a 12-h light/dark, full-spectrum light cycle with no twilight. Research was conducted in a facility accredited by the Association for Assessment and Accreditation of Laboratory Animal Care-International (AAALAC-I). All procedures involving animals were reviewed and approved by the Armed Forces Radiobiology Research Institute (AFRRI) Institutional Animal Care and Use Committee.

2.3. Radiation exposure

2.3.1. Cell culture

Radiation exposures were conducted using AFRRI's ^{60}Co source. Cells suspended in growth medium were placed in 6-well plates (5 x 10^6 cells/ml; 2 ml per well) and exposed to ^{60}Co gamma-radiation at various total doses using a dose rate of 0.6 Gy/min. Cells were then returned to the incubator in a 5% CO2 atmosphere at 37 °C for the specified time.

2.3.2. *Animals*

For survival experiments, mice (n=16 per group) received 9.25 Gy (equivalent to $LD_{90/30}$) of total-body ^{60}Co gamma-photon radiation administered at a dose rate of 0.6 Gy/min. Sham-treated mice were handled identically but received no radiation. After treatment, mice were returned to their original cages and survival was monitored for 30 d. Body weight and facial dropsy were assessed. Mean survival times (ST_{50}) were observed. The moribund mice found during the observation period were euthanized in accordance with recommendations [39, 40] and guidelines [41]. For mechanistic experiments, mice also received 9.25 Gy (n=6 per group). At specified time points after irradiation, mice were euthanized in accordance with recommendations [39, 40] and guidelines [41]. Interested tissues were harvested 1 and 7 d after irradiation and stored at -80°C until use for biochemical assays and western blots. Ileum was also prepared for immunofluorescence assessment.

2.4. Western blots

To investigate amounts of caspase-3 and LC3 proteins, ileum was minced, mixed in 100 μL Na^+ Hanks' solution containing protease inhibitors, sonicated, and centrifuged at 8000 x g for 10 min. The supernatant was collected and total protein was determined with Bio-Rad reagent (Bio-Rad, Richmond, CA, USA). Aliquots containing 20 μg of protein in tris buffer (pH=6.8) containing 1% sodium dodecyl sulfate (SDS) and 1% 2-mercaptoethanol were resolved on SDS-polyacrylamide slab gels (Novex precast 4-20 % gel; Invitrogen, Grand Island, NY, USA). After electrophoresis, proteins were blotted onto a PDVF nitrocellulose membrane (type NC, 0.45 μm; Invitrogen), using a Novex blotting apparatus and the manufacturer's protocol. After blocking the nitrocellulose membrane by incubation in tris-buffered saline-0.5% tween20 (TBST) containing 3% nonfat dried milk for 90 min at room temperature, the blot was incubated for 60 min at room temperature with monoclonal antibodies directed against caspase-3 and LC3 at a concentration of 1 μg/ml in TBST - 3% dry milk. The blot was then washed 3 times (10 min each) with TBST before incubating the blot for 60 min at room temperature with a 1000X dilution of species-specific IgG peroxidase conjugate (Santa Cruz Biotechnology) in TBST. The blot was washed 6 times (5 min each) in TBST before detection of peroxidase activity using the Enhanced Chemiluminescence Plus kit (Amersham Life Science Inc., Arlington Heights, IL, USA). IgG levels were not altered by radiation; we therefore used IgG as a control for protein loading. Protein bands of interest were quantitated densitometrically and normalized to IgG.

2.5. Nitric oxide measurements

Nitric oxide (NO) production was measured under acidic conditions as nitrite, using a commercial kit (Biomedical Research Service, School of Medicine and Biomedical Sciences, State University of New York at Buffalo, NY, USA; www.bmrservice.com).

2.6. Lipid peroxidation measurements

Malondialdehyde (MDA), a lipid peroxidation end product, was measured colorimetrically using a commercial lipid peroxidation assay kit (CalBiochem, San Diego, CA).

2.7. Detection and analysis of caspase-3/7 activity by confocal microscopy

The Magic Red® Caspase Detection kit (MP Biomedicals; Solon, OH, USA) was used for the detection of caspase-3/7 activity, following the manufacturer's protocol. Briefly, about 2×10^5 cells were stained in the presence of up to 300 µl of OPTI-MEM I medium (Invitrogen). Cells were seeded onto #1 borosilicate glass slides with 4-well chambers (Fisher Science Education, Hanover Park, IL). An LSM 5 PASCAL Zeiss laser scanning confocal microscope (Carl Zeiss MicroImaging; Thornwood, NY, USA) with a 100×/1.3 NA Plan Apochromat oil objective was used to scan the signals. Each resulting image was provided with a simultaneous scan of differential interference contrast (DIC).

2.8. Immunofluorescence staining and image analysis

Small intestine specimens (5 per each of animal groups) collected at necropsy were processed for the immunofluorescence analysis and analyzed using fluorescence confocal microscopy [42]. Donkey normal serum and antibody were diluted in phosphate buffered saline (PBS) containing 0.5% BSA and 0.15% glycine. Any nonspecific binding was blocked by incubating the samples with purified donkey normal serum (Santa Cruz Biotechnology, Inc., Santa Cruz, CA, USA) diluted 1:20. The primary antibodies were raised against CD15 (mouse monoclonal biotin conjugated IgM from eBioscience), MAP LC3and AD4 (vendors indicated above). This was followed by incubation with secondary fluorochrome-conjugated antibody and/or streptavidin-AlexaFluor 610 conjugate (Molecular Probes, Inc., Eugene, OR, USA), and with Heochst 33342 (Molecular Probes) diluted 1:3000. The secondary antibodies used were AlexaFluor 488 and AlexaFluor 594 conjugated donkey IgG (Molecular Probes Inc.) Negative controls for nonspecific binding included normal goat serum without primary antibody or with secondary antibody alone. Five confocal fluorescence and DIC images of crypts (per specimen) were captured with Zeiss LSM 7100 microscope. Immunofluorescence image analysis was conducted as described previously [43]. The index of spatial correlation (r) of proteins was determined by multiple pixel analysis for pairwise signal interaction of green and red channels. Paneth cell identification was conducted by i) their spatial localization in crypts; ii) presence of immunoreactivity to CD15 and AD4, which are specific to this epithelial phenotype; iii) spatial appearance of the immunoreactivity to CD15, AD4, and the FISH reactivity to AD4 mRNA (see above) in morphologically identified Paneth cells.

2.9. Solutions

Na+ Hanks' solution contained in mM: 145 NaCl, 4.5 KCl, 1 .3 MgCl₂, 1.6 CaCl₂, and 10 HEPES (pH 7.40 at 24 °C). Na+ Hanks' stop buffer contained in mM: 50 tris-HCl, 1% NP-40, 0 .25% Na+-deoxycholate, 150 NaCl, 1 EDTA, 1 phenylmethanesulfonyl fluoride, 1 Na₃VO₄, 1 NaF, along with aprotinin, leupeptin, and pepstatin (10 µg/mL each). Na+ Hanks' wash buffer contained in mM: 1 EDTA, 1 phenylmethanesulfonyl fluoride, 1 DTT, 1 Na₃VO₄, 1 NaF, along with aprotinin, leupeptin, and pepstatin (10 µg/mL each).

2.10. Statistical analysis

Results represent the mean ± s.e.m. One-way ANOVA, Studentized-range test, Bonferroni's inequality, and Student's t-test were used for comparison of groups with 5% as a significant level.

3. Response of human T cells to irradiation

3.1. Gamma radiation increased lipid peroxidation production

Cells were irradiated with 2, 4, 6, or 8 Gy; the lipid peroxidation marker MDA was measured in these cells at 4, 24, 48, and 72 h postirradiation. Figure 1 shows MDA levels increased in a radiation dose- and postirradiation time-dependent manner in cells receiving 2, 4, and 6 Gy (Fig. 1A-C). Cells receiving 8 Gy (Fig. 1D) had lipid peroxidation levels above the baseline at all times tested. But in cells receiving 6 Gy and 8 Gy, by 24 h postirradiation the levels were significantly lower than those observed in cells receiving 4 Gy (Fig. 1C vs. 1B and Fig. 1D vs. 1B). This observation may be a reflection of the drop in cell viability previously observed in cells after doses greater than 4 Gy.

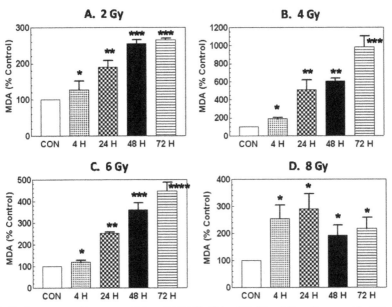

Jurkat T cells were exposed to gamma radiation at 2 (A), 4 (B), 6 (C), or 8 Gy (D) and allowed to respond for 4, 24, 48, or 72 h (n=3) before preparation of cell lysates. Lipid peroxidation as indicated by MDA was measured. Data are expressed relative to that of unirradiated controls. For panels A: *P<0.05 vs. control, 24 H, 48 H, and 72 H; **P<0.05 vs. control, 4 H, 48 H, and 72 H; ***P<0.05 vs. control, 4 H, and 24 H. For Panel B: *P<0.05 vs. control, 24 H, 48 H, and 72 H; **P<0.05 vs. control, 4 H, and 72 H; ***P<0.05 vs. control, 4 H, 24 H, and 48 H. For panels C: *P<0.05 vs. control, 24 H, 48 H, and 72 H; **P<0.05 vs. control, 4 H, 48 H, and 72 H; ***P<0.05 vs. control, 4 H, 24 H, and 72 H; ****P<0.05 vs. control, 4 H, 24 H, and 48 H. For panel D: *P<0.05 vs. control, determined by one way ANOVA and studentized-range test. CON: unirradiated controls

Figure 1. Gamma radiation increased lipid peroxidation in T cells.

3.2. Gamma radiation increased NO production

Because NO is known to react with O_2- to form peroxynitrite (ONOO-) that can oncrease lipid peroxidation [23], we measured NO production levels in irradiated Jurkat cells. Figure 2 shows NO production in irradiated cells. By 4 h postirradiation cells receiving 2, 4, or 6 Gy showed NO levels statistically lower than non-irradiated controls, but cells receiving 8 Gy showed an increase. By 24 and 48 h, all irradiated cells exhibited increased NO production, generally in a radiation dose-dependent manner. By 72 h postirradiation NO level returned to baseline in cells receiving 4, or 8 Gy, while NO levels in cells receiving 2 or 6 Gy remained above the baseline and had dropped below baseline, respectively.

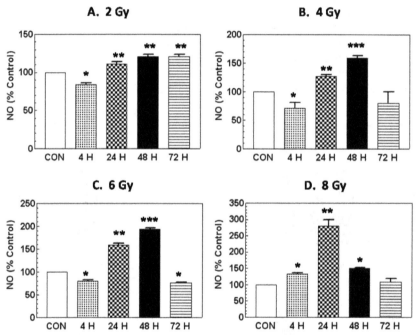

Jurkat T cells were exposed to gamma radiation at 2 (A), 4 (B), 6 (C), or 8 Gy (D) and allowed to respond for 4, 24, 48, or 72 h (n=3) before preparation of cell lysates. NO production was measured. Data are expressed relative to that of unirradiated controls. For panel A: *P<0.05 vs. control, 24 H, 48 H, and 72 H; **P<0.05 vs. control and 4 H. For Panel B: *P<0.05 vs. control, 24 H, and 48 H; **P<0.05 vs. control, 4 H, 48 H, and 72 H; ***P<0.05 vs. control, 4 H, 24 H, and 72 H. For panel C: *P<0.05 vs. control, 24 H, and 48 H; **P<0.05 vs. control, 4 H, 48 H, and 72 H. ***P<0.05 vs. control, 4 H, 24 H, and 72 H. For panel D: *P<0.05 vs. control, 24 H, and 72 H; **P<0.05 vs. control, 4 H, 48 H, and 72 H, determined by one way ANOVA and studentized-range test. CON: unirradiated controls

Figure 2. Gamma radiation increased NO production in T cells.

3.3. Gamma radiation increased apoptosis

Lipid peroxidation occurring within the cell and mitochondrial membranes can trigger apoptosis [28]. Because increased caspase-3 and -7 are indicators of cells undergoing apoptosis [16-18], we measured caspase-3 and -7 in irradiated Jurkat cells. As shown in Fig. 3, cells receiving 8 Gy displayed significantly increased immunofluorescence of caspase-3/-7, compared to non-irradiated cells. The percentages of non-irradiated cells and irradiated cells presenting the caspase-3/-7 immunofluorescence were 23 and 85 %, respectively, suggesting that gamma irradiation increases apoptosis.

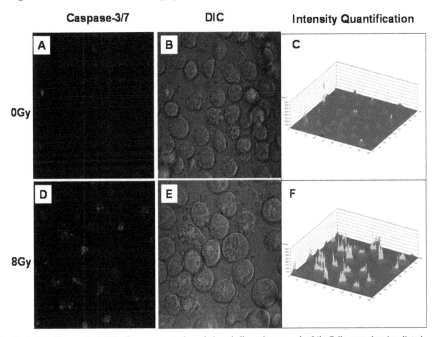

Freshly isolated human Jurkat T cells were counted, seeded, and allowed to grow for 24h. Cells were then irradiated at 8 Gy and returned to incubator. Live cells were studied 24 h postirradiation for internal caspase-3/7 activities by Magic Red® staining.

Figure 3. Gamma radiation increased apoptosis in T cells.

4. Response of mice to irradiation

The organs most sensitive to radiation are the hematopoietic, lymphoid, gastrointestinal, reproductive, vascular, and cutaneous systems [44]. We used mice irradiated with 9.25 Gy ^{60}Co-gamma photons to determine if events similar to those observed *in vitro* also occurred *in vivo*. Thirty-day survival, body weight, facial dropsy, and ileum were also assessed.

4.1. Gamma radiation decreased mouse survival and body weight

Irradiated mice first demonstrated mortality 9 d after irradiation and showed no survival 23 d after irradiation. The mean survival time (ST$_{50}$) was 12 d. All non-irradiated mice survived

(Fig. 4A). Irradiated mice demonstrated significant body weight loss 10 d after irradiation and lost over 20% by the end of the experiment; non-irradiated mice gained weight daily. The rate of weight change was -0.6 g/d for irradiated mice and 0.17 g/d for non-irradiated mice. The observations are in agreement with those obtained in previous experiments using irradiated B6D2F1/J mice [45]

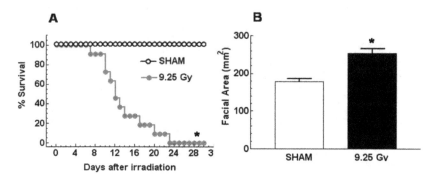

Mice (n=16 per group) received 9.25 Gy. (A) Radiation significantly reduced 30-d survival after irradiation. *P<0.01 vs. sham group, determined by one-way ANOVA and Student's t-test. (B) Radiation induced facial dropsy. Facial dropsy was assessed by measuring increase in facial area. Facial area was approximated by multiplying width between outer edges of ears and distance from ear-to-ear midpoint to tip of snout and then dividing by 2 (area of a triangle). Facial dropsy in each mouse was calculated approximately 10 d after 9.25 Gy (9-12 d, point of maximal swelling) and averaged. *P<0.05 vs. sham group, determined by Student's t-test.

Figure 4. Gamma radiation decreased survival and increased facial dropsy.

4.2. Gamma radiation induced facial dropsy

Irradiated mice began to show facial dropsy around 10 d postirradiation (Fig. 4B). The overall increase in dropsy was approximately 43%. Facial dropsy did not occur in previous experiments using irradiated B6D2F1/J mice [44] or in humans [46], suggesting the response is mouse strain- and species-specific.

4.3. Gamma radiation induced lipid peroxidation and NO production

Ileum lysate was obtained for MDA measurement from mice exposed to 9.25 Gy. As shown in Fig. 5, irradiation increased MDA levels (Fig. 5A) and NO production (Fig. 5B). The results were consistent with *in vitro* findings using human T cells.

4.4. Gamma radiation induced increases in caspase-3 and LC3

To measure apoptosis in ileum from irradiated mice, we performed an immunoblot analysis of ileum caspase-3, a biomarker for apoptosis [29-31]. Radiation induced an approximate 2.5-fold increase in caspase-3 (Fig. 6A-B). The result was consistent with the *in vitro* findings with Human T cells.

To measure autophagy in ileum from irradiated mice, we performed an immunoblot analysis of ileum LC3 protein, a biomarker for Autophagy [13]. Radiation induced an approximate 3.5-fold increase in LC3 (Fig. 6A and C).

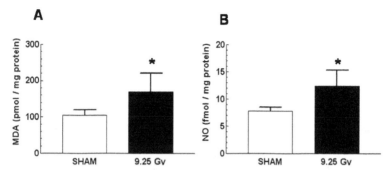

Mice (n=6 per group) received 9.25 Gy. Ileum tissues were collected 1 d postirradiation and cell lysates prepared. (A) Lipid peroxidation as indicated by MDA was measured. (B) NO production was measured. *P<0.05 vs. sham group, determined by Student's t-test.

Figure 5. Gamma increased lipid peroxidation and NO production in ileum.

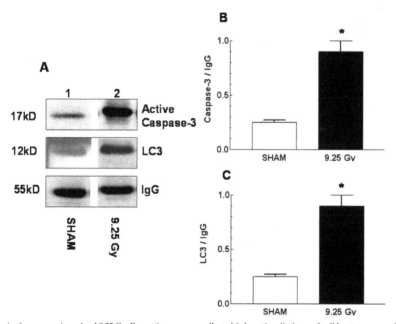

Mice (n=6 per group) received 9.25 Gy. Ileum tissues were collected 1 d postirradiation and cell lysates prepared. (A) Representative Western blots. (B) Caspase-3 and LC3 were quantitated densitometrically and normalized to IgG. *P<0.05 vs. sham, determined by Student's t-test.

Figure 6. Gamma radiation increased caspase-3 and LC3 in ileum.

4.5. Gamma radiation induced apoptosis in mouse ileal villi

The immunofluoresence images in Fig. 7 show that there was little immunofluorescence present in ileal villi of non-irradiated mice (Fig. A and C), whereas there was a significant increase in immunofluorescence in ileal villi of irradiated mice (Fig. B and D). It is known that ileal epithelial cells regularly slough off after undergoing apoptosis and are then replenished within 7 d [47]. It is not clear if irradiation accelerated the rate of apoptosis, but ionizing radiation-induced increases in apoptosis were observed in these studies (Fig. B vs. A). Similar results were observed in earlier studies using irradiated B6D2F1/J mice [48].

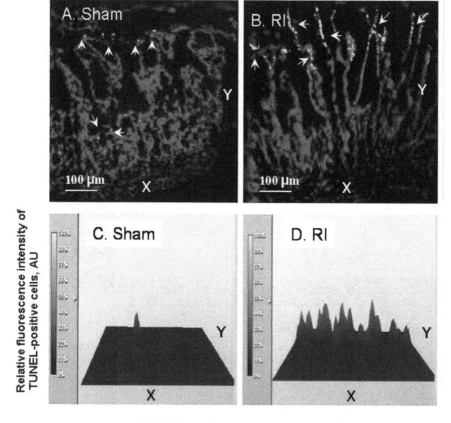

TUNEL - Green; Nuclei - Blue

Mice (n=6 per group) received 9.25 Gy. Ileum tissues were collected 7 d postirradiation and slide preparations were stained using TUNEL assay to detect apoptosis. (A-B) Confocal microscopy fluorescent images: green fluorescence indicates apoptotic cells; blue Hoechst 33342 fluorescence indicates all nuclei. (C-D) Quantitation of fluorescence intensities.

Figure 7. Gamma radiation induced apoptosis in ileum.

4.6. Gamma radiation induced autophagy in mouse ileal crypts

In ileum, stem cells anchored in the crypts gives rise to proliferating progenitor cells that exit the cell cycle as they migrate and differentiate into 4 different cell types, including enterocytes, goblet cells, enteroendocrine cells, and Paneth cells [see reviews 49, 50]. Paneth cells produce defensins to protect against bacterial entry via the gut barrier [49, 50]. However, stem cells have also been shown to inhibit bacterial entry [51]. The immunofluoresence images in Fig. 8 show that there was little immunofluorescence present in ileal crypts of non-irradiated mice (A and C), whereas there was significant increase in immunofluorescence in ileal crypts of irradiated mice (B and D). Autophagy occurred in both stem cells and Paneth cells (using CD-15 as a biomarker). Ionizing radiation-induced increases in autophagy (B vs. A) have also been observed in B6D2F1/J mice [44].

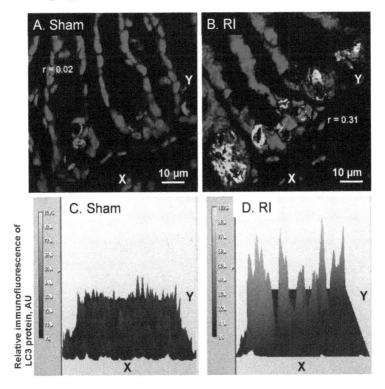

Paneth cells – Red; LC3 – Green; Nuclei - Blue

Mice (n=6 per group) received 9.25 Gy irradiation. Ileum tissues were collected 7 d postirradiation and slide preparations were stained with anti-LC3 antibody to detect autophagy indicator LC-3. (A-B) Confocal microscopy fluorescent images: green fluorescence indicates anti-LC3 antibody; blue Hoechst 33342 fluorescence indicates all nuclei. (C-D) Quantitation of fluorescence intensities.

Figure 8. Gamma radiation induced autophagy in ileum.

5. Conclusion

The complexity of the cellular response to ionizing radiation complicates efforts to design approaches to treat or prevent injury resulting from ionizing radiation. Ionizing radiation activates many signal transduction pathways [44, 52], including the one involving lipid peroxidation. It is known that Bcl-2 (an anti-apoptotic protein) decreases lipid peroxidation [53]. Although in cultured cells, silencing of the iNOS gene by iNOS siRNA inhibits lipid peroxidation and the subsequent production of apoptosis-related proteins [12], ionizing radiation increases lipid peroxidation even in iNOS knockout mice (Lu and Kiang, unpublished data). This suggests that radiation-induced activation of iNOS pathway is not all for its occurrence. Nevertheless, total body ionizing irradiation causes lipid peroxidation, caspase-3 activation, LC3 increases, and apoptosis and autophagy in ileum, suggesting the ileal cell death may contribute to ionizing radiation-induced mortality and body weight loss (Fig. 4). The mechanism underlying radiation-induced facial dropsy is unclear and warrants further investigation.

Figure 9 shows a schematic representation of lipid peroxidation-mediated induction of apoptosis and autophagy after exposure to gamma-radiation. Mitochondrial membrane lipid peroxidation is induced either directly by radiation or indirectly by increased NO production as a result of radiation-induced iNOS upregulation. Lipid peroxidation in the mitochondrial membrane leads to release of cytochrome c into the cytoplasm. Cytochrome c complexes with Apf-1 and seven molecules of caspase-9 to form an apoptosome. Apoptosomes then activate caspase-3 and -7. Active caspase-3 subsequently activates other caspases in the cytoplasm, which leads to apoptosis. It is not known what role plasma membrane lipid peroxidation plays in the apoptosis and autophagy processes. Lipid peroxidation also increases LC3 to cause autophagy, a process whose poorly understood molecular mechanisms warrant further investigation.

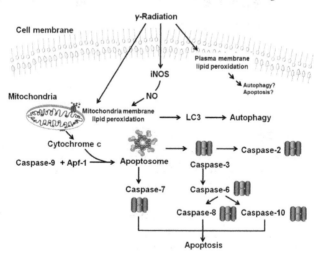

Figure 9. Schematic representation of lipid peroxidation-mediated induction of apoptosis and autophagy after exposure to gamma-radiation. iNOS: inducible nitric oxide synthase; NO: nitric oxide; LC3: light chain 3; Apf-1: ATP-dependent proteolysis factor 1

6. Perspective

Cell death resulting from ionizing radiation is a common scenario in the clinical practice of medicine and probably occurs in virtually all organ systems. Its hallmarks are relatively consistent across species and probably organ systems. Blockade of lipid peroxidation could be a useful approach to prevent radiation injury. In the murine model [14, 17-19, 21], radiation-induced lipid peroxidation occurs in fetal brain cultures [14], liver [17-19], and serum [21] as well as in the ileum, as reported in this chapter. A wide range of agents has been shown to effectively inhibit radiation-induced lipid peroxidation, including N-acetylcysteine [15], melatonin [16], vitamin E [20, 22], MnSOD-Plasmid Liposomes [14], leaf extract of Moringa oleifera [17], cinnamon extract [18], green tea polyphenol [21], hesperidin [19], and 17-DMAG [Lu and Kiang, unpublished data]. Since the response to radiation involves many signal transduction pathways, a combination of drugs targeting different signaling pathways may be a useful approach to address the difficult problem of protecting from ionizing radiation injury.

Author details

Juliann G. Kiang*, Risaku Fukumoto and Nikolai V. Gorbunov
*Radiation Combined Injury Program, Armed Forces Radiobiology Research Institute,
Uniformed Services University of the Health Sciences, Bethesda, Maryland,
USA*

Acknowledgement

The authors thank HM1 Neil Agravante and Ms. Joan T. Smith for their technical support and Dr. David E. McClain for his scientific discussion and editorial assistance.

Grants

This work was supported by AFRRI Intramural RAB2CF (to JGK) and DTRA CBM.RAD.01.10.AR.010 (to JGK). There are no ethical and financial conflicts in the presented work.

Disclaimer

The opinions or assertions contained herein are the authors' private views and are not to be construed as official or reflecting the views of the United States Department of Defense.

7. References

[1] Jay A, LaVerne JA. OH Radicals and Oxidizing Products in the Gamma Radiolysis of Water. Radiation Research. 2000; 153:196-200.

---

* Corresponding Author

[2] Nakagawa K, Nishio T. Electron paramagnetic resonance investigation of sucrose irradiated with heavy ions. Radiat Res. 2000; 153:835-9.

[3] Nauser T, Koppenol WH, Gebicki JM. The kinetics of oxidation of GSH by protein radicals. Biochem J. 2005; 392(Pt 3):693-701.

[4] Gebicki JM, Nauser T, Domazou A, Steinmann D, Bounds PL, Koppenol WH. Reduction of protein radicals by GSH and ascorbate: potential biological significance. Amino Acids. 2010; 39:1131-7.

[5] Østdal H, Davies MJ, Andersen HJ. Reaction between protein radicals and other biomolecules. Free Radic Biol Med. 2002; 33:201-9.

[6] Tappel AL. Vitamin E and selenium protection from in vivo lipid peroxidation. Ann N Y Acad Sci. 1980; 355:18-31.

[7] Davies MJ, Forni LG, Willson RL. Vitamin E analogue Trolox C. E.s.r. and pulse-radiolysis studies of free-radical reactions. Biochem J. 1988; 255:513-22.

[8] Bild W, Ciobica A, Padurariu M, Bild V. The interdependence of the reactive species of oxygen, nitrogen, and carbon. J Physiol Biochem. 2012 Mar 29. [Epub ahead of print]

[9] Azzam EI, Jay-Gerin JP, Pain D. Ionizing radiation-induced metabolic oxidative stress and prolonged cell injury. Cancer Lett. 2011 Dec 17. [Epub ahead of print]

[10]. Leach JK, Van Tuyle G, Lin PS, Schmidt-Ullrich R, Mikkelsen RB. Ionizing radiation-induced, mitochondria-dependent generation of reactive oxygen/nitrogen. Cancer Res. 2001; 61:3894-901.

[11] Tyurina YY, Tyurin VA, Epperly MW, Greenberger JS, Kagan VE. Oxidative lipidomics of gamma-irradiation-induced intestinal injury. Free Radic Biol Med. 2008; 44:299-314.

[12] Kiang JG, Smith JA, and Agravante NG. Geldanamycin analog 17-DMAG inhibits iNOS and caspases in gamma irradiated human T cells. Radiat Res. 2009; 172: 321-30.

[13] Gorbunov NV, Kiang JG. Up-regulation of Autophagy in the Small Intestine Paneth Cell in Response to Total-Body γ-Irradiation. J Pathol. 2009; 219: 242-52.

[14] Bernard ME, Kim H, Rwigema JC, Epperly MW, Kelley EE, Murdoch GH, Dixon T, Wang H, Greenberger JS. Role of the esophageal vagus neural pathway in ionizing irradiation-induced seizures in nitric oxide synthase-1 homologous recombinant negative NOS1-/- mice. In Vivo. 2011; 25:861-9.

[15] Mansour HH, Hafez HF, Fahmy NM, Hanafi N. Protective effect of N-acetylcysteine against radiation induced DNA damage and hepatic toxicity in rats. Biochem Pharmacol. 2008; 75:773-89.

[16] El-Missiry MA, Fayed TA, El-Sawy MR, El-Sayed AA. Ameliorative effect of melatonin against gamma-irradiation-induced oxidative stress and tissue injury. Ecotoxicol Environ Saf. 2007; 66:278-86.

[17] Sinha M, Das DK, Bhattacharjee S, Majumdar S, Dey S. Leaf extract of Moringa oleifera prevents ionizing radition-induced oxidative stress in mice. J Med Food. 2011; 14:1167-72.

[18] Azab KSh, Mostafa AH, Ali EM, Abdel-Aziz MA. Cinnamon extract ameliorates ionizing radiation-induced cellular injury in rats. Ecotoxicol Environ Saf. 2011; 74:2324-9.

[19] Kalpana KB, Devipriya N, Srinivasan M, Vishwanathan P, Thayalan K, Menon VP. Evaluating the radioprotective effect of hesperidin in the liver of Swiss albino mice. Eur J Pharmacol. 2011; 658:206-12.

[20] Karslioglu I, Ertekin MV, Kocer I. Protective role of intramuscularly administered vitamin E on the levels of lipid peroxidation and the activities of antioxidant enzymes in the lens of rats made cataractous with gamma radiation. Eur J Ophthalmol. 2004; 14:478-95.

[21] Hu Y, Guo DH, Liu P, Cao JJ, Wang YP, Yin J, Zhu Y, Rahman K. Bioactive components from the tea polyphenols influence on endogenous antioxidant defense system and modulate inflammatory cytokines after total-body irradiation in mice. Phytomedicine. 2011; 18:970-5.

[22] Yilmaz S, Yilmaz E. Effects of melatonin and vitamin E on oxidative-antioxidative status in rats exposed to irradiation. Toxicology 2006; 222:1-7.

[23] Zhou X, Taghizadeh K, Dedon PC. Chemical and biological evidence for base propenals as the major source of the endogenous M1dG adduct in cellular DNA. J Biol Chem 2005; 280:25377-82.

[24] Bentzen SM. Preventing or reducing late side effects of radiation therapy: radiobiology meets molecular pathology. Nat Rev Cancer. 2006; 6:702-13.

[25] Buonanno M, de Toledo SM, Pain D, Azzam EI. Long-term consequence of radiation-induced bystander effects depend on radiation quality and dose and correlate with oxidative stress. Radiat Res. 2011; 175:405-15.

[26] Marnett LJ. Lipid peroxidation-DNA damage by malondialdehyde. Mutation Research. 1999; 424:83-95.

[27] Muller FL, Lustgarten M S, Jang Y, Richardson A, Van Remmen H. Trends in oxidative aging theories. Free Radic Biol Med. 2007; 43:477-503.

[28] Choi AO, Cho SJ, Desbarats J, Lovrić J, Maysinger D.Quantum dot-induced cell death involves Fas upregulation and lipid peroxidation in human neuroblastoma cells. J Nanobiotechnology. 2007; 5:1.

[29] Hill MM, Adrain C, Martin SJ. Portrait of a killer: The mitochondrial apoptosome emerges from the shadow. Mol Interventions. 2003; 3:19-25.

[30] Jiang X, Wang X. Cytochrome-c-mediated apoptosis. Annu Rev biochem. 2004; 73:87-106.

[31] Kiang JG, Tsen KT. Biology of hypoxia. Chin J Physiol. 2006; 49:223-33.

[32] Mizushima N, Levine B, Cuervo AM, Klionsky DJ. Autophagy fights disease through cellular self-digestion. Nature. 2008; 451:1069–75.

[33] Yang Z, Klionsky DJ. Eaten alive: a history of macroautophagy. Nat Cell Biol. 2010; 12:814-22.

[34] Klionsky DJ. The Autophagy Connection. Dev Cell. Author manuscript; available in PMC 2011 July 20.

[35] Tooze SA, Yoshimori T. The origin of the autophagosomal membrane. Nat Cell Biol. 2010; 12:831-5.

[36] Weidberg H, Shvets E, Elazar Z. Biogenesis and cargo selectivity of autophagosomes. Annu Rev Biochem. 2011; 80:125-56.

[37] Eskelinen EL. New insights into the mechanisms of macroautophagy in mammalian cells. Int Rev Cell Mol Biol. 2008; 266:207-47.

[38] Sridhar S, Botbol Y, Macian F, Cuervo AM. Autophagy and Disease: always two sidesto a problem. J Pathol. 2011 Oct 12. doi: 10.1002/path.3025. [Epub ahead of print]

[39] Montgomery CA. Oncologic and toxicologic research: Alleviation and control of pain and distress in laboratory animals. Cancer Bulletin. 1990; 42:230-37.

[40] Tomasivic SP, Coghlan LG, Gray KN, Mastromarino AJ, Travis EL. IACUC evaluation of experiments requiring death as an end point: A cancer center's recommendations. Lab Animal January/February, 31-34 (1988).

[41] American Veterinary Medical Association, Report of the AVMA Panel on Euthanasia. J Am Veterinary Medical Association. 2001; 218:669-96.

[42] Müller CA, Autenrieth IB, Peschel A. Innate defenses of the intestinal epithelial barrier. Cell Mol Life Sci. 2005; 62:1297-307.

[43] Gorbunov NV, Garrison BR, Kiang JG. Response of crypt Paneth cells in the small intestine following total-body gamma-irradiation. Int J Immunopathol Pharmacol. 2010; 23:1111-23.

[44] Kiang JG, Garrison BR, Gorbunov NV. Radiation combined injury: DNA damage, apoptosis, and autophagy. Adapt Med. 2010; 2:1-10.

[45] Kiang JG, Garrison BR, Burns TM, Zhai M, Fukumoto R, Cary LH, Elliott TB, Ledney GD. Radiation combined injury complicates biodosimetric assessment. Radiat Res. 2012; In review.

[46] http://en.wikipedia.org/wiki/Poisoning_of_Alexander_Litvinenko.

[47] AFRRI Handbook. 2009 QuickSeries Publishing, Canada.

[48] Kiang JG, Jiao W, Cary L, Mog SR, Elliott TB, Pellmar TC, Ledney GD. Wound trauma increases radiation-induced mortality by increasing iNOS, cytokine concentrations, and bacterial infections. Radiate Res. 2010; 173: 319-332.

[49] Haegebarth A, Clevers H. Wnt signaling, lgr5, and stem cells in the intestine and skin. Am J Pathol. 2009; 174:715-21.

[50] Van der Flier LG, Clevers H. Stem cells, self-renewal, and differentiation in the intestinal epithelium. Annu Rev Physiol. 2009; 71:241-60.

[51] Gorbunov NV, Garrison BR, Zhai M, McDaniel DP, Ledney GD, Elliott TB, Kiang JG. Autophagy-mediated defense response of mouse mesenchymal stromal cells (MSCs) to challenge with Escherichia coli. In: Protein Interaction / Book 1; ISBN 979-953-307-577-7. Eds.: Cai J. InTech Open Access Publisher; 2011; www.intechweb.org. pp. 23-44.

[52] Haegebarth A, Perekatt AO, Bie W, Gierut JJ, Tyner AL. Induction of protein tyrosine kinase 6 in mouse intestinal crypt epithelial cells promotes DNA damage-induced apoptosis. Gastroenterology. 2009; 137:945-54.

[53] Wilkins HM, Marquardt K, Lash LH, Linseman DA. Bcl-2 is a novel interacting partner for the 2-oxoglutarate carrier and a key regulator of mitochondrial glutathione. Free Radical Biol & Med. 2012; 52:410-19.

Chapter 2

The Role of Physical Exercise on Lipid Peroxidation in Diabetic Complications

Yaşar Gül Özkaya

Additional information is available at the end of the chapter

1. Introduction

Diabetes mellitus is a group of metabolic disorder characterized by hyperglicemia and insufficiency of action or secretion of insulin. More than 346 million people worldwide have diabetes. 80 per cent of diabetes-induced deaths ocur in low- and middle-income countries. Most people with diabetes are above the age of retirement in developed countries, whereas in developing countries those most frequently affected are aged between 35 and 64 [1]. Although the etiology of this disease is not well defined, viral infections, autoimmunity, genetic and environmental factors have been implicated [2-5]. Four major types of diabetes have been defined by the American Diabetes Association (ADA): type 1 diabetes, type 2 diabetes, other spesific types of diabetes and gestational diabetes mellitus (GDM) [6].

Type 1 diabetes (T1D) usually develops in childhood and adolescence and the cause of the disease is an absolute deficiency of insulin secretion. Individuals at increased risk of developing this type of diabetes can often be identified by serological evidence of an autoimmune pathologic process occurring in the pancreatic islets and by genetic markers [6].

Type 2 diabetes (T2D) usually develops in adulthood and is related to obesity, lack of physical activity, and unhealthy diets. This is the more common type of diabetes (representing 90% of diabetic cases worldwide) and the cause is a combination of resistance to insulin action and an inadequate compensatory insulin secretory response. In Type 2 diabetes, a degree of hyperglycemia sufficient to cause pathologic and functional changes in various target tissues, but without clinical symptoms, may be present for a long period of time before diabetes is detected. During this asymptomatic period, it is possible to demonstrate an abnormality in carbohydrate metabolism by measurement of plasma glucose in the fasting state or after a challenge with an oral glucose load.

The third category "other spesific types of diabetes" includes diabetes caused by a spesific and identified underlying defect, such as genetic syndromes, acquired processes such as pancreatitis, diseases such as cystic fibrosis, exposure to certain drugs, viruses, and unknown causes. Gestational diabetes is a state of hyperglicemia which develops during pregnancy [6].

Currently, ADA recommends the use of any of the following four criteria for diagnosing diabetes: 1) glycated hemoglobin (A1c) value of 6.5% or higher, 2) fasting plasma glucose ≥ 126 mg.dL^{-1} (7.0 mmol.L^{-1}), 3) 2-h plasma glucose ≥ 200 mg.dL^{-1} (11.1 mmol.L^{-1}) during an oral glucose tolerance test using 75 g of glucose, and/or 4) classic symptoms of hyperglycemia (e.g., polyuria, polydipsia, and unexplained weight loss) or hyperglycemic crisis with a random plasma glucose of 200 mg.dL^{-1} (11.1 mmol.L^{-1}) or higher. In the absence of unequivocal hyperglycemia, the first three criteria should be confirmed by repeat testing [6].

Hyperglycaemia and hyperlipidaemia are key promoters of diabetes dysmetabolism, namely, through the formation of reactive oxygen species (ROS) and advanced glycation end products (AGEs), which causes cell damage and insulin resistance [7-9]. Moreover, both of them stimulate proinflammatory cytokines, thus contributing to β-cell degradation, particularly due to apoptosis pathways [10].

Increased oxidative stress is a widely accepted participant in the development and progression of diabetes and its complications [11-13]. Diabetes is usually accompanied by increased production of free radicals [14,15] or impaired antioxidant defenses [16]. Mechanisms by which increased oxidative stress is involved in the diabetic complications are partly known, including activation of transcription factors, advanced glycated end products (AGEs) [2], and protein kinase C [17].

Modern medical care uses a vast array of lifestyle and pharmaceutical interventions aimed at preventing and controlling hyperglycemia. In addition to ensuring the adequate delivery of glucose to the tissues of the body, treatment of diabetes attempts to decrease the likelihood that the tissues of the body are harmed by hyperglycemia. The importance of protecting the body from hyperglycemia cannot be overstated; the direct and indirect effects on the human vascular tree are the major source of morbidity and mortality in both type 1 and type 2 diabetes. Generally, the injurious effects of hyperglycemia are separated into macrovascular complications (coronary artery disease, peripheral arterial disease, and stroke) and microvascular complications (diabetic nephropathy, neuropathy, and retinopathy) [11].

Physical activity (PA) and diet are cornerstones of diabetes therapy [19]. Physical activity is a multifaceted behavior of which exercise is just one component. PA is defined as "bodily movement produced by the contraction of skeletal muscle that substantially increases energy expenditure" and exercise is defined as "a subset of PA done with the intention of developing physical fitness (i.e., cardiovascular, strength, and flexibility training)." [19]. In this chapter, PA and exercise is used interchangeably.

In last decades, an impressive body of research has accumulated that demonstrates the varied benefits of regular physical activity for people with type 1 or type 2 diabetes [20]. Notably, exercise has been shown to improve glycemic control, reduce the need for insulin and oral hypoglycemic agents, and improve body weight control. Exercise has been shown to promote beneficial effects on insulin resistance, both in humans and in rodent models of T2DM [21, 22]. Moreover, exercise has myriad benefits for all people beyond those relating to diabetes alone. It can work wonders for the heart, improving the lipid profile, reducing risk for heart disease, restoring function after a heart attack, and moderating blood pressure. It helps in maintaining bone health regardless of age, it can significantly relieve depression and anxiety, and it appears to help maintain cognitive function in old age [23, 24]. A correlation between the effects of acute and chronic aerobic exercise upon oxidative stress and inflammation and the diabetic dysmetabolism has been previously described [25-27].

This chapter focuses on recent clinical and experimental studies of diabetes and exercise interventions done within the context of lipid peroxidation.

2. Lipid peroxidation and diabetic complications

2.1. Overview of lipid peroxidation and diabetic complications

Excessively high levels of free radicals cause damage to cellular proteins, membrane lipids and nucleic acids, and eventually cell death [2]. Various mechanisms have been suggested to contribute to the formation of these reactive oxygen-free radicals in diabetic state. Glucose oxidation is believed to be the main source of free radicals. In its enediol form, glucose is oxidized in a transition-metal dependent reaction to an enediol radical anion that is converted into reactive ketoaldehydes and to superoxide anion radicals. The superoxide anion radicals undergo dismutation to hydrogen peroxide, which if not degraded by catalase or glutathione peroxidase, and in the presence of transition metals, can lead to production of extremely reactive hydroxyl radicals [28, 29]. Superoxide anion radicals can also react with nitric oxide to form reactive peroxynitrite radicals [30, 31]. Hyperglycemia is also found to promote lipid peroxidation of low density lipoprotein (LDL) by a superoxide-dependent pathway resulting in the generation of free radicals [32, 33]. Another important source of free radicals in diabetes is the interaction of glucose with proteins leading to the formation of an Amadori product and then advanced glycation endproducts (AGEs) [34, 35]. These AGEs, via their receptors (RAGEs), inactivate enzymes and alter their structures and functions [36], promote free radical formation [37, 38], and quench and block antiproliferative effects of nitric oxide [39, 40]. By increasing intracellular oxidative stress, AGEs activate the transcription factor NF-κB, thus promoting up-regulation of various NF-κB controlled target genes [41]. NF-κB enhances production of nitric oxide, which is believed to be a mediator of islet beta cell damage. Considerable evidence also implicates activation of the sorbitol pathway by glucose as a component in the pathogenesis of diabetic complications, for example, in lens cataract formation or peripheral neuropathy [42-44]. Efforts to understand cataract formation have provoked various hypotheses. In the aldose reductase osmotic hypothesis, accumulation of polyols initiates lenticular osmotic changes.

In addition, oxidative stress is linked to decreased glutathione levels and depletion of NADPH levels [45, 46]. Alternatively, increased sorbitol dehydrogenase activity is associated with altered NAD+ levels, which results in protein modification by nonenzymatic glycosylation of lens proteins [47, 48]. Mechanisms linking the changes in diabetic neuropathy and induced sorbitol pathway are not well delineated. One possible mechanism, metabolic imbalances in the neural tissues, has been implicated in impaired neurotrophism [49, 50], neurotransmission changes [51, 52], Schwann cell injury [53, 54], and axonopathy [55, 56].

2.2. Overview of antioxidants

While on the one hand hyperglycemia engenders free radicals, on the other hand it also impairs the endogenous antioxidant defense system in many ways during diabetes [57]. Antioxidant defense mechanisms involve both enzymatic and nonenzymatic strategies. Common antioxidants include the vitamins A, C, and E, glutathione, and the enzymes superoxide dismutase, catalase, glutathione peroxidase, and glutathione reductase. Other antioxidants include α-lipoic acid, mixed carotenoids, coenzyme Q10, several bioflavonoids, antioxidant minerals (copper, zinc, manganese, and selenium), and the cofactors (folic acid, vitamins B1, B2, B6, B12). They work in synergy with each other and against different types of free radicals. Vitamin E suppresses the propagation of lipid peroxidation; vitamin C, with vitamin E, inhibits hydroperoxide formation; metal complexing agents, such as penicillamine, bind transition metals involved in some reactions in lipid peroxidation [58] and inhibit Fenton and Haber- Weiss-type reactions; vitamins A and E scavenge free radicals [30, 37].

Several discrepancies observed in the activities of SOD, catalase, and glutathione peroxidase in experimentally diabetic animals. Decreased levels of glutathione and elevated concentrations of thiobarbituric acid reactants are consistently observed in diabetes [59, 60]. In addition, changes in nitric oxide and glycated proteins are also seen in diabetes.

3. Biomarkers of lipid peroxidation

Since the initial discoveries of Dilliard and colleagues [61], several commercial assay kits have been made available for the measurement of oxidative stress, with many new kits emerging each year. Furthermore, the discovery and utilization of F2-isoprostanes, a prostaglandin like compound, measured via gas chromotomography mass spectrometry has emerged as a substantially more reliable and valid measure of lipid peroxidation [62]. Newly developed ELISA kits for both isoprostanes as well as protein carbonyls are also now available, proving an opportunity for a more widespread use of these biomarkers. In regards to measurement of oxidative stress, due to the high reactivity and relatively short half lives (e.g., 10^{-5}, 10^{-9} seconds for superoxide radical and hydroxyl radical, respectively) of reactive oxygen and nitrogen species (RONS), direct measurement is extremely difficult to employ. However, direct assessment of free radical production is possible via electron spin resonance spectroscopy (ESR) involving spin traps, as well as two other less common techniques such as radiolysis and laser flash photolysis [63]. ESR works by recording the energy changes that occur as unpaired electrons align in response to a magnetic field [64]. Due to the high cost of such equipment and the high degree of labor associated with each direct method, the majority of free radial research related to exercise has utilized indirect methods for the assessment of resultant oxidative stress. Indirect assessment of oxidative stress involves the measurement of the more stable molecular products formed via the reaction of RONS with certain biomolecules. Common molecular products include stable metabolites (e.g., nitrate/nitrite), and/or concentrations of oxidation target products, including lipid peroxidation end products [isoprostanes, malondialdehyde (MDA), thiobarbituric acid reactive substances (TBARS), lipid hydroperoxides (LOOH), conjugated dienes (CD), oxidized low density lipoprotein (oxLDL)], oxidized proteins [protein carbonyls (PC), individual oxidized amino acids, nitrotyrosine (NT), and nucleic acids [8-hydroxy-2-deoxyguanosine (8-OHdG), oxidized DNA bases (via the Comet Assay), strand breaks] [65]. Additionally, oxidative stress can be measured by observing alterations in the body's antioxidant defense system. This is typically done by measuring the redox changes in the major endogenous antioxidant glutathione, as well as circulating levels of vitamin E, and vitamin C. Moreover, the activity of certain antioxidant enzymes [e.g., superoxide dismutase (SOD), glutathione peroxidase (GPx), catalase (CAT), glutathione reductase (GR)] can be assessed as indicators of the oxidative stress imposed on the tissue. Numerous antioxidant capacity assays also exist and include: Trolox Equivalent Antioxidant Capacity (TEAC), Total Antioxidant Status (TAS), Ferric Reducing Ability of Plasma (FRAP), Total RadicalTrapping Antioxidant Parameter (TRAP), and Oxygen Radical Absorbance Capacity (ORAC) [66].

4. Exercise and lipid peroxidation

4.1. Lipid peroxidation and antioxidant status in acute exercise research

Numerous studies have reported an increase in several lipid peroxidation markers following both maximal [67-69] and submaximal [70, 71] exercise. In opposition to these findings, a few studies have reported no increase in lipid peroxidation despite the use of similar maximal [72-74] and submaximal [75, 76] protocols. Increased lipid peroxidation seems to be a result of increased mitochondrial oxidative enyzme activation during aerobic exercise. However, studies reporting conflictiong findings for lipid peroxidation may be partially related to the timing of sampling, in addition to the trained status of the subjects or an insufficient intensity of exercise.

In response to conditions of strenuous physical work the body's antioxidant capacity may be temporarily decreased as its components are used to quench the harmful radicals produced. It appears that the antioxidant capacity may be temporarily reduced during and immediately post exercise [77, 78], after which time levels typically increase above basal conditions during the recovery period [79, 80]. However, conflicting findings have been reported for each of the four main enzymes, with investigators noting increases in GPx [81, 82], SOD [82, 83], and CAT [70, 84, 85], as well as decreases in GPx [86], GR [81], SOD [78].

Furthermore, no change has also been reported for GPx [69], GR [87], SOD (Tauler et al., 2006), CAT [87] activity following exercise. Clearly, these results are mixed and likely depend on the time of sampling, as well as the duration and intensity of exercise, which has varied considerably across studies.

During low-intensity and duration protocols, antioxidant defenses appear sufficient to meet the RONS production, but as intensity and/or duration of exercise increases, these defenses are no longer adequate, potentially resulting in oxidative damage to surrounding tissues [67]. Other factors appear to impact the degree of antioxidant defenses present, including age, training status [81, 88], and dietary intake [80].

It has been shown that anaerobic exercise results in increased RONS production [89]. The mechanisms responsible for the exercise-induced increases in RONS have been suggested to be largely a function of radical generating enzymes (activated in response to ischemia followed by reperfusion) and/or phagocytic immune response following muscle damaging exercise. In the literature, there are fewer data on the markers of lipid peroxidation after anaerobic exercise. It currently remains to be elucidate whether increased RONS formation observed during anaerobic exercise represents a necessary or harmful event.

4.2. Lipid peroxidation and antioxidant status after exercise training

Regular physical exercise exerts numereous adaptive responses in several tissues. In the context of lipid peroxidation, repeated exposure of RONS production appears to induced to maintain the optimal health. Literature data demonstrated that regular moderate exercise is strengthening the endogenous antioxidant defense system [90, 91,92], and in some animal studies, it has decreased lipid peroxidation. On the other hand, exercise training – both endurance and interval type - appears to protects against exercise induced oxidative stress [93, 94, 95].

5. Exercise and diabetes

The therapeutic use of physical exercise for diabetes treatment has been promoted since 600 B.C. before the discovery of insulin in 1922. Some investigators highlighted the interaction between this hormone and regular physical activity, with possible beneficial results in diabetes treatment [96]. Recent guidelines provide exercise recommendations for people with diabetes based on the strong and convincing epidemiologic association of aerobic exercise with lower cardiovascular disease risk in people with diabetes. The recent 2010 ADA/American College of Sports Medicine (ACSM) exercise guidelines recommend 150 minutes of weekly aerobic exercise (i.e. brisk walking or an equivalent activity with intensity $\geq 40\%$ VO2max); and resistance exercise of major muscle groups two to three times weekly on non-consecutive days (ACSM evidence category B, ADA B level recommendation). The ADA/ACSM guidelines also suggest adding unstructured physical activity as much as possible. Before undertaking exercise more intense than brisk walking, sedentary people with T2D should be evaluated by a physician and an exercise trainer [19].

Exercise has been shown to promote beneficial effects on insulin resistance, both in humans and in rodent models of diabetes [97, 98]. Regular physical exercise may prevent diabetes complications through beneficial effects on glycemic control, insulin sensitivity, blood pressure, lipid profile, and endothelial function. Moderate exercise training has been demonstrated to decrease the plasma glucose concentration in STZ-induced diabetic rats [99]. Hypoglycemic effect of exercise can be explained by exercise induced increase in uptake of glucose of muscle which induces increase of GLUT 4 expression and translocation from intracellular pool [100, 101]. Increase in glucose uptake seems to be related to the increased number of GLUT-4 glucose transporters, although the type of training, strain, age and sex of the animals seem to affect significantly the expression of GLUT-4 [100]. On the other hand, Etgen et al. [101] found that exercise training of normal rats results in an elevated maximal insulin-stimulated hindlimb glucose uptake. They suggested that this increase was only partially explained by an increase in total muscle GLUT-4 protein content. A recent study showed that physical training improves in vivo mitochondrial function concomitantly with increased insulin sensitivity in type 2 diabetes patients and control participants [102].

6. Exercise and chronical complications of diabetes

6.1. Exercise and cardiovascular disease (CVD)

Regular exercise has beneficial effects on glucose control and cardiovascular disease (CVD) risk factors. Exercise improves and maintains cardiorespiratory fitness, muscular strength, endurance, and body composition [103]. Exercise has a favorable effect on cardiovascular risk factors. In particular, it has specific beneficial effects on the reduction of hypertension, hyperlipidemia, and obesity and the improvement in blood lipid profile [104] even when combined with a rigorous calorie-restricted diet in obese patients with T2DM [105].

The effects of exercise training on abnormal vascular structure and function (including endothelial dysfunction and vascular distensibility) associated with diabetes are yet to be fully understood [106].

Oxidative stress has been suggested to play a role in either the primary or secondary etiology of both congestive heart failure (CHF) and coronary arter disease (CAD) [107, 108] evident by increased oxidative stress biomarkers and/or decreased antioxidant defenses at rest in diseased compared to healthy controls [109]. Increased TBARS [110, 111] and GSSG [110] have been reported following submaximal aerobic exercise in type 1 diabetic subjects. In regards to maximal exercise, direct production of RONS via electron spin resonance spectroscopy has been reported following a graded exercise testing. However, it is important to note that significance was only achieved when data for both type 1 diabetic and healthy control subjects were pooled [112]. Despite the observation of increased levels of exercise-induced oxidative stress biomarkers in studies involving type 1 diabetics, when compared to healthy individuals, the relative magnitude of increase does not differ; rather the group differences at rest are merely maintained during the post exercise period. Other investigators have reported no changes in MDA [112], total glutathione (TGSH), antioxidant enzyme activity or circulating antioxidants [111] in response to acute exercise in type 1 diabetics .

6.2. Diabetic nephropathy

Diabetic nephropathy is the most feared complication of diabetes, due to its substantial co-morbidity (need for dialysis, blindness, amputations, etc.), cost, and mortality (the annual mortality rate of diabetic patients with kidney failure on dialysis is about 25%) [113, 114]. The major determinants of kidney disease and its progression to end-stage kidney failure in diabetes are uncontrolled blood glucose, blood pressure and albuminuria [115, 116].

Diabetic nephropathy is an important complication of diabetes since it can lead to end-stage renal failure and also it is a risk factor of cardiovascular disease. The clinical problems caused by diabetic nephropathy are proteinuria and decreased renal function. Diabetic nephropathy is defined by proteinuria > 500 mg in 24 hours in the setting of diabetes, but this is preceded by lower degrees of proteinuria, or "microalbuminuria.". Microalbuminuria is defined as albumin excretion of 30–299 mg/24 hours [11]. Without intervention, diabetic patients with microalbuminuria typically progress to proteinuria and overt diabetic nephropathy.

In vitro studies indicate that hyperglycemia directly enhances oxidative stress in cultured endothelial and mesangial cells, which are targets for injury in diabetes [38, 117]. Several different antioxidants, including vitamin E (VE), vitamin C (VC), taurine, and α-lipoic acid (LA), have been reported to ameliorate renal injury in experimental diabetes [118-120]. In human diabetes, there is evidence that short-term (3 to 4 mo), high-dose (1600 to 1800 IU/d) VE supplementation reduces proteinuria in type 1 and 2 patients with overt nephropathy and decreases hyperfiltration in type 1 patients without overt nephropathy [121].

Urinary albumin excretion occurs normally after exercise [122, 123]. Post-exercise urinary albumin excretion is explained by increased glomerular capillary membrane permeability as a result of increased filtration pressure with increased filtered protein load, and decreased tubular absorption [122, 123]. In normal subjects, proteinuria is better related to exercise intensity and lactate production than to exercise duration [124], it diminishes after 1 h [125] and returns to baseline within 24 h [122]. In diabetes mellitus, the kidneys are more sensitive to the haemodynamic exercise stress [122]. Under exercise, patients with Type 1 diabetes show a partial depletion of negative charges on the glomerular capillary wall [126] that permits the increase of urinary albumin excretion [127]. Reports indicate that urinary albumin excretion increases after exercise, without correlation with glycaemic control, renal function, disease evolution or resting urinary albumin excretion [128, 129]. In contrast, post-exercise albuminuria has been found to be associated with HbA1c [127, 129].

On the other hand, recent data demonstrated that exercise might protect the diabetic renal function [130]. Kutlu et al. demonstrated that moderate exercise with combined vitamin E and C supplement was strengthen the antioxidant defense system and reduced the lipid peroxidation in STZ-induced diabetic rat kidney [131].

6.3. Diabetic retinopathy

Diabetic retinopathy may be the most common microvascular complication of diabetes. The risk of developing diabetic retinopathy or other microvascular complications of diabetes depends on both the duration and the severity of hyperglycemia. Development of diabetic retinopathy in patients with type 2 diabetes was found to be related to both severity of hyperglycemia and presence of hypertension in the U.K Prospective Diabetes Study (UKPDS) [132] and most patients with type 1 diabetes develop evidence of retinopathy within 20 years of diagnosis [133]. Retinopathy may begin to develop as early as 7 years before the diagnosis of diabetes in patients with type 2 diabetes [134]. There are several proposed pathological mechanisms by which diabetes may lead to development of retinopathy.

Oxidative stress may also play an important role in cellular injury from hyperglycemia. High glucose levels can stimulate free radical production and reactive oxygen species formation. Animal studies have suggested that treatment with antioxidants, such as vitamin E, may attenuate some vascular dysfunction associated with diabetes, but treatment with antioxidants has not yet been shown to alter the development or progression of retinopathy or other microvascular complications of diabetes [11].

6.4. Diabetic neuropathy

Diabetic polyneuropathy affects 30% of the hospital-based population and 20% of community based samples of diabetic patients [135]. There is a growing body of evidence to support the notion that oxidative stress is the biochemical trigger for nerve dysfunction. Various disturbances such as reduced endoneurial blood flow, altered electroconductive properties of the myelin sheath, impaired incorporation of acetate and glucose into the neuron cells should also be mentioned in diabetic condition. It has been shown [136] that superoxide dismutase activity is decreased in nerves from streptozotocin-induced diabetic rats. Glutathione content and glutathione peroxidase activity are also diminished in sciatic nerves from diabetic rats [137, 138]. Nerves of diabetic rats show lower amounts of vitamin E compared to control animals [139]. Lipid peroxidation products such as malondialdehydes or conjugated dienes are elevated in diabetic sciatic nerves [136, 139]. Treatment of diabetic rats with insulin or antioxidants is associated with improved nerve function [51, 140].

Sensory, visual and auditory neural conduction deficits are well documented both in diabetic animals and human studies. As an early marker of visual system deficits observed in diabetic state, visual evoked potential (VEP) latencies were measured in STZ-induced diabetic rats in our laboratories. The results of the previous studies were demonstrated that visual evoked potential (VEP) latencies were prolonged in STZ-induced diabetic rats whereas the latencies were restored by moderate physical exercise [60, 141, 142]. The VEP alterations were found to be accompanied with the increased TBARS concentration in the brain tissues of the diabetic rats.

The impact of diabetes on nervous system is complex and poorly elucidated. The brain is particularly vulnerable to oxidative damage because of its high rate of oxygen consumption, intense production of reactive radicals, and high levels of transition metals, such as iron, that catalyze the production of reactive radicals [143]. Moreover, neuronal membranes are rich in poly unsaturated fatty acids, which are a source of lipid peroxidation [37]. Free radicals are formed disproportionately in diabetes by glucose oxidation, non-enzymatic glycation of proteins, and the subsequent oxidative degradation of glycated proteins. Abnormally high levels of free radicals and the simultaneous decline of antioxidant defense mechanisms can lead to damage of cellular organelles and enzymes, increased lipid peroxidation, and development of insulin resistance. These consequences of oxidative stress can promote the development of complications of diabetes mellitus [143].

Previous experimental studies demonstrated that diabetes resulted increased lipid peroxidation and decreased antioxidant enzymes in several brain regions such as hypocampus, striatum and cerebral cortex as well as in whole brain tissue homogenates [59]. The lipids oxidation in the CNS usually demonstrates different concentrations at different regions of the brain, and it can be attributed to regional differences in the O2 consumption [144, 145].

Nervous system complications of diabetes mellitus can become one of the most debilitating complications and affect sensitive and cognitive functions that modulates memory function, resulting in significant functional impairment and dementia. Oxidative stress forms the foundation for the induction of multiple cellular pathways that can ultimately lead to both the onset and subsequent complications of DM [146]. Defects in hippocampal synaptic plasticity and transmission resulting in impairment of learning and memory is one the central nervous system complications of diabetes mellitus [147, 148]. Increasing evidence in both experimental and clinical studies suggests that oxidative stress plays a central role in the onset and subsequent complications of diabetes mellitus [149].

Physical exercise has been demonstrated to induce several neurobiological changes in the brain and to prevent diabetes-induced cognitive decline. The neurobiological changes induced by physical exercise have been demonstrated to facilitate the acquisition of a spatial memory task in rats. Exercise has also been demonstrated to increase the cognitive function both in healthy and diabetic people [150, 151, 152]. However, intense exercise has been shown to impaired the cognitive function in murine model that was prevented by vitamin C and E supplementation [153].

7. Conclusion

Literature results emphasize the beneficial role of physical exercise in the promotion of in diabetic complications probably by decreasing hyperglicemia, increasing insulin sensitivity and enhancing antioxidant status of the several systems. The type, duration and the intensity of the exercise as well as the degree of the diabetic complications should be determined before the exercise prescription in diabetic person. For future research, the effects of the different exercise protocols for maintaining the optimum health and stimulating the cellular processes for decreasing the hyperglicemia-induced complications in diabetes in children and older people remains to be explored.

Author details

Yaşar Gül Özkaya

School of Physical Education and Sports, Akdeniz University, Antalya, Turkey

8. References

[1] World Health Organization. Media Centre. Diabetes. http://www.who.int/mediacentre/factsheets/fs312/en/index.html (accessed 12 May 2008).

[2] Maritim, A. C., R. A. Sanders & J. B. Watkins. Diabetes, oxidative stress, and antioxidants: A review. Journal of Biochemical and Molecular Toxicology. 2003;17:24-38.

[3] Paik SG, Blue ML, Fleischer N, Shin SI. Diabetes susceptibility of balb/cbom mice treated with streptozotocin - inhibition by lethal irradiation and restoration by splenic lymphocytes. Diabetes. 1982;31(9):808-15.

[4] Sandler S, Andersson AK, Barbu A, Hellerstrom C, Holstad M, Karlsson E, Sandberg JO, Strandell E, Saldeen J, Sternesjo J, Tillmar L, Eizirik DL, Flodstrom M & Welsh N. Novel experimental strategies to prevent the development of type 1 diabetes mellitus. Upsala Journal of Medical Sciences. 2000;105(2):17–34.

[5] Shewad Y, Tirth S & Bhonde RR. Pancreatic islet-cell viability, functionality and oxidative status remain unaffected at pharmacological concentrations of commonly used antibiotics in vitro. Journal of Biosciences. 2001;26(3):349–355.

[6] American Diabetes Association (ADA) Position Statement (2010). Diagnosis and Classification of Diabetes Mellitus. Diabetes Care.2010;33(Supplement 1): S62-S69.

[7] de Lemos ET, Pinto Ri Oliveira J, Garrido P, Sereno J, Mascarenhas-Melo F, Pascoa Pinheiro J, Teixeira F & Reis F. (2011). Differential effects of acute (extenuating) and chronic (training) exercise on inflammation and oxidative stress status in an animal model of Type 2 diabetes mellitus. Mediators of Inflammation. 2011;253061:1-8.

[8] P´erez-Matute P, Zulet MA & Mart´ınez JA (2009). Reactive species and diabetes: Counteracting oxidative stress to improve health. Current Opinion in Pharmacology. 2009;9(6):771–779.

[9] Brunner Y, Schvartz D, Priego-Capote F, Cout´e Y & Sanchez JC (2009). Glucotoxicity and pancreatic proteomics. Journal of Proteomics.2009;71(6):576–591.

[10] Donath MY, Schumann DM, Faulenbach M, Ellingsgaard H, Perren A & Ehses JA. Islet inflammation in type 2 diabetes: from metabolic stress to therapy. Diabetes Care. 2008;31(Supplement 2):S161–164.

[11] Fowler MJ. Microvascular and Macrovascular Complications of Diabetes. Clinical Diabetes. 2008;26(2):77-82.

[12] Ballas LM, Jirousek MR, Umeda F, Nawata H, King GL. Vitamin E prevents diabetes induced abnormal retinal blood flow via the diacylglycerolprotein kinase C pathway. American Journal of Physiology. 1995;269: E239– E246.

[13] Watkins PJ. Retinopathy. BMJ.2003;326:924–926.

[14] Gross JL, de Azevedo MJ, Silveiro SP, Canani LH, Caramori ML & Zelmanovitz T. Diabetic nephropathy: diagnosis, prevention, and treatment. Diabetes Care. 2005;28:164-176.

[15] Chaturvedi N, Bandinelli S, Mangili R, Penno G, Rottiers RE.& Fuller JH. Microalbuminuria in type 1 diabetes: rates, risk factors and glycemic threshold. Kidney International. 2001;60: 219– 227.

[16] Adler AI, Stevens RJ, Manley SE, Bilous RW, Cull CA & Holman RR (2003). Development and progression of nephropathy in type 2 diabetes: the United Kingdom Prospective Diabetes Study (UKPDS 64). Kidney International. 2003;63:225–232.

[17] Kunisaki M, Bursell SE, Clermont AC, Ishii H, Ballas LM, Jirousek MR, Umeda F, Nawata H, King GL. Vitamin E prevents diabetes-induced abnormal retinal blood flow via the diacylglycerol-protein kinase C pathway. American Journal of Physiology-Endocrinology and Metabolism. 1995;269(2):E239-E46.

[18] Sigal RJ, Kenny GP, Wasserman DH, Castaneda-Sceppa C & White RD. Physical activity/exercise and type 2 diabetes: a consensus statement from the American Diabetes Association. Diabetes Care. 2006;29(6):1433-8.

[19] Colberg SR, Albright AL, Blissmer BJ, Braun B, Chasan-Taber L, Fernhall B, Regensteiner JG, Rubin RR & Sigal RJ. Exercise and Type 2 Diabetes ACSM ADA Joint Position Statement. Medicine & Science In Sports & Exercise. 2010;42(12): 2282-2303.

[20] Marrero DG. Time to get moving: helping patients with diabetes adopt exercise as part of a healthy lifestyle. Clinical Diabetes. 2005;23 (4):154-159.

[21] Hevener AL, Reichart D & Olefsky J. Exercise and thiazolidinedione therapy normalize insulin action in the obese Zucker fatty rat. Diabetes. 2000;49(12):2154– 2159.

[22] Pold R, Jensen LS, Jessen N et al. Long-term AICAR administration and exercise prevents diabetes in ZDF rats. Diabetes. 2005;54(4):928–934.

[23] Klein S, Sheard NF, Pi-Sunyer X, Daly A, Wylie-Rosett J, Kulkarni K & Clark NG. American Diabetes Association, American Association for the Study of Obesity, American Society for Clinical Nutrition: Weight management through lifestyle modification for the prevention and management of type 2 diabetes: rationale and strategies. Diabetes Care.2004;27:2067–2073.

[24] Boule N, Haddad E, Kenny G, Wells G & Sigal R. Effects of exercise on glycemic control and body mass in type 2 diabetes: a meta-analysis of controlled clinical trials. JAMA.2001;286:1218 1227.

[25] de Lemos ET, Reis F, Baptista S. et al. Exercise training is associated with improved levels of C-reactive protein and adiponectin in ZDF (type 2) diabetic rats. Medical Science Monitor. 2007;13(8):BR168–BR174.

[26] Connor TJ, Brewer C, Kelly JP, Harkin A. Acute stress suppresses pro-inflammatory cytokines TNF-alpha and IL-1 beta independent of a catecholamine-driven increase in IL-10 production. Journal of Neuroimmunology. 2005;159(1-2):119-28.

[27] Giraldo E, Garcia JJ, Hinchado MD, Ortega E. Exercise Intensity-Dependent Changes in the Inflammatory Response in Sedentary Women: Role of Neuroendocrine Parameters in the Neutrophil Phagocytic Process and the Pro-/Anti-Inflammatory Cytokine Balance. Neuroimmunomodulation. 2009;16(4):237-44.

[28] Jiang ZY, Woollard ACS, Wolff SP. Hydrogen-peroxide production during experimental protein glycation. Febs Letters. 1990;268(1):69-71.

[29] Wolff SP, Dean RT. Glucose autoxidation and protein modification - the potential role of autoxidative glycosylation in diabetes. Biochemical Journal. 1987;245(1):243-50.

[30] Halliwell B, Gutteridge JMC. Role of free-radicals and catalytic metal-ions in human-disease - an overview. Methods in Enzymology. 1990;186:1-85.

[31] Hogg N, Kalyanaraman B, Joseph J, Struck A, Parthasarathy S. Inhibition of low-density-lipoprotein oxidation by nitric-oxide - potential role in atherogenesis. Febs Letters. 1993;334(2):170-4.

[32] Tsai EC, Hirsch IB, Brunzell JD, Chait A. Reduced plasma peroxyl radical trapping capacity and increased susceptibility of LDL to oxidation in poorly controlled IDDM. Diabetes. 1994;43(8):1010-4.

[33] Kawamura M, Heinecke JW, Chait A. Pathophysiological concentrations of glucose promote oxidative modfication of low-density-lipoprotein by a superoxide-dependent pathway. Journal of Clinical Investigation. 1994;94(2):771-8.

[34] Hori O, Yan SD, Ogawa S, Kuwabara K, Matsumoto M, Stern D, et al. The receptor for advanced glycation end-products has a central role in mediating the effects of advanced glycation end-products on the development of vascular disease in diabetes mellitus. Nephrology Dialysis Transplantation. 1996;11:13-6.

[35] Mullarkey CJ, Edelstein D, Brownlee M. Free-radical generation by early glycation products - a mechanism for accelerated atherogenesis in diabetes. Biochemical and Biophysical Research Communications. 1990;173(3):932-9.

[36] McCarthy AD, Etcheverry SB, Cortizo AM. Effect of advanced glycation endproducts on the secretion of insulin-like growth factor-I and its binding proteins: role in osteoblast development. Acta Diabetologica. 2001;38(3):113-22.

[37] Baynes JW. Role of oxidative stress in development of complications in diabetes. Diabetes. 1991;40(4):405-12.

[38] Baynes JW, Thorpe SR. Role of oxidative stress in diabetic complications - A new perspective on an old paradigm. Diabetes. 1999;48(1):1-9.

[39] Vlassara H. Recent progress in advanced glycation end products and diabetic complications. Diabetes. 1997;46:S19-S25.

[40] Wautier JL, Wautier MP, Schmidt AM, Anderson GM, Hori O, Zoukourian C, et al. Advanced glycation end-products (ages) on the surface of diabetic erythrocytes bind to the vessel wall via a specific receptor inducing oxidant stress in the vasculature - a link

between surface-associated ages and diabetic complications. Proceedings of the National Academy of Sciences of the United States of America. 1994;91(16):7742-6.

[41] Mohamed AK, Bierhaus A, Schiekofer S, Tritschler H, Ziegler R, Nawroth PP. The role of oxidative stress and NF-kappa B activation in late diabetic complications. Biofactors. 1999;10(2-3):157-67.

[42] Kador PF, Kinoshita JH. Diabetic and galactosemic cataracts. Ciba Foundation Symposia. 1984;106:110-31.

[43] Greene DA, Sima AAF, Stevens MJ, Feldman EL, Lattimer SA. Complications - neuropathy, pathogenetic considerations. Diabetes Care. 1992;15(12):1902-25.

[44] Obrosova I, Faller A, Burgan J, Ostrow E, Williamson JR. Glycolytic pathway, redox state of NAD(P)-couples and energy metabolism in lens in galactose-fed rats: Effect of an aldose reductase inhibitor. Current Eye Research. 1997;16(1):34-43.

[45] Gonzalez AM, Sochor M, Hothersall JS, Mclean P. Effect of aldose reductase inhibitor (sorbinil) on integration of polyol pathway, pentose-phosphate pathway, and glycolytic route in diabetic rat lens. Diabetes. 1986;35(11):1200-5.

[46] Cheng HM, Gonzalez RG. The effect of high glucose and oxidative stress on lens metabolism, aldose reductase, and senile cataractogenesis. Metabolism-Clinical and Experimental. 1986;35(4):10-4.

[47] Yano M, Matsuda S, Bando Y, Shima K. Lens protein glycation and the subsequent degree of opacity in streptozotocin-diabetic rats. Diabetes Research and Clinical Practice. 1989;7(4):259-62.

[48] Ramalho JS, Marques C, Pereira PC & Mota MC. Role of glycation in human lens protein structure change. European Journal of Ophthalmology.1996;6(2):155–161.

[49] Mizisin AP, Bache M, distefano PS, Acheson A, Lindsay RM, Calcutt NA. BDNF attenuates functional and structural disorders in nerves of galactose-fed rats. Journal of Neuropathology and Experimental Neurology. 1997;56(12):1290-301.

[50] Hounsom L, Horrobin DF, Tritschler H, Corder R, Tomlinson DR. A lipoic acid gamma linolenic acid conjugate is effective against multiple indices of experimental diabetic neuropathy. Diabetologia. 1998;41(7):839-43.

[51] Stevens MJ, Obrosova I, Cao XH, Van Huysen C, Greene DA. Effects of DL-alpha-lipoic acid on peripheral nerve conduction, blood flow, energy metabolism, and oxidative stress in experimental diabetic neuropathy. Diabetes. 2000;49(6):1006-15.

[52] Ralevic V, Belai A, Burnstock G. Effects of streptozotocin-diabetes on sympathetic-nerve, endothelial and smooth-muscle function in the rat mesenteric arterial bed. European Journal of Pharmacology. 1995;286(2):193-9.

[53] Mizisin AP, Kalichman MW, Bache M, Dines KC, distefano PS. NT-3 attenuates functional and structural disorders in sensory nerves of galactose-fed rats. Journal of Neuropathology and Experimental Neurology. 1998;57(9):803-13.

[54] Kalichman MW, Powell HC, Mizisin AP. Reactive, degenerative, and proliferative Schwann cell responses in experimental galactose and human diabetic neuropathy. Acta Neuropathologica. 1998;95(1):47-56.

[55] Chokroverty S, Seiden D, Navidad P & Cody R. Distal axonopathy in streptozotocin diabetes in rats. Experientia. 1988;44(5):444–446.

[56] Fernyhough P, Gallagher A, Averill SA, Priestley JV, Hounsom L, Patel J, et al. Aberrant neurofilament phosphorylation in sensory neurons of rats with diabetic neuropathy. Diabetes. 1999;48(4):881-9.

[57] Saxena AK, Srivastava P, Kale RK, Baquer NZ. Impaired antioxidant status in diabetic rat-liver - effect of vanadate. Biochemical Pharmacology. 1993;45(3):539-42.

[58] Feher J, Cosmos G & Vereckei A. Free Radical Reactions in Medicine. Berlin:Springer-Verlag;1987.

[59] Ozkaya YG, Agar A, Yargicoglu P, Hacioglu G, Bilmen-Sarikcioglu S, Ozen I, et al. The effect of exercise on brain antioxidant status of diabetic rats. Diabetes & Metabolism. 2002;28(5):377-84.

[60] Ozkaya YG, Hacioglu G, Kucukatay V, Yargicoglu P, Agar A. The Effect of Chronic N(G)-Nitro-L-arginine Methyl Ester (L-NAME) Administration on Visual Evoked Potentials and Oxidative Stress in Streptozotocin Induced Diabetic Rats. Journal of Neurological Sciences-Turkish. 2011;28(2):132-41.

[61] Dillard CJ, Litov RE, Savin WM, Dumelin EE, Tappel AL. Effects of exercise, vitamin-E, and ozone on pulmonary-function and lipid peroxidation. Journal of Applied Physiology. 1978;45(6):927-32.

[62] Morrow JD, Roberts LJ. Mass spectrometric quantification of F-2-isoprostanes in biological fluids and tissues as measure of oxidant stress. Oxidants and Antioxidants, Pt B. 1999;300:3-12.

[63] Knight JA. Free Radicals, antioxidants, aging, and disease. Washington: American Association for Clinical Chemistry Press, 1999.

[64] Halliwell B, Cross CE. Oxygen-derived species - their relation to human-disease and environmental-stress. Environmental Health Perspectives. 1994;102:5-12.

[65] Dalle-Donne I, Rossi R, Colombo R, Giustarini D, Milzani A. Biomarkers of oxidative damage in human disease. Clinical Chemistry. 2006;52(4):601-23.

[66] Fisher-Wellman K & Bloomer RJ. Acute exercise and oxidative stress: a 30 year history. Dynamic Medicine. 2009;8(1):1-8.

[67] Knez WL, Jenkins DG, Coombes JS. Oxidative stress in half and full ironman triathletes. Medicine and Science in Sports and Exercise. 2007;39(2):283-8.

[68] Steinberg JG, Ba A, Bregeon F, Delliaux S, Jammes Y. Cytokine and oxidative responses to maximal cycling exercise in sedentary subjects. Medicine and Science in Sports and Exercise. 2007;39(6):964-8.

[69] Miyazaki H, Oh-ishi S, Ookawara T, Kizaki T, Toshinai K, Ha S, et al. Strenuous endurance training in humans reduces oxidative stress following exhausting exercise. European Journal of Applied Physiology. 2001;84(1-2):1-6.

[70] Nikolaidis MG, Kyparos A, Hadziioannou M, Panou N, Samaras L, Jamurtas AZ, et al. Acute exercise markedly increases blood oxidative stress in boys and girls. Applied

Physiology Nutrition and Metabolism-Physiologie Appliquee Nutrition Et Metabolisme. 2007;32(2):197-205.

[71] Meijer EP, Goris AHC, van Dongen JLJ, Bast A, Westerterp KR. Exercise-induced oxidative stress in older adults as a function of habitual activity level. Journal of the American Geriatrics Society. 2002;50(2):349-53.

[72] Rahnama N, Gaeini AA, Hamedinia MR. Oxidative stress responses in physical education students during 8 weeks aerobic training. Journal of Sports Medicine and Physical Fitness. 2007;47(1):119-23.

[73] Silvestro A, Scopacasa F, Oliva G, de Cristofaro T, Iuliano L, Brevetti G. Vitamin C prevents endothelial dysfunction induced by acute exercise in patients with intermittent claudication. Atherosclerosis. 2002;165(2):277-83.

[74] Gaeini AA, Rahnama N, Hamedinia MR. Effects of vitamin E supplementation on oxidative stress at rest and after exercise to exhaustion in athletic students. Journal of Sports Medicine and Physical Fitness. 2006;46(3):458-61.

[75] Morillas-Ruiz J, Zafrilla P, Almar M, Cuevas M, Lopez FJ, Abellan P, et al. The effects of an antioxidant-supplemented beverage on exercise-induced oxidative stress: results from a placebo-controlled double-blind study in cyclists. European Journal of Applied Physiology. 2005;95(5-6):543-9.

[76] Goldfarb AH, Patrick SW, Bryer S, You TJ. Vitamin C supplementation affects oxidative-stress blood markers in response to a 30-minute run at 75% VO2max. International Journal of Sport Nutrition and Exercise Metabolism. 2005;15(3):279-90.

[77] Di Massimo C, Scarpelli P, Tozzi-Ciancarelli MG. Possible involvement of oxidative stress in exercise-mediated platelet activation. Clinical Hemorheology and Microcirculation. 2004;30(3-4):313-6.

[78] Tozzi-Ciancarelli MG, Penco M, Di Massimo C. Influence of acute exercise on human platelet responsiveness: possible involvement of exercise-induced oxidative stress. European Journal of Applied Physiology. 2002;86(3):266-72.

[79] Alessio HM, Hagerman AE, Fulkerson BK, Ambrose J, Rice RE, Wiley RL. Generation of reactive oxygen species after exhaustive aerobic and isometric exercise. Medicine and Science in Sports and Exercise. 2000;32(9):1576-81.

[80] Watson TA, Callister R, Taylor RD, Sibbritt DW, macdonald-Wicks LK, Garg ML. Antioxidant restriction and oxidative stress in short-duration exhaustive exercise. Medicine and Science in Sports and Exercise. 2005;37(1):63-71.

[81] Elosua R, Molina L, Fito M, Arquer A, Sanchez-Quesada JL, Covas MI, et al. Response of oxidative stress biomarkers to a 16-week aerobic physical activity program, and to acute physical activity, in healthy young men and women. Atherosclerosis. 2008;197(2):967.

[82] Buczynski A, Kedziora J, Tkaczewski W, Wachowicz B. Effect of submaximal physical exercise on antioxidative protection of human blood-platelets. International Journal of Sports Medicine. 1991;12(1):52-4.

[83] Chen MF, Hsu HC, Lee YT. Effects of acute exercise on the changes of lipid profiles and peroxides, prostanoids, and platelet activation in hypercholesterolemic patients before and after treatment. Prostaglandins. 1994;48(3):157-74.

[84] Michailidis Y, Jamurtas AZ, Nikolaidis MG, Fatouros IG, Koutedakis Y, Papassotiriou I, et al. Sampling time is crucial for measurement of aerobic exercise-induced oxidative stress. Medicine and Science in Sports and Exercise. 2007;39(7):1107-13.

[85] Vider J, Lehtmaa J, Kullisaar T, Vihalemm T, Zilmer K, Kairane C, Landor A, Karu T &, Zilmer M. Acute immune response in respect to exercise-induced oxidative stress. Pathophysiology.2001;7(4):263-270.

[86] Akova B, Surmen-Gur E, Gur H, Dirican M, Sarandol E, Kucukoglu S. Exercise-induced oxidative stress and muscle performance in healthy women: role of vitamin E supplementation and endogenous oestradiol. European Journal of Applied Physiology. 2001;84(1-2):141-7.

[87] Tauler P, Aguilo A, Gimeno I, Fuentespina E, Tur JA, Pons A. Response of blood cell antioxidant enzyme defences to antioxidant diet supplementation and to intense exercise. European Journal of Nutrition. 2006;45(4):187-95.

[88] Fatouros IG, Jamurtas AZ, Villiotou V, Pouliopoulou S, Fotinakis P, Taxildaris K, et al. Oxidative stress responses in older men during endurance training and detraining. Medicine and Science in Sports and Exercise. 2004;36(12):2065-72.

[89] Guzel NA, Hazar S, Erbas D. Effects of different resistance exercise protocols on nitric oxide, lipid peroxidation and creatine kinase activity in sedentary males. Journal of Sports Science and Medicine. 2007;6(4):417-22.

[90] Chung HY, Cesari M, Anton S, Marzetti E, Giovannini S, Seo AY, et al. Molecular inflammation: Underpinnings of aging and age-related diseases. Ageing Research Reviews. 2009;8(1):18-30.

[91] Radak Z, Chung HY, Koltai E, Taylor AW & Goto S. Exercise, oxidative stress and hormesis. Ageing Research Reviews. 2008;7(1):34-42.

[92] Gocmen AY, Ozkaya YG, Yazar H, Agar A, Gunaydin I & Gumuslu S. (2011b). Effect of exercise on paraoxonase-1 activity and lipid peroxidation in diabetes. Bozok Medical Journal. 2011;1(2):13-21.

[93] Atalay M, Marnila P, Lilius EM, Hanninen O, Sen CK. Glutathione-dependent modulation of exhausting exercise induced changes in neutrophil function of rats. European Journal of Applied Physiology and Occupational Physiology. 1996;74(4):342-7.

[94] Atalay M, Seene T, Hanninen O, Sen CK. Skeletal muscle and heart antioxidant defences in response to sprint training. Acta Physiologica Scandinavica. 1996;158(2):129-34.

[95] Powers SK, Ji LL, Leeuwenburgh C. Exercise training-induced alterations in skeletal muscle antioxidant capacity: a brief review. Medicine and Science in Sports and Exercise. 1999;31(7):987-97.

[96] Steppel TH & Horton ES. (2005). Exercise in patients with diabetes mellitus. In: Kahn CR, Weir GC, King GL, Jacobson AM, Moses AC, Smith RJ, Editors: Joslins' Diabetes Mellitus. Philadelphia: Lippincott Williams and Wilkins, 2005. pp: 649-658.

[97] Hughes VA, Fiatrone MA, Fielding RA, Kahn BB, Ferrara CM, Shepherd P, Fisher EC, Wolfe RR, Elahi D & Evans WJ. Exercise increases muscle GLUT-4 levels and insulin action in subjects with impaired glucose tolerance. American Journal of Physiology.1993;264:E855–E862.

[98] Cox JH, Cortright RN, Dohm GL, Houmard JA. Effect of aging on response to exercise training in humans: skeletal muscle GLUT-4 and insulin sensitivity. Journal of Applied Physiology. 1999;86(6):2019-25.

[99] Giacca A, Groenewoud Y, Tsui E, McClean P, Zinman B. Glucose production, utilization, and cycling in response to moderate exercise in obese subjects with type 2 diabetes and mild hyperglycemia. Diabetes. 1998 Nov;47(11):1763-70.

[100] Kainulainen H, Komulainen J, Joost HG, Vihko V. Dissociation of the effects of training on oxidative-metabolism, glucose-utilization and GLUT4 levels in skeletal-muscle of streptozotocin-diabetic rats. Pflugers Archiv-European Journal of Physiology. 1994;427(5-6):444-9.

[101] Etgen GJ, Brozinick JT, Kang HY, Ivy JL. Effects of exercise training on skeletal-muscle glucose-uptake and transport. American Journal of Physiology. 1993;264(3):C727-C33.

[102] Wohaieb SA, Godin DV. Alterations in free-radical tissue-defense mechanisms in streptozocin-induced diabetes in rat - effects of insulin-treatment. Diabetes. 1987;36(9):1014-8.

[103] Marwick TH, Hordern MD, Miller T, Chyun DA, Bertoni AG, Blumenthal RS, et al. Exercise Training for Type 2 Diabetes Mellitus Impact on Cardiovascular Risk A Scientific Statement From the American Heart Association. Circulation. 2009;119(25):3244-62.

[104] Wilmore JH, Green JS, Stanforth PR, Gagnon J, Rankinen T, Leon AS, et al. Relationship of changes in maximal and submaximal aerobic fitness to changes in cardiovascular disease and non-insulin-dependent diabetes mellitus risk factors with endurance training: The Heritage Family Study. Metabolism-Clinical and Experimental. 2001;50(11):1255-63.

[105] Pi-Sunyer, X., Blackburn, G., Brancati, F.L., et al. (2007). Look AHEAD Research Group. Reduction in weight and cardiovascular disease risk factors in individuals with type 2 diabetes: one-year results of the Look AHEAD trial. Diabetes Care. Volume 30, pp:1374–1383.

[106] Simpson SH, Corabian P, Jacobs P, Johnson JA. The cost of major comorbidity in people with diabetes mellitus. Canadian Medical Association Journal. 2003;168(13):1661-7.

[107] Fisher-Wellman K, Bell HK, Bloomer RJ. Oxidative stress and antioxidant defense mechanisms linked to exercise during cardiopulmonary and metabolic disorders. Oxidative Medicine and Cellular Longevity. 2009;2(1):43-51.

[108] Soccio M, Toniato E, Evangelista V, Carluccio M, De Caterina R. Oxidative stress and cardiovascular risk: the role of vascular NAD(P)H oxidase and its genetic variants. European Journal of Clinical Investigation. 2005;35(5):305-14.

[109] Tsutsui H. Oxidative stress in heart failure: The role of mitochondria. Internal Medicine. 2001;40(12):1177-82.

[110] Laaksonen DE, Uusitupa M, Atalay M, Hanninen O, Niskanen L, Sen CK. Increased resting and exercise-induced oxidative stress in young IDDM men. Diabetes Care. 1996;19(6):569-74.

[111] Atalay M, Laaksonen DE, Niskanen L, Uusitupa M, Hanninen O, Sen CK. Altered antioxidant enzyme defences in insulin-dependent diabetic men with increased resting and exercise-induced oxidative stress. Acta Physiologica Scandinavica.1997;161:195-201.

[112] Davison GW, George L, Jackson SK, Young IS, Davies B, Bailey DM, et al. Exercise, free radicals, and lipid peroxidation in type 1 diabetes mellitus. Free Radical Biology and Medicine. 2002;33(11):1543-51.

[113] Leehey DJ, Moinuddin I, Bast JP, Qureshi S, Jelinek CS, Cooper C, et al. Aerobic exercise in obese diabetic patients with chronic kidney disease: a randomized and controlled pilot study. Cardiovascular Diabetology. 2009;8:62.

[114] US Renal Data System: USRDS 2008 Annual Data Report Bethesda: National Institutes of Health, National Institute of Diabetes and Digestive and Kidney Diseases; 2008.

[115] Stratton IM, Cull CA, Adler AI, Matthews DR, Neil HAW, Holman RR. Additive effects of glycaemia and blood pressure exposure on risk of complications in type 2 diabetes: a prospective observational study (UKPDS 75). Diabetologia. 2006;49(8):1761-9.

[116] Rossing K, Christensen PK, Hovind P, Parving HH. Remission of nephrotic-range albuminuria reduces risk of end-stage renal disease and improves survival in type 2 diabetic patients. Diabetologia. 2005;48(11):2241-7.

[117] Giugliano D, Ceriello A, Paolisso G. Oxidative stress and diabetic vascular complications. Diabetes Care. 1996;19(3):257-67.

[118] Lee,EY, Lee MY, Hong SW, Choon Hee Chung CH, Hong SY. Blockade of oxidative stress by vitamin C ameliorates albuminuria and renal sclerosis in experimental diabetic rats. Yonsei Medical Journal. 2007; 31; 48(5): 847–855.

[119] Trachtman H, Futterweit S, Maesaka J, Ma C, Valderrama E, Fuchs A, et al. Taurine ameliorates chronic streptozocin-induced diabetic nephropathy in rats. American Journal of Physiology-Renal Fluid and Electrolyte Physiology. 1995;269(3):F429-F38.

[120] Craven PA, derubertis FR, Kagan VE, Melhem M, Studer RK. Effects of supplementation with vitamin C or E on albuminuria, glomerular TGF-beta, and glomerular size in diabetes. Journal of the American Society of Nephrology. 1997;8(9):1405-14.

[121] Bursell SE, Clermont AC, Aiello LP, Aiello LM, Schlossman DK, Feener EP, et al. High-dose vitamin E supplementation normalizes retinal blood flow and creatinine clearance in patients with type 1 Diabetes. Diabetes Care. 1999;22(8):1245-51.

[122] Poortmans JR, Labilloy D. The influence of work intensity on postexercise proteinuria. European Journal of Applied Physiology and Occupational Physiology. 1988;57(2):260-3.

[123] Heathcote KL, Wilson MP, Quest DW, Wilson TW. Prevalence and duration of exercise induced albuminuria in healthy people. Clinical and Investigative Medicine. 2009;32(4):E261-E5.

[124] Poortmans JR, Geudvert C, Schorokoff K, deplaen P. Postexercise proteinuria in childhood and adolescence. International Journal of Sports Medicine. 1996;17(6):448-51.

[125] Poortmans JR, Vanderstraeten J. Kidney-function during exercise in healthy and diseased humans - an update. Sports Medicine. 1994;18(6):419-37.

[126] Ala-Houhala I. Effects of exercise on glomerular passage of macromolecules in patients with diabetic nephropathy and in healthy subjects. Scandinavian Journal of Clinical and Laboratory Investigation.1990;50:27–33.

[127] Huttunen NP, Kaar ML, Puukka R, Akerblom HK. Exercise-induced proteinuria in children and adolescents with type-1 (insulin dependent) diabetes. Diabetologia. 1981;21(5):495-7.

[128] Feldtrasmussen B, Baker L, Deckert T. Exercise as a provocative test in early renal-disease in type-1 (insulin-dependent) diabetes - albuminuric, systemic and renal hemodynamic-responses. Diabetologia. 1985;28(7):389-96.

[129] Kruger M, Gordjani N, Burghard R. Postexercise albuminuria in children with different duration of type-1 diabetes mellitus. Pediatric Nephrology. 1996;10(5):594-7.

[130] Kurdak H, Sandikci S, Ergen N, Dogan A, Kurdak SS. The effects of regular aerobic exercise on renal functions in streptozotocin induced diabetic rats. Journal of Sports Science and Medicine. 2010;9(2):294-9.

[131] Kutlu M, Naziroglu M, Simsek H, Yilmaz T, Kukner AA. Moderate exercise combined with dietary vitamins C and E counteracts oxidative stress in the kidney and lens of streptozotocin-induced diabetic rat. International Journal for Vitamin and Nutrition Research. 2005;75(1):71-80.

[132] Turner RC, Holman RR, Cull CA, Stratton IM, Matthews DR, Frighi V, et al. Intensive blood-glucose control with sulphonylureas or insulin compared with conventional treatment and risk of complications in patients with type 2 diabetes (UKPDS 33). Lancet. 1998;352(9131):837-53.

[133] Keenan HA, Costacou T, Sun JK, Doria A, Cavellerano J, Coney J, et al. Clinical factors associated with resistance to microvascular complications in diabetic patients of extreme disease duration - The 50-year medalist study. Diabetes Care. 2007;30(8):1995-7.

[134] Fong DS, Aiello LP, Ferris FL, Klein R. Diabetic retinopathy. Diabetes Care. 2004;27(10):2540-53.

[135] Ziegler D, Sohr CGH, Nourooz-Zadeh J. Oxidative stress and antioxidant defense in relation to the severity of diabetic polyneuropathy and cardiovascular autonomic neuropathy. Diabetes Care. 2004;27(9):2178-83.

[136] Low PA, Nickander KK. Oxygen free-radical effects in sciatic-nerve in experimental diabetes. Diabetes. 1991;40(7):873-7.

[137] Stevens MJ, Lattimer SA, Kamijo M, Vanhuysen C, Sima AAF, Greene DA. Osmotically-induced nerve taurine depletion and the compatible osmolyte hypothesis in experimental diabetic neuropathy in the rat. Diabetologia. 1993;36(7):608-14.

[138] Hermenegildo C, Raya A, Roma J, Romero FJ. Decreased glutathione-peroxidase activity in sciatic-nerve of alloxan-induced diabetic mice and its correlation with blood-glucose levels. Neurochemical Research. 1993;18(8):893-6.

[139] Nickander KK, mcphee BR, Low PA, Tritschler H. Alpha-lipoic acid: Antioxidant potency against lipid peroxidation of neural tissues in vitro and implications for diabetic neuropathy. Free Radical Biology and Medicine. 1996;21(5):631-9.

[140] Low PA, Nickander KK, Tritschler HJ. The roles of oxidative stress and antioxidant treatment in experimental diabetic neuropathy. Diabetes. 1997;46:S38-S42.

[141] Ozkaya YG, Agar A, Hacioglu G, Yargicoglu P. Exercise improves visual deficits tested by visual evoked potentials in streptozotocin-induced diabetic rats. Tohoku Journal of Experimental Medicine. 2007;213(4):313-21.

[142] Ozkaya YG, Agar A, Hacioglu G, Yargicoglu P, Abidin I, Senturk UK. Training induced alterations of visual evoked potentials are not related to body temperature. International Journal of Sports Medicine. 2003;24(5):359-62.

[143] Kumar JSS, Menon VP. Effect of diabetes on levels of lipid peroxides and glycolipids in rat-brain. Metabolism-Clinical and Experimental. 1993;42(11):1435-9.

[144] Aguiar, A.S. & Pinho, R.A. (2007). Effects of physical exercise over the redox brain state. The Revista Brasileira de Medicina do Esporte. 2007;13(5): 355-360.

[145] Floyd RA, Carney JM. Age influence on oxidative events during brain ischemia reperfusion. Archives of Gerontology and Geriatrics. 1991;12(2-3):155-77.

[146] Maiese K, Chong ZZ, Shang YC. Mechanistic insights into diabetes mellitus and oxidative stress. Current Medicinal Chemistry. 2007;14(16):1729-38.

[147] Alipour M, Salehi I & Ghadiri Soufi F. (2012). Effect of Exercise on Diabetes-Induced Oxidative Stress in the Rat Hippocampus. Iran Red Crescent Medical Journal. 2012;14(4): 222-228.

[148] Kucukatay V, Hacioglu G, Ozkaya G, Agar A, Yargicoglu P. The effect of diabetes mellitus on active avoidance learning in rats: The role of nitric oxide. Medical Science Monitor. 2009;15(3):BR88-BR93.

[149] Tahirovic I, Sofic E, Sapcanin A, Gavrankapetanovic I, Bach-Rojecky L, Salkovic-Petrisic M, et al. Brain antioxidant capacity in rat models of betacytotoxic-induced experimental sporadic Alzheimer's disease and diabetes mellitus. Journal of Neural Transmission-Supplement. 2007;(72):235-40.

[150] Ozkaya GY, Aydin H, Toraman NF, Kizilay F, Ozdemir O, Cetinkaya V. Effect of strength and endurance training on cognition in older people. Journal of Sports Science and Medicine. 2005;4(3):300-13.

[151] Cetin E, Top EC, Sahin G, Ozkaya YG, Aydin H, Toraman F. Effect of vitamin e supplementation with exercise on cognitive functions and total antioxidant capacity in older people. Journal of Nutrition Health & Aging. 2010;14(9):763-9.

[152] Gomez-Pinilla F, Ying Z, Roy RR, Molteni R, Edgerton VR. Voluntary exercise induces a BDNF-mediated mechanism that promotes neuroplasticity. Journal of Neurophysiology. 2002;88(5):2187-95.

[153] Rosa EF, Takahashi S, Aboulafia J, Nouailhetas VLA, Oliveira MGM. Oxidative stress induced by intense and exhaustive exercise impairs murine cognitive function. Journal of Neurophysiology. 2007;98(3):1820-6.

Chapter 3

Reactive Oxygen Species Act as Signaling Molecules in Liver Carcinogenesis

María Cristina Carrillo, María de Luján Alvarez, Juan Pablo Parody, Ariel Darío Quiroga and María Paula Ceballos

Additional information is available at the end of the chapter

1. Introduction

Reactive Oxygen Species (ROS) were viewed as the "bad" molecules of cells for a long time, but in the recent years, several lines of evidence indicate the contrary: ROS are essential participants in cell signaling and regulation depending on their concentration.

At present it is well established that ROS signaling is an important factor of many gene- and enzyme-catalyzed processes. ROS signaling is responsible for activation or inhibition of numerous processes catalyzed by protein kinases, phosphatases, and many other enzymes although these reactions proceed by heterolytic (non-free radical) mechanisms [1]. Therefore, ROS signaling can initiate both inhibition and activation of tumor formation. This fact might be of utmost importance for the development of anticancer treatment by the drugs possessing both prooxidant and antioxidant properties.

In this chapter, we summarize a series of experiments that have allowed us to establish the role of oxidative stress in the early development of liver cancer process and the effects of cytokines on the modulation of this process.

Through a series of *in vivo* and *in vitro* experiments we are able to describe:

- The oxidative stress status of a preneoplastic liver
- The modulating effect of Interferon α-2b (IFN α-2b) on this oxidative status that triggers the apoptotic mechanism in hepatic cells
- The role of TGFβ1 in the whole process
- The participation of FOXO transcription family proteins in the programmed cell death activated by IFN α-2b and TGFβ1

2. Experimental models of liver cancer development

Hepatocellular carcinoma (HCC) is a malignant solid tumor that arises from the major cell type in the liver: the hepatocyte. HCC is the most common type of primary hepatic tumor; it represents approximately 6% of all malignancies and is the fifth most common tumor worldwide [2].

Nearly all types of primary liver tumors known to occur in humans can be reproduced by chemicals in laboratory animals, especially in rats [3]. In experimental carcinogenesis, preneoplastic foci of altered hepatocytes (AHF) emerge weeks or months before the appearance of hepatocellular adenomas and HCCs [4,5] and this has also been discovered in human with hepatocellular neoplasms and/or cirrhosis [6]. This fact has led to the development of a number of *in vivo* systems for the study of early neoplasia in rat liver [7,8]. The initiation-promotion or two-stage model of cancer development mimics the early events of the latent period of human carcinogenesis. Several two stages models have been developed, including the protocols of Solt and Farber [9], Ito *et al.* [10] and Rao *et al.* [11], that comprise necrogenic doses of carcinogens or other models such as the protocols of Peraino et al. [12] and Pitot et al. [13] that use low, non toxic doses of carcinogens.

In this context, the initiation stage of cancer development can be produced in rat liver by the administration of diethylnitrosamine (DEN) [9–11], a complete carcinogen that produces DNA ethylation and mutagenesis [13]. Necrogenic doses of DEN cause massive hepatic necrosis followed by regeneration [14] and would be expected to cause not only increased gene expression related to regeneration, but also increased expression related to oncogene mutation. Administration of promoting agents causes selective enhancement of the proliferation of initiated cell populations over non-initiated cells in the target tissue [5].

Accordingly, we have developed a two-phase model of liver preneoplasia in rat: basically, the animals are initiated with two necrogenic doses of DEN and subsequently 2-acetylaminofluorene (2-AAF) is administered as promoting agent. The experimental protocol takes six weeks, and at the end of the treatment animals show 5% of liver tissue occupied by microscopic preneoplastic foci. A diagram of the experimental model is shown in Figure 1.

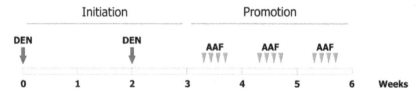

Figure 1. Two-phase or initiation-promotion (IP) model of rat chemical hepatocarcinogenesis.
Initiation stage is performed by the administration of 2 necrogenic doses of diethylnitrosamine (DEN, 150 mg / kg body weight, intraperitoneal), separated by 2 weeks. A week after the last injection of DEN, the promotion phase begins by the administration of 2-acetylaminofluorene (2-AAF, 20 mg / kg body weight) by gavage, 4 days per week during 3 weeks. At the end of the sixth week, rat livers show microscopic preneoplastic foci.

The presence of preneoplastic foci in this two-phase (initiation-promotion, IP) hepatocarcinogenic model was determined using rat Pi class isoenzyme of glutathione S-transferase (GST) as a foci marker [15]. This isoenzyme has been described as the most effective single marker of hepatic preneoplasia in the rat [16], and immunohistochemical detection of Pi class GST is the most widely used method for identification, quantitation and assessment of rat AHF [17].

3. GSTs and liver preneoplasia: Our first studies

GSTs are a family of multifunctional dimeric enzymes with an important role in detoxification processes of several xenobiotics, including anticancer drugs, carcinogens and mutagens [18–20]. These enzymes catalyze the nucleophilic attack of reduced glutathione (GSH) on electrophilic compounds [19,21].

Thus, GSTs are part of a cellular defense system which also includes GSH levels (and enzymes related to its biosynthesis) and proteins involved in the uptake of drugs and in the excretion of glutathione conjugates [22]. In the liver, among the several cytosolic classes of GSTs, Pi class GST (GST P), is particularly interesting because its expression in the adult tissue is associated with preneoplastic and neoplastic development [23]. In addition, increased expression of GST P was found to be associated with resistance of tumor tissues to several cytostatic drugs [24,25].

There is a significant increase of GST P in preneoplastic livers. This enzyme has shown to be the more efficient isoenzyme in the catalysis of conjugation of ethacrynic acid (EA) with GSH. How does this enzyme act in the preneoplastic condition?

EA, an electrophilic loop diuretic drug, causes hepatotoxicity through lipid peroxidation mediated by its oxidative metabolism [26,27]. This drug has a preferential conjugation with GSH either spontaneous or GST catalyzed, reducing its intracellular levels and consequently favoring oxidative stress in isolated hepatocytes [27]. The glutathione conjugate of EA (EA-SG) is a substrate of human multidrug-resistance protein 2 (MRP2) and probably of rat Mrp2 [28]. Thus, it has been suggested that EA-SG is excreted through this active canalicular transport protein into bile [29]. In addition, EA and EA-SG (as well as many others α,α-unsaturated carbonyl derivatives and their glutathione conjugates) are important *in vivo* and *in vitro* inhibitors of several human and rat GSTs activities [19,30,31].

As was stated above, at the inactivation step, GSTs are playing a major role by catalyzing the conjugation reaction of GSH with the drug and leading to the inactivation of the therapeutic agent. EA and EA-SG have been proved to be good inhibitors of GSTs activities [19,30,31]. For this reason, we evaluated the enzymatic and cellular *in vitro* response to EA in isolated hepatocytes from preneoplastic rat livers, which present high levels of GST P, and analyzed the role of the GSTs/GSH system and Mrp2 (as a measure of the multidrug resistance) in these cells [15].

Results showed that hepatocytes from IP animals presented higher levels of cell viability than control hepatocytes in the presence of EA. In accordance with this data, IP hepatocytes showed lower levels of release of lactate dehydrogenase and alanine aminotransferase than control cells throughout the incubation time, indicating higher levels of cellular integrity. This suggests that hepatocytes from IP animals are more resistant to the cytotoxic effects of EA than control cells.

Control hepatocytes suspensions showed increased levels of lipid peroxidation measured through the quantification of TBARS (Thiobarbituric Acid Reactive Substances, [32]) production in a time- and dose-dependent manner in the presence of EA. This data was expected since oxidative metabolism of the drug and a subsequent lipid peroxidation was already described as part of the EA-induced toxicity [27]. However, IP hepatocytes suspensions did not show increased levels of lipid peroxidation during incubation at all times and EA-doses evaluated.

The higher basal levels of TBARS observed in preneoplastic hepatocytes could be attributed to the hepatocarcinogenic treatment, since it was described that lipid hydroperoxides are produced by some promotion regimens [33]. The unchanged levels during EA incubation are probably a consequence of both GST P activities: lipid peroxidase [34] and catalysis of EA conjugation with GSH, avoiding its oxidative metabolism.

Although intracellular total GSH (tGSH) levels decreased in both hepatocytes suspensions without EA, initial tGSH levels showed a mild although significantly higher value in hepatocytes from IP animals. This fact suggests that the small extra pool of tGSH is certainly an advantageous factor to prevent susceptibility to oxidative stress.

GST P has shown to be the more efficient isoenzyme in the catalysis of conjugation of EA with GSH [35,36] and may have a role in the detoxification of lipid hydroperoxides through its selenium-independent peroxidase activity [34]. We observed decreased levels of both Alpha and Mu class GSTs in preneoplastic hepatocytes. This fact, and the presence of GST P in hepatocytes from IP rats, gives to this induced isoenzyme a compensatory role in these cells. Based on the measurement of total GST activity and in data obtained from other publications [37,38], we have proposed that GST P could be playing a major role in the defense system against the cytotoxic effects of EA in our preneoplastic model. At high EA concentrations this resistance is overwhelmed over time, probably as a consequence of inhibition of GST P activity by EA-SG acummulation and depleted levels of intracellular tGSH. In the absence of GST P and GSH, EA may alkylate cell proteins thiols, which might be the major determinants of the cytotoxic effects observed with higher EA concentrations.

It has been demonstrated that MRP2 expression, the canalicular membrane protein reported to be the transporter of GSH and GSH conjugates, is higher in human HCCs than in normal cells [39]. MRP2 increased expression could suggest accelerated GSH depletion and hence, enhanced toxicity of cytotoxic compounds. On the other hand, diminution of MRP2 expression could indicate a preservation of GSH intracellular pool. In spite of the data demonstrated in human HCC, in our preneoplastic model, we observed a 75–85% decrease in the expression of Mrp2 in freshly isolated hepatocytes compared to control cells. Thus, for this reason, reduced levels of Mrp2 in preneoplastic liver cells could contribute to the preservation of intracellular GSH and would result, in addition, in an accumulation of EA-SG and a consequent inhibition of GSTs activities suppressing more GSH consumption.

We also saw increased levels of Mrp2 in intracellular microsomal membrane fraction in a EA-dose dependent manner in both control and IP hepatocytes. This internalization phenomenon was already described [40] and could imply a process which takes place as a feedback mechanism under GSH-consumption conditions at the highest dose of EA. Our results showed that the rate of internalization of Mrp2 with increasing doses of EA was markedly higher in IP hepatocytes, although the initial basal values were significantly lower. To our knowledge, this was the first study evaluating this accelerated internalization process in isolated hepatocytes from preneoplastic rat livers.

In conclusion, hepatocytes of IP rats showed an intrinsic resistance to the cytotoxic effects of low doses of EA and it seems likely that the presence of GST P, the higher levels of GSH, and the lower expression of Mrp2 in the cellular membrane are closely related to this phenotype.

4. Interferon α-2b gets into scene

Human lymphoblastoid IFN α has been shown to have a powerful antiproliferative effect on human hepatoma cell line PLC/PRF/5 in a dose-dependent manner, both *in vitro* and *in vivo*, after implantation in nude mice. Moreover, IFN α inhibits liver regeneration by decreasing DNA and total protein synthesis [41,42].

Considerable expectations in reducing the incidence of HCC were connected with the use of IFN α in antiviral treatment of hepatitis B or C. By now, clinical trials have indeed confirmed a reduced incidence of HCC in IFN α–treated patients with chronic hepatitis B or C [43,44]. In contrast, the benefit derived from IFN α treatment of established HCC remains controversial [45,46]. It is important to deepen the understanding of the action of IFN α on HCC cells, because some patients with hepatitis B– or hepatitis C–related liver diseases may already have small, clinically undetectable preneoplastic foci during IFN α therapy. Experimental studies have shown that IFN α exerts its antiproliferative effects against HCC cell lines *in vitro* by inducing apoptosis and inhibiting cell-cycle progression [47–49]. However, the sensitivity of early-stage HCC to IFN α could not be estimated from the sensitivity of the cell lines that have a larger number of gene abnormalities and higher proliferation capability, whereas the activity of IFN is expected to be minimal [50]. However, it was unknown whether IFN α prevents *in vivo* oncogenesis by expressing these effects in the very-early-stage, clinically undetectable cancer cells.

In this context, we have demonstrated that administration of IFN α-2b during the development of rat liver preneoplasia significantly decreased both number and volume percentage of GST P–positive foci [14]. Particularly, these reductions where observed when IFN α-2b was administered during the initiation phase or during the entire experimental protocol. However, when IFN α-2b was administered during the promotion phase no effect on these parameters could be observed. Thus, the use of IFN α-2b as an antitumor agent was lost when it was administered only at the 2-AAF phase. Nevertheless, we cannot discount that the lack of IFN α-2b effect during the 2-AAF phase reflects some interaction between 2-AAF and the cytokine. Administration of IFN α-2b during the initiation stage seems to be essential to exert inhibitory effects against DEN-initiated hepatic carcinogenesis in the rat.

Contrary to our expectations, the proliferation index (measured by immunohistochemical detection of proliferating cell nuclear antigen or PCNA) in preneoplastic foci was not reduced by treatment with IFN α-2b. On the other hand, the apoptotic index (measured by TUNEL technique) in AHF was significantly increased in the groups that received IFN α-2b. The number of apoptotic cells and bodies in AHF after treatment with IFN α-2b was higher than for control rats. Then, the reduction of both number and volume percentage of AHF in IFN α-2b–treated animals is explained by a greater programmed cell death within the foci.

In regard to the effects of IFN α on the cell cycle progression of various normal and tumor cell lines, most studies have observed inhibitory effects on G1 to S phase transition [51,52]; other studies have shown S phase accumulation in response to treatment with IFN α [48,49]. In our studies, the animals with liver preneoplasia that were treated with IFN α-2b showed a diminution in the percentage of preneoplastic hepatocytes in S phase and an accumulation in the G1 phase. Although apoptosis may be initiated in any phase of the cell cycle, most cells undergo apoptosis primarily in the G1 phase of cycling cells, and there is a positive relationship between apoptosis and cell proliferation [53]. This relationship is explained by the presence of many cell cycle regulators/apoptosis inducers such as p53, which operates at the G1/S checkpoint [54].

In this connection, we examined whether p53 and 3 members of the Bcl-2 family (Bax, Bcl-2, and Bcl-xL), which are important regulators of apoptosis [53] were involved in IFN α-2b–mediated programmed cell death. It is known that p53 down-regulates Bcl-2 [55] and up-regulates Bax genes [56]. The role of the Bcl-2 family in IFN α–induced apoptosis still remains controversial. For example, IFN α–induced apoptosis in cells of hematopoietic and hepatic origins can occur without involvement of the Bcl-2 family [48,57] whereas transfection of IFN α–sensitive cell lines with a Bcl-2 expression vector conferred partial resistance to cell death mediated by IFN α [58]. Our results showed that members of the Bcl-2 family were involved in the apoptotic elimination of preneoplastic hepatocytes after treatment with IFN α-2b. Specifically, treatment with IFN α-2b increased levels of the proapoptotic protein Bax, in parallel with increases of p53 protein levels. In addition, there were decreases in the levels of Bcl-2 and Bcl-xL proteins, which are known to promote cell survival through homodimerization. Bax protein promotes cell death via homodimerization, whereas heterodimerization with either Bcl-2 or Bcl-xL results in cell survival [59,60]. The relative prevalence of Bax and Bcl-xL protein are critical factors influencing cell fate, promoting either survival or death, whose ultimate outcome largely depends on the Bax/Bcl-xL ratio. Thus, apoptosis pathways can be activated under conditions in which Bax protein expression is elevated and/or Bcl-xL protein expression is decreased.

We also observed increased Bax protein translocation into the mitochondria in the animals that received IFN α-2b. It has been established that subcellular localization of Bax protein is an important regulator of apoptosis. Bax is localized in the cytoplasm and translocates to the mitochondria at the early stage of apoptosis. Bax mediates its proapoptotic effects through a channel-forming activity of the mitochondrial membrane, resulting in disruption of mitochondrial function, release of cytochrome c, and apoptosis [61].

In brief, our experimental observations led us conclude that preneoplastic hepatocytes in the IFN α-2b–treated rats are "primed" for apoptosis and undergo programmed cell death as a primary result of a substantial increase in the level of mitochondrial Bax protein, producing a further increase in the Bax/Bcl-xL protein ratio.

5. Has TGFβ1 any role in this scenario?

Given its antiproliferative, proapoptotic role in the liver, TGFβ1 could be expected to act as a tumor suppressor. However, various types of neoplastic liver cells respond quite differently to TGFβ1. Whereas some human and rat hepatoma cell lines are sensitive to TGFβ1 [62–64], resistance has been reported for other hepatoma cells [64,65]. In addition, TGFβ1 overexpression seems to be a hallmark of human liver cancer [66]. Thus, the relationship between TGFβ1 and cancer is complex: TGFβ1 may stimulate malignant progression itself; conversely, it can have tumor suppressor activity [67]. The escape of certain hepatoma cells from TGFβ1–induced apoptosis seems to be an important and essential step in malignant progression [68,69]. Moreover, it has been suggested that TGFβ1 overexpression is a late event in human hepatocarcinogenesis [66]. These data indicate that loss of TGFβ1 responsiveness is not an initiating or strongly predisposing event, but rather a late event in carcinogenesis [67,70].

Therefore, it was of interest to study if liver preneoplasia as an early stage of cancer development is still sensitive toward TGFβ1 actions.

Given that the changes of pro- and anti-apoptotic proteins induced by IFN α-2b in rats with liver preneoplasia were similar to those attributed to TGFβ1 in other experimental models [62,63,71], we studied the possibility that TGFβ1 could be involved in the programmed cell death induced by IFN α-2b [72]. Primary, we observed that serum TGFβ1 levels in the animals treated with IFN α-2b were significantly increased. In accordance with this, immunohistochemical studies showed that IFN α-2b treatment significantly augmented the quantity of TGFβ1–positive hepatocytes in preneoplastic livers. At first sight, these findings seemed to indicate that administration of IFN α-2b increased serum TGFβ1 production and the number of TGFβ1–positive hepatocytes. Although the mechanisms by which IFN α-2b treatment induced TGFβ1 in the preneoplastic livers were not completely explored, we observed, using Western blot analysis, that preneoplastic livers expressed higher levels of IFN α receptors than control livers. In addition, IFN α-2b administration in animals subjected to the preneoplastic protocol induced elevated levels of phosphorylated Stat1, indicating activation of the IFN α pathway.

Recent investigations have reported that the induction of apoptosis by endogenous TGFβ1 does not require an overall increase in its hepatic concentration [70]. In view of the fact that TGFβ1 hepatic content may not reflect the induction of apoptosis by this cytokine, we determined the nuclear content of p-Smads-2/3 (critical intracellular transducers of TGFβ1 signaling). We observed high levels of p-Smads-2/3 proteins in the nuclear extracts of IFN α-2b–treated animals. These results corresponded with the increased number of TGFβ1–positive hepatocytes, indicating increased TGFβ1 activation in rats with liver preneoplasia that received IFN α-2b.

Nonparenchymal cells, including Kupffer cells and peritoneal macrophages, are the main source of hepatic TGFβ1 [73,74]. Hepatocytes, however, may synthesize TGFβ1 *in vitro* [75] as well as during hepatocarcinogenesis [66]. During liver preneoplasia, neither peritoneal macrophages nor Kupffer cells secreted detectable levels of TGFβ1 when they were stimulated with IFN α-2b. Conversely, hepatocytes from normal, untreated livers did not secrete TGFβ1 in the absence or presence of IFN α-2b. Nevertheless, hepatocytes from preneoplastic livers produced and secreted detectable levels of TGFβ1 when they were cultured without IFN α-2b stimulus, and IFN α-2b presence in the culture media induced several-fold increases of TGFβ1 production.

In vitro studies with isolated hepatocytes have allowed us to demonstrate that IFN α-2b induces apoptosis in hepatocytes from preneoplastic livers, measured by fluorescence microscopy and caspase-3 activity. These cells also had higher nuclear accumulation of p-Smads-2/3, indicating increased TGFβ1 activation. When anti–TGFβ1 was added to the culture media, TGFβ1 activation and apoptosis induced by IFN α-2b were completely blocked. Therefore, the apoptotic effect of IFN α-2b is mediated by the production of TGFβ1 from hepatocytes.

Thus, our work determined for the first time that endogenous TGFβ1 is implicated in the increased apoptosis into the AHF of IFN α-2b-treated rats. Taken together, these data clearly showed that TGFβ1, which is produced and secreted by hepatocytes from preneoplastic liver under IFN α-2b treatment, stimulates hepatocytes apoptotic cell death in an autocrine/paracrine fashion. This postulated mode of action is in agreement with data published previously [70,76,77]. The reduction of preneoplastic foci by endogenous TGFβ1 early in the carcinogenesis process would likewise protect against tumor formation.

6. Participation of ROS

In a new series of *in vitro* experiments, we proved that IFN α-2b induces the production of TGFβ1 in hepatocytes from preneoplastic livers by activation of NADPH oxidase complex (superoxide-producing enzyme consisting of membrane (gp91phox and p22phox) and cytosolic (p47phox, p67phox, and p40phox) components [78]), and TGFβ1 induces apoptosis through a mechanism linked to the production of ROS by the same oxidase [79]. In order to confirm that the induction of NADPH oxidase activity was the main pathway producing ROS, additional experiments were made using IFN α-2b plus an inhibitor of NADPH oxidase activity, diphenyleneiodonium (DPI). Presence of DPI in the culture media totally blocked the activity of NADPH oxidase, the production of ROS and the subsequent apoptosis induced by IFN α-2b.

ROS production induced by IFN α-2b showed a singular pattern of two peaks: one peak in ROS generation at 1 hour of culture, and another peak at 9 hours. The addition of anti-TGFβ1 to the culture media did not block the production of the first peak of ROS whereas totally blocked the appearance of the second one. On the other hand, when ASC was added to the culture media the production of both peaks was abolished. Based on these findings, the postulated mechanism by which ROS act as signaling molecules in liver preneoplasia is as follow: IFN α-2b induces, via NADPH oxidase activation, an early ROS production that serves as a messenger, promoting TGFβ1 production and secretion. This growth factor triggers the production of more reactive oxygen intermediates, as a late event, by inducing the same enzyme complex. It was demonstrated that synthesis of new protein is required for NADPH activation and subsequent apoptosis [80]. This event shows an additive response in ROS production and imposes the final onset of the apoptotic effect. The presence of ASC in the culture media totally blocked the increase in the activity of the NADPH oxidase complex, ROS production and the final apoptotic effect induced by IFN α-2b.

Once the source of ROS was assessed, we analyzed the cellular antioxidant defenses and their behavior during the studied times. We observed a reduction in tGSH levels from 7 hours of culture onwards. For that reason we studied if any form of glutathione was being exported out of the cell, and whether the biosynthetic GSH capacity was altered. We found an increase in oxidize glutathione (GSSG) levels probably due to the oxidation of the reduced form within the cytosol, and its exportation to the culture media, possibly in order to protect cells from a shift in the redox equilibrium. IFN α-2b treatment resulted in the loss of GSH biosynthetic capacity since glutamate cysteine ligase (GCL) activity was decreased at 7 hours of culture and a rapid decrease of the mRNA expression of the catalytic subunit of GLC (GCLC) through a mechanism mediated by TGFβ1 was also observed. Moreover, it was found that IFN α-2b-induced apoptosis in hepatocytes from rat preneoplastic livers is accompanied by the cleavage and loss of GCLC protein, through a mechanism mediated by TGFβ1.

A decrease in the antioxidant enzymes catalase (CAT) and superoxide dismutase (SOD) activities was observed when hepatocytes were treated with IFN α-2b. On the other hand, treatment with anti-TGFβ1 or ASC totally blocked the decrease in CAT and SOD enzymatic activities. These findings indicate that IFN α-2b induced the decrease in enzymatic CAT and SOD activities by a mechanism mediated by ROS and TGFβ1. These enzymes probably protect hepatocytes from the initial IFN α-2b-induced burst of ROS and this may be the reason for the rapid decrease of the first peak of ROS.

These results confirmed that the perturbation of the redox status produced by the IFN α-2b induction of NADPH oxidase complex triggered TGFβ1 synthesis and secretion and assessed the downregulation of antioxidative systems. Similar data have been reported by Herrera et al. [80] when they treated fetal rat hepatocytes with TGFβ1.

Since ASC abolished all the apoptotic effects induced *in vitro* by IFN α-2b, we determined the relevance of ROS on the onset of the apoptotic process *in vivo*, in the whole preneoplastic liver. IFN α-2b plus ASC treatment of rats with liver preneoplasia abrogated the apoptotic effect induced by IFN α-2b, leading to no reduction on size/number of foci. Interestingly, foci volume was almost twice higher in the animals that received IFN α-2b plus ASC than in IFN α-2b-treated rats. This result highlights the importance of ROS signaling during the beneficial effects of IFN α-2b treatment of hepatic preneoplasia. In this regard, it was found that ASC at low concentrations stimulates growth of malignant cells [81], while inhibits their growth at high doses [82]. At the present time, many cancer patients combine some forms of complementary and alternative medicine therapies with their conventional therapies. The most common choice of these therapies is the use of antioxidants such as vitamin C. It must be assumed that any antioxidant, used to reduce toxicity of tumor therapy on healthy tissue, has the potential to decrease effectiveness of cancer therapy on malignant cells [83]. Some data suggest that antioxidants can ameliorate toxic side effects of therapy without affecting treatment efficacy, whereas other data suggest that antioxidants interfere with radiotherapy or chemotherapy [83].

In summary, we demonstrated that increase in ROS levels turns on the process of programmed hepatocytes death, leading to the elimination of these malignant cells. The inhibition of ROS production with an antioxidant such as ASC in the co-treatment with IFN α-2b may be not a beneficial therapy for the prevention of preneoplastic foci.

7. Is p38 MAPK implied in the process?

p38 MAPK pathway has been implicated in a wide range of cellular functions. However, it is now well established that p38 MAPK activation and its role depends on the cellular context, on the specific stimuli, and on the specific p38 MAPK activated isoform [84]. There are controversies about the role of p38 MAPK in apoptosis. It has been shown that p38 MAPK signaling promotes cell death [85,86], whereas it has also been shown that p38 MAPK cascades enhance survival [87,88], cell growth [89], and differentiation [90]. Furthermore, it has been reported that p38 MAPK participates on the estradiol-mediated inhibition of apoptosis in endothelial cells [91], while participates on the apoptosis induced by thrombospondin-1 [92], or by high leves of D-glucose in the same cells [93]. It is believed that p38 MAPK mediates its apoptotic effects through the phosphorylation of proteins of the apoptotic pathways [94].

Previous reports in hematopoietic cells have shown that IFN α and TGFβ1 play their growth inhibitory effects through activation of the p38 MAPK pathway via phosphorylation (activated p38 MAPK or p-p38 MAPK) [95]. However, these effects are primarily ascribed to G1 cell cycle arrest and not to induction of apoptosis. Others have suggested that during the TGFβ1-induced apoptosis in fetal rat hepatocytes, ROS activates p38 MAPK not by induction of apoptosis, but mediating ROS regulation of TGFβ1-gene expression [96]. On the other hand, it was demonstrated that inactivation of p38 MAPK pathway in cultured mice fibroblasts promotes tumor development [97]. Moreover, it was demonstrated that treatment with an inhibitor of p38 MAPK activation, induced carcinogenesis in mice resistant to tumor development, indicating the leading role of p38 MAPK in the regulation of tumor growth [98].

Using *in vivo* studies we could demonstrate that rats subjected to a 2-phase model of chemical hepatocarcinogenesis have less hepatic p38 MAPK activation than control rats, determined as p-p38 MAPK levels [79]. This is in agreement with Honmo et al. [99] that showed that 2-AAF administration induces a decrement of p38 MAPK activation promoting tumor development.

Another important finding of the *in vivo* studies was the effect of IFN α-2b on the activation of p38 MAPK in rat preneoplastic livers. Preneoplastic animals treated with IFN α-2b showed similar p-p38 MAPK levels to those in controls. In this connection, cultured hepatocytes from preneoplastic livers treated with IFN α-2b plus SB-203580 (inhibitor of α and β isoforms of p38 MAPK), totally blocked the IFN α-2b-induced apoptosis. It is clear that activation of p38 MAPK pathway plays a key role in promoting apoptosis after IFN α-2b treatment in our model of experimental preneoplasia. It was previously reported that IFN α suppresses the growth of leukemia cell progenitors through activation of p38 MAPK, which leads to cell cycle arrest in different phases [100].

We demonstrated that IFN α-2b induces an early production of ROS (first peak), in hepatocytes from preneoplastic livers. Then, ROS stimulate the production and secretion of TGFβ1 from hepatocytes, which in turn, generates a new burst of ROS (second peak). These oxygen radicals act as signaling mediators of the onset of the IFN α-2b-induced apoptosis.

Activation of p38 MAPK after IFN α-2b stimulus occurred preceding each increment in ROS generation and so, the particular pattern of two peaks was also functioning for p38 MAPK activation. Interestingly, treatment with ASC was able to block only the second peak, indicating that early activation of the pathway was independent of ROS, while late activation depended on ROS produced by endogenous TGF-β1. Treatment with anti-TGFβ1 completely blocked the second p38 MAPK, demonstrating that TGF β1 induces activation of p38 MAPK through ROS, as previously reported in fetal rat hepatocytes [96].

Another relevant issue is the activation of transcription factors by p38 MAPK. Cell signaling pathway activation could be transmitted to the nucleus in different ways, depending on the stimulus. To assess whether activation of p38 MAPK transmitted the IFN α-2b stimulus to the nucleus, we analysed phosphorylation status of specific p38 MAPK transcription factors CREB/ATF-1 and ATF-2. Our findings documented that early p38 MAPK activation under IFN α-2b stimulus mainly activates the transcription of ATF-2-regulated genes, whereas the late signal of p38 MAPK activation is transmitted to the nucleus mainly by the phosphorylation of CREB/ATF-1. Moreover, it can be also inferred that early phosphorylation of ATF-2 may be dependent on activation of p38 MAPK by IFN α-2b, while late phosphorylation of CREB/ATF-1 may be dependent on activation of p38 MAPK by TGFβ1.

8. Relationship between p38 and NADPH oxidase

We inferred that p38 MAPK activation is essential for NADPH oxidase to function in preneoplastic hepatocytes treated with IFN α-2b, because the presence of p38 MAPK inhibitor SB-203580 totally blocked the activation of the enzyme [101]. Cytosolic component of NADPH oxidase complex, p47phox got phosphorylated following the same pattern as p38 MAPK induction and ROS generation: an early, first increment and a late, second increase. The first increase of p47phox phosphorylation by IFN α-2b was independent of ROS, since ASC did not block such phosphorylation. However, it was dependent of p38 MAPK activation, since it was blocked by SB203580. This is a very interesting finding since it suggests that p-p38 MAPK phosphorylates p47phox, initiating the activation of NADPH oxidase in cells from preneoplastic livers. Analysis at higher times demonstrated that late phosphorylation of p47phox was completely blocked by anti-TGFβ1 or ASC, evidencing the participation of TGF β1 and ROS in this process. Studies of p47phox translocation from cytosol to plasma membrane were consistent with the phosphorylation findings.

It is clear that in liver preneoplasia there is a positive cross-talk between IFN α-2b, TGFβ1 and p38 MAPK pathways. Taken altogether, evidence indicates that p38 MAPK pathway plays a critical role in the generation of the suppressive effects of IFN α-2b, as well as TGFβ1 in the very early stages of hepatic neoplasia. There is strong indication that this signaling cascade acts as a converging signaling point for signaling pathways activated by different cytokines to mediate apoptotic or suppressive signals. These findings may have important clinical implications, as improving the pharmacological development of better drugs for the prevention and treatment of hepatic illness such as cancer.

9. How are IFN α and TGFβ1 signaling pathways connected?

Interactions between TGFβ and other cytokines signaling pathways have been extensively studied, particularly the cross-talk between TGFβ/Smad and IFN γ/Stat signaling in their antagonistic role on collagen deposition and fibrosis [102–107]. However, despite the fact that TGFβ plays a crucial role in cancer, little is known about TGFβ signaling interactions during this process. An investigation in hepatoma cells have described a cross-talk between Il-6 and TGFβ signaling [108] and another study in a melanoma cell line normally resistant to IFN α, have demonstrated that co-stimulation with IFN α and TGFβ induces antiproliferative activity [109].

As was stated above, the relationship between TGFβ and cancer is complex: it functions as a tumor suppressor in early epithelial carcinogenesis, but often becomes prooncogenic in late stages of tumor progression [110]. Autocrine TGFβ1 is known to suppress tumorigenesis and tumor progression in normal and early transformed cells, but it can also promote the survival of various cancer cells [111]. Besides, dysregulation of the downstream effectors of TGFβ has been described in late steps of promotion stage, indicating that may contribute to the progression of preneoplastic lesions [112].

We demonstrated that during liver preneoplasia TGFβ1 has a beneficial role, promoting apoptotic death of AHF. Therefore, we attempted to get more insight into the relationship between IFN α-2b and autocrine TGFβ1 in preneoplastic rat livers. Many *in vitro* cell systems are good tools to explain related actions of distinct types of cytokines in various biological signaling pathways, but they are not physiological. However, the study of IFN α-2b and TGFβ1 signals interactions in hepatocytes derived from the whole preneoplastic liver may be relevant for understanding the mechanisms operating in patients with chronic hepatitis B or C treated with IFN α-2b, who already have small, clinically undetectable preneoplastic liver foci during therapy.

The obtained results provided evidence for the integration of TGFβ1 and IFN α-2b signaling pathways during the development of liver carcinogenesis. IFN α-2b treatment of hepatocytes from preneoplastic livers produced a rapid activation of IFN α signaling, with increased p-Stat1 levels. Subsequently, autocrine TGFβ1 produced under IFN α-2b stimulus was able to induce the activation of TGFβ1/Smad signaling pathway, determined by nuclear content of p-Smad2/3 and confirmed by the use of specific TGF β1 signaling inhibitors (anti-TGFβ1 and SB-431542) [113].

A critical mechanism for regulating the cellular response to cytokines resides at the level of receptor expression. TGFβRII plays a key role in receptor activation and subsequent TGFβ1 signal propagation, functioning both to bind ligand and to activate TGFβRI. Disorders of TGFβRII expression lead to various diseases. For example, reduction of TGFβRII levels contributes to the resistance of tumor cells to TGFβ [114].

We observed that TGFβRII was up-regulated at mRNA and protein levels. This induction was mediated by autocrine TGFβ1, since it was blocked by inhibitors of TGFβ1 signaling. This is an outstanding finding, since TGFβ1-dependent regulation of TGFβRII has not been previously reported.

Inhibitory Smad7 is a key component of TGFβ1 signals. Its expression is not only induced by TGFβ, but also controlled by, for example, IFN γ [102,107]. Therefore, Smad7 is considered as a protein involved in the fine-tuning of the cellular responses to the TGFβ family by integrating various signaling pathways. However, in our model, Smad7 did not show changes in its protein levels, at least during the studied times. Furthermore, Smad7 protein levels in hepatocytes from preneoplastic livers were significantly reduced with respect to their levels in hepatocytes from normal livers. So, additional experiments of Smad7 induction by phorbol 12-myristate 13-acetate (PMA) were performed in order to evaluate if the decreased Smad7 levels showed in preneoplastic livers may contribute in TGFβ1 signaling activation. Results showed that this possibility seems unlikely; provided that Smad7 protein reached similar levels to those in normal hepatocytes, and TGFβ1 signaling continued activated. These experiments indicated that Smad7 protein is not directly related with TGFβ1 and IFN α signals interaction in hepatocytes from preneoplastic livers.

Another decisive aspect in signaling pathways relationships is the availability of certain co-activators for interacting with specific transcription factors. The cofactor p300 is an important component of the transcriptional machinery that integrates TGFβ/ IFN γ-induced signals [115].

In normal fibroblasts exposed to IFN γ and TGFβ simultaneously, activated Stat1 and activated Smad2/3 compete each other for limiting p300. IFN γ-activated Stat1 appears to sequester p300, thereby disrupting TGFβ-induced interaction of p300 with Smad2/3. Ectopic p300 rescues stimulation in the presence of IFN γ, suggesting that p300 acts as an integrator of IFN γ/Stat1 and TGFβ/Smad2/3 signals [103]. In addition, Inagaki et al. [116] have demonstrated that IFN α antagonizes TGFβ/Smad-induced hepatic fibrosis by competition between Stat1 and Smad3 for binding to p300 protein.

In our study, we found that IFN α-2b induced a direct interaction between activated Stat1 and p300 in hepatocytes from preneoplastic livers. Furthermore, activated Smad2/3 induced by autocrine TGFβ1 were able to physically associate with p300. In addition, levels of p300 in hepatocytes from preneoplastic livers were significantly higher than in normal hepatocytes. Together, these findings suggested that in hepatocytes from preneoplastic livers, the intracellular signals triggered by TGFβ1 and IFN α-2b are integrated at the nuclear level, where p-Stat1 and p-Smad2/3 are capable of interact with p300, present in no restrictive cellular amounts.

It was recently found that TGFβ signals potentiate Il-6 signaling in hepatoma cells. This cross-talk occurs by physical interactions between Stat3 and Smad3, bridged by p300 [108]. In our model of liver preneoplasia we did not observe physical interaction between Stat1 and Smad3, but it seems to be enough p300 protein available to interact with p-Stat1 on one hand, and with p-Smad2/3 on the other, leading to the activation of TGFβ1 and IFN α signaling simultaneously.

In fact, we have described for the first time a positive cross-talk between IFN α and TGFβ1 signaling.

10. Summary # 1

In these series of experiments, it was demonstrated that NADPH oxidase complex is activated when IFN α-2b binds to type I receptor. This binding produces early amounts of ROS. ROS, in turn, trigger TGFβ1 production and secretion. TGFβ1, when binding to its receptor, also induces NADPH oxidase complex activation, and, besides, decreases the antioxidant defenses of the cell. Moreover, we demonstrated that p38 MAPK activation is essential for NADPH oxidase to function.

Furthermore, ROS initiate mitochondrial apoptosis directly and/or acting by the Bcl-2 family proteins inducing a mitochondrial permeability transition pore (MPTP), releasing cytochrome c and activating caspase 3. TGFβ1 could induce, as a late event, the activation of caspase 8, which, in turn, induces a higher MPTP through activation of Bid, another Bcl-2 family member [117]. A graphic outline of these concerns is shown in Figure 2.

Altogether, our results demonstrate that the oxidative stress induced in preneoplastic liver by IFN α-2b is able to trigger the apoptotic mechanism and brings into the play another key cytokine in the cancer process: TGFβ1.

11. Targeting the Wnt/β-catenin signaling pathway

Among the growth factor signaling cascades dysregulated in HCC, evidences suggest that the Wnt/Frizzled-mediated signaling pathway plays a key role in hepatic carcinogenesis. Aberrant activation of the signaling in HCC is mostly due to dysregulated expression of the Wnt/β-catenin signaling components. This leads to the activation of the β-catenin/TCF dependent target genes, which control cell proliferation, cell cycle, apoptosis or motility. It has been shown that disruption of the Wnt/β-catenin signaling cascade displayed anti-cancer properties in HCC [118].

For this reason, we determined the status of the Wnt/β-catenin/TCF pathway in the preneoplastic stage and evaluated the possible effects of IFN α-2b on this pathway.

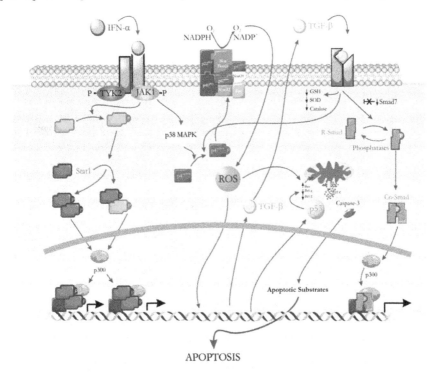

APOPTOSIS

Figure 2. Graphic outline of IFN α-2b, TGFβ1, p38 MAPK, NADPH oxidase and ROS interactions in liver preneoplasia.

The major findings of our studies were related to the impairment of the canonical Wnt/β-catenin/TCF pathway in a very early stage of hepatic carcinogenesis. In addition, we demonstrated that *in vivo* IFN α-2b treatment produces an attenuation of TCF transcriptional activity and enhances FOXO transcriptional activity in preneoplastic livers.

The common denominator of an abnormal Wnt signaling is the stabilization and accumulation of unphosphorylated β-catenin in the cytoplasm of a cell. Eventually, this allows entry of unphosphorylated β-catenin into the nucleus where it promotes the transcription of a subset of genes implicated in cellular proliferation. This β-catenin stabilization was demonstrated in our two-phase carcinogenic model, where plasma membrane delocalization and cytoplasmic accumulation of β-catenin were observed [119]. Moreover, significant reductions of phosphorylated β-catenin levels were found in IP animals. Since total β-catenin (phosphorylated and unphosphorylated) protein levels were preserved in all studied groups, these results indicate a lower phosphorylation rate of cytoplasmic β-catenin in IP rats.

We have also found up-regulation of TCF target genes Cyclin D1, MMP-7, Axin 2, and SP5 in preneoplastic livers. Up-regulation of Cyclin D1 was predicted since this protein is an important regulator of cell cycle progression, and its activity is required for G1 to S-phase transition. Overexpression of this gene has been associated with the development and progression of several cancers [120]. In addition, it has been reported that overexpression of Cyclin D1 in tumor cells contributes with their resistance to cytotoxic drugs [121]. In fact, inhibition of Cyclin D1 enhances the effects of several chemotherapeutic agents [121]. In agreement with these results, we have previously described (see *GSTs and liver preneoplasia: our first studies*) a drug-resistance phenotype in isolated hepatocytes obtained from rat preneoplastic livers. Thus, it is possible that the overexpression of Cyclin D1 could play a role in the drug-resistance phenotype of this model. MMP-7, a member of the matrix metalloproteinase family, acts as a specific proteolytic enzyme for degradation of certain components of the extracellular matrix. This protein was already shown to be important for the growth of early adenomas [122] and its function is essential in more advanced stages such as tumor progression and metastasis, where an invasive growth is a highlight of these steps [123,124]. Hence, enhanced MMP-7 expression could be proposed as an indicator of potential tumor progression, invasiveness, and metastatic ability at a very early stage of hepatocarcinogenic development. It has been reported that the tumor suppressor Axin 2 is a target of Wnt signaling [125,126]. The up-regulation of Axin 2 showed in IP rats, which is known to be a negative regulator of free β-catenin [127,128], could be an expression of a feedback preservation mechanism of the preneoplastic tissue, and might not be sufficient to prevent cytoplasmic β-catenin accumulation. SP5, a member of the SP1 transcription factor family and known target of Wnt signaling [129] was also over-expressed. This protein seems to work as a transcriptional repressor, preventing the expression of genes involved in cell cycle G1 phase arrest such as p21 [129].

In order to determine the involvement of a mutated β-catenin protein in the activation of this pathway as was described for HCC [130–133], we performed a direct sequencing of amplicons encoding a region of exon 2 of rat liver β-catenin gene. Our results demonstrated that this sequence had no deletion or point mutations in any of the studied groups.

Even with a wild-type β-catenin, the pathway can also be triggered because of alterations in other components of the cascade signaling. The Frizzled protein family acts as a seven-span transmembrane receptor for Wnt proteins. It was recently reported an up-regulation of the Frizzled-7 receptor in the presence of wild-type β-catenin in four murine transgenic models of hepatocarcinogenesis [134] and in human HCC [135] with activation of the Wnt/β-catenin/TCF pathway. Therefore, it was suggested that overexpression of Frizzled-7 could lead or contribute to activation of Wnt signaling. The obtained data showed a marked increase of this receptor in preneoplastic livers at mRNA and protein levels. Since it was reported that Frizzled-7 is also a target gene of the Wnt/β-catenin/TCF pathway [136], we presume that overexpression is rather a consequence than a cause of abnormal activation of the Wnt/β-catenin/TCF pathway.

Once we demonstrated that the Wnt/β-catenin/TCF pathway is activated in preneoplastic rat livers, we analyzed the effects of IFN α-2b treatment. Results showed that *in vivo* IFN α-2b administration did not prevent β-catenin delocalization and cytoplasmic accumulation; however, it certainly attenuates activation of the canonical Wnt/β-catenin/TCF pathway as measured by four TCF target genes. The transcription levels of these genes were similar to controls in IP animals that received IFN α-2b.

In addition, IFN α-2b-treated IP rats showed that Frizzled-7 levels remained unchanged compared to control animals. These results reinforced our hypothesis that Frizzled-7 up-regulation occurs as a result of the abnormal activation of the studied pathway.

In an attempt to get more insight into the regulation of Wnt/β-catenin/TCF pathway, FOXO transcription family has come into scene. Recent studies reported that FOXO interacts with β-catenin in a competitive manner with TCF, particularly under cellular oxidative stress conditions [137,138]. Taking this into consideration and the fact that *in vivo* IFN α-2b treatment induces endogenous ROS formation in preneoplastic livers, we analyzed interactions between β-catenin with TCF4 and FoxO3a and association of these transcription factors with their corresponding target gene promoters. Co-immunoprecipitation assays showed that β-catenin/TCF4 interaction effectively occurs in preneoplastic livers and administration of IFN α-2b not only attenuates this interaction but also promotes β-catenin/FoxO3a association. Using ChIP assay, we verified that interaction of FoxO3a with the promoter region of its target gene is enhanced in preneoplastic livers treated with IFN α-2b. On the other hand, TCF4 remains associated with SP5 gene promoter region in all studied groups. It is known that TCF4 contains a conserved domain that binds DNA irrespective of its interaction with β-catenin; however, the transcriptional activity is blocked by the presence of a family of transcriptional repressors [139,140]. TCF4 must bind β-catenin for its transactivation and this interaction was verified by co-immunoprecipitation assays. In addition, it has been demonstrated that interaction of β-catenin with FOXO enhances its transcriptional activity [137,138], so we measured the expression of p130, a FOXO target gene whose main function is related to the maintenance of cell cycle arrest. Furthermore, it was suggested that p130 may exert a proapoptotic effect on certain tumor samples [141]. We found up-regulation of p130 transcript in preneoplastic livers treated with IFN α-2b. These findings suggest that IFN α-2b treatment in preneoplastic livers decreases β-catenin/TCF interaction and consequently reduces TCF transcriptional activity probably via ROS induction. Furthermore, IFN α-2b-induced ROS production could stimulate β-catenin/FOXO interaction, thereby favoring cell cycle arrest and apoptosis. In agreement with this proposal, recent unpublished results from our group demonstrate the participation of ROS in these events.

Collectively, our data demonstrate that the canonical Wnt/β-catenin/TCF signaling pathway is activated at a very early stage of the development of the hepatocarcinogenic process, even with a wild-type β-catenin. More importantly, *in vivo* IFN α-2b treatment could be an efficient therapy to attenuate Wnt/β-catenin/TCF signaling promoting diminution of preneoplastic foci by an apoptotic process. A graphic outline of these concerns is shown in Figure 3.

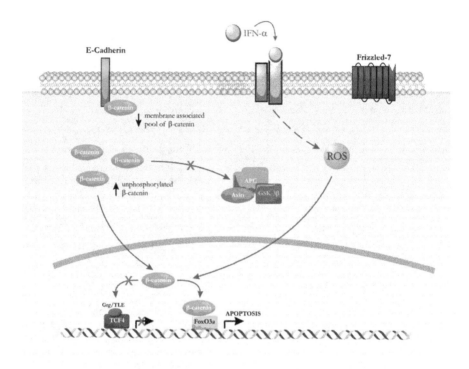

Figure 3. Graphic outline of IFN α-2b, Wnt/β-catenin pathway and ROS/FOXO interactions in liver preneoplasia.

12. Studies in HCC cell lines

The elucidation of the signals induced by IFN α and TGFβ in human liver tumor cells, and their possible cross-talks with other intracellular signals, would have relevance in the future design of therapeutic tools to balance the cellular responses in favor of liver tumor suppression. To gain mechanistic insights into these cooperative signals, we analyzed the effects of IFN α-2b and TGFβ1 on Wnt/β-catenin pathway and Smads intermediates in HepG2/C3A and Huh7 HCC cell lines. We could demonstrate that IFN α-2b or TGFβ1 stimulations not only decreased cellular proliferation but also increased apoptotic cell death [142]. The apoptotic and anti-proliferative effects of both cytokines separately have already been reported in HepG2 and Huh7 [143–145]. More interestingly, we demonstrated that the combined treatment increased these effects. Until now, combined treatment with both cytokines has only been used to analyze their impact on proliferation in human melanoma cell lines [109]. Treatments impact on Wnt/β-catenin pathway was analyzed, together with the analysis of the effects of IFN α-2b and TGFβ1 on Smads proteins. Insufficient information is available concerning TCF4/Smads association and their impact on carcinogenesis in HCC cell lines. Labbé et al. [146] and Letamendia et al. [147] reported the interaction between Smads 2, 3 and 4 and TCF/LEF in HepG2. Additionally, treatment with TGFβ1 in HepG2 reduced the amount of Smad4 protein bound to TCF/LEF and this was associated with the capacity of TGFβ1 of inhibiting cell proliferation [148]. To date, no study on IFN α and Smads has been carried out. More insight could be gained by analyzing the amount of each Smad protein inside the β-catenin/TCF4 nuclear complex, since all Smads coexist in this complex and the balance between them could contribute to the overall cell response by differently regulating gene expression as suggested by Edlund et al. [149].

Our findings clearly showed a negative modulation of IFN α-2b and TGFβ1 on Wnt/β-catenin pathway. This attenuation was evidenced by a decrease in β-catenin and Frizzled-7 receptor proteins levels in C3A and Huh7 and by a diminution in the amount of β-catenin bound to TCF4. Stimulation with both cytokines also caused a decrease in Smads protein contents and their association with TCF4. This effect on Smads proteins seems to be linked to the decrease of β-catenin. Finally, the inhibition of β-catenin/TCF4/Smads complexes formation may have a critical role in slowing down oncogenesis, since the overall action of IFN α-2b and/or TGFβ1 treatments on both HCC cell lines was the diminution in cellular proliferation and the increase in apoptotic cell death. In conclusion, our results support the efficacy of inhibiting Wnt/β-catenin pathway in HCC cell lines through an IFN α-2b and TGFβ1 combined treatment, proving that is effective against either wild-type or truncated β-catenin. These findings open a wide therapeutic option for patients with HCC.

13. Summary # 2

The presented data suggest a model in which IFN α-2b provides a link between TGFβ1 and Wnt signaling pathways and the oxidative stress/FOXO pathway. The stress caused by IFN α-2b treatment might strengthen the interaction between FOXO and β-catenin and potentially inhibit the interaction with TCF and Smads. The inhibition of β-catenin/TCF4/Smads complexes formation may have a critical role in slowing down oncogenesis. These findings may have important clinical implications, since β-catenin, Smads, TCF, and FOXO arise as molecular targets for novel therapies that can modify their interactions favoring cellular apoptosis over proliferation in patients that underwent a potential carcinogenic hepatic injury.

14. Concluding remarks: Oxidative stress as a critical factor in cancer therapy

Preneoplastic hepatocytes are more resistant to oxidative stress than normal ones. Nevertheless, we demonstrated that increase in ROS levels triggered by IFN α-2b enhances the process of programmed hepatocytes death, leading to the elimination of malignant cells. The study of the mechanism of IFN α-2b-induced apoptosis led to demonstrate a link between TGFβ1 and Wnt signaling pathways and the oxidative stress/FOXO pathway.

In conclusion, reactive oxygen species emerge as key mediators in the context of using cytokines as therapeutic agents in the treatment of human liver diseases, so the use of antioxidants could have the potential to decrease effectiveness of the therapy.

Author details

María Cristina Carrillo*, María de Luján Alvarez, Juan Pablo Parody,
Ariel Darío Quiroga and María Paula Ceballos
*Institute of Experimental Physiology (IFISE-CONICET), Faculty of Biochemistry and
Pharmacological Sciences, National University of Rosario, Rosario, Argentina*

Acknowledgement

The authors kindly thank BioSidus Laboratory and Cassará Laboratory for the generous gift of recombinant IFN α-2b.

The work presented in this chapter was supported by research grants from Consejo Nacional de Investigaciones Científicas y Técnicas (CONICET) and from Agencia Nacional de Promoción Científica y Tecnológica (ANPCyT).

15. References

[1] Afanas'ev IB (2009) Signaling Mechanisms of Oxygen and Nitrogen Free Radicals. Boca Raton, FL, USA: CRC Press, Taylor & Francis Group. 212 p.

[2] Motola-Kuba D, Zamora-Valdés D, Uribe M, Méndez-Sánchez N (2006) Hepatocellular carcinoma. An overview. Ann Hepatol. 5(1):16–24.

[3] Stewart H. (1975) Comparative aspects of certain cancers. In: Becker F, editor. Cancer. A Comprehensive Treatise. New York: Plenum Press. pp. 303–74.

[4] Williams GM (1989) The significance of chemically-induced hepatocellular altered foci in rat liver and application to carcinogen detection. Toxicol Pathol.17(4 Pt 1):663–72; discussion 673–4.

[5] Pitot HC, Campbell HA, Maronpot R, Bawa N, Rizvi TA, Xu YH, Sargent L, Dragan Y, Pyron M (1989) Critical parameters in the quantitation of the stages of initiation, promotion, and progression in one model of hepatocarcinogenesis in the rat. Toxicol Pathol. 17(4 Pt 1):594–611; discussion 611–2.

[6] Altmann HW. (1994) Hepatic neoformations. Pathol Res Pract. 190(6):513–77.

[7] Farber E, Sarma DS (1986) Chemical carcinogenesis: the liver as a model. Pathol Immunopathol Res. 5(1):1–28.

[8] Goldsworthy TL, Hanigan MH, Pitot HC (1986) Models of hepatocarcinogenesis in the rat--contrasts and comparisons. Crit Rev Toxicol.17(1):61–89.

[9] Solt D, Farber E. (1976) New principle for the analysis of chemical carcinogenesis. Nature. 263(5579):701–3.

---

* Corresponding Author

[10] Ito N, Tsuda H, Tatematsu M, Inoue T, Tagawa Y, Aoki T, Uwagawa S, Kagawa M, Ogiso T, Masui T, et al. (1988) Enhancing effect of various hepatocarcinogens on induction of preneoplastic glutathione S-transferase placental form positive foci in rats-- an approach for a new medium-term bioassay system. Carcinogenesis. 9(3):387–94.

[11] Rao PM, Nagamine Y, Roomi MW, Rajalakshmi S, Sarma DS (1984) Orotic acid, a new promoter for experimental liver carcinogenesis. Toxicol Pathol. 12(2):173–8.

[12] Peraino C, Staffeldt EF, Carnes BA, Ludeman VA, Blomquist JA, Vesselinovitch SD (1984) Characterization of histochemically detectable altered hepatocyte foci and their relationship to hepatic tumorigenesis in rats treated once with diethylnitrosamine or benzo(a)pyrene within one day after birth. Cancer Res. 44(8):3340–7.

[13] Pitot HC, Barsness L, Goldsworthy T, Kitagawa T (1978) Biochemical characterisation of stages of hepatocarcinogenesis after a single dose of diethylnitrosamine. Nature. 271(5644):456–8.

[14] de Luján Alvarez M, Cerliani JP, Monti J, Carnovale C, Ronco MT, Pisani G, Lugano MC, Carrillo MC. (2002) The in vivo apoptotic effect of interferon alfa-2b on rat preneoplastic liver involves Bax protein. Hepatology. 35(4):824–33.

[15] Parody JP, Alvarez M de L, Quiroga A, Ronco MT, Francés D, Carnovale C, Carrillo MC (2007) Hepatocytes isolated from preneoplastic rat livers are resistant to ethacrynic acid cytotoxicity. Arch Toxicol. 81(8):565–73.

[16] Imai T, Masui T, Ichinose M, Nakanishi H, Yanai T, Masegi T, Muramatsu M, Tatematsu M (1997) Reduction of glutathione S-transferase P-form mRNA expression in remodeling nodules in rat liver revealed by in situ hybridization. Carcinogenesis. 18(3):545–51.

[17] Pitot HC (1990) Altered hepatic foci: their role in murine hepatocarcinogenesis. Annu Rev Pharmacol Toxicol. 30:465–500.

[18] Coles B, Ketterer B (1990) The role of glutathione and glutathione transferases in chemical carcinogenesis. Crit Rev Biochem Mol Biol. 25(1):47–70.

[19] Mannervik B, Danielson UH (1988) Glutathione transferases--structure and catalytic activity. CRC Crit Rev Biochem. 23(3):283–337.

[20] Morrow CS, Cowan KH (1990) Glutathione S-transferases and drug resistance. Cancer Cells. 2(1):15–22.

[21] Keen JH, Jakoby WB (1978) Glutathione transferases. Catalysis of nucleophilic reactions of glutathione. J Biol Chem. 253(16):5654–7.

[22] Hayes JD, McLellan LI (1999) Glutathione and glutathione-dependent enzymes represent a co-ordinately regulated defence against oxidative stress. Free Radic Res. 31(4):273–300.

[23] Satoh K, Kitahara A, Soma Y, Inaba Y, Hatayama I, Sato K (1985) Purification, induction, and distribution of placental glutathione transferase: a new marker enzyme for preneoplastic cells in the rat chemical hepatocarcinogenesis. Proc Natl Acad Sci U S A. 82(12):3964–8.

[24] Armstrong DK, Gordon GB, Hilton J, Streeper RT, Colvin OM, Davidson NE (1992) Hepsulfam sensitivity in human breast cancer cell lines: the role of glutathione and glutathione S-transferase in resistance. Cancer Res. 52(6):1416–21.

[25] Batist G, Tulpule A, Sinha BK, Katki AG, Myers CE, Cowan KH (1986) Overexpression of a novel anionic glutathione transferase in multidrug-resistant human breast cancer cells. J Biol Chem. 261(33):15544–9.

[26] Tolman KG, Gray PD, el Masry S, Luther RR, Janicki RS (1989) Toxicity of uricosuric diuretics in rat hepatocyte culture. Biochem Pharmacol. 38(7):1181–4.

[27] Yamamoto K, Masubuchi Y, Narimatsu S, Kobayashi S, Horie T (2002) Toxicity of ethacrynic acid in isolated rat hepatocytes. Toxicol In Vitro. 16(2):151–8.

[28] Evers R, Kool M, van Deemter L, Janssen H, Calafat J, Oomen LC, Paulusma CC, Oude Elferink RP, Baas F, Schinkel AH, Borst P (1998) Drug export activity of the human canalicular multispecific organic anion transporter in polarized kidney MDCK cells expressing cMOAT (MRP2) cDNA. J Clin Invest.101(7):1310–9.

[29] Tirona RG, Tan E, Meier G, Pang KS (1999) Uptake and glutathione conjugation of ethacrynic acid and efflux of the glutathione adduct by periportal and perivenous rat hepatocytes. J Pharmacol Exp Ther. 291(3):1210–9.

[30] Iersel ML, Ploemen JP, Struik I, van Amersfoort C, Keyzer AE, Schefferlie JG, van Bladeren PJ (1996) Inhibition of glutathione S-transferase activity in human melanoma cells by alpha,beta-unsaturated carbonyl derivatives. Effects of acrolein, cinnamaldehyde, citral, crotonaldehyde, curcumin, ethacrynic acid, and trans-2-hexenal. Chem Biol Interact. 102(2):117–32.

[31] Ploemen JH, van Ommen B, van Bladeren PJ (1990) Inhibition of rat and human glutathione S-transferase isoenzymes by ethacrynic acid and its glutathione conjugate. Biochem Pharmacol. 40(7):1631–5.

[32] Ohkawa H, Ohishi N, Yagi K (1979) Assay for lipid peroxides in animal tissues by thiobarbituric acid reaction. Anal Biochem. 95(2):351–8.

[33] Sato K (1989) Glutathione transferases as markers of preneoplasia and neoplasia. Adv Cancer Res. 52:205–55.

[34] Meyer DJ, Beale D, Tan KH, Coles B, Ketterer B (1985) Glutathione transferases in primary rat hepatomas: the isolation of a form with GSH peroxidase activity. FEBS Lett. 184(1):139–43.

[35] Mannervik B, Alin P, Guthenberg C, Jensson H, Tahir MK, Warholm M, Jörnvall H (1985) Identification of three classes of cytosolic glutathione transferase common to several mammalian species: correlation between structural data and enzymatic properties. Proc Natl Acad Sci U S A. 82(21):7202–6.

[36] Tahir MK, Guthenberg C, Mannervik B (1989) Glutathione transferases in rat hepatoma cells. Effects of ascites cells on the isoenzyme pattern in liver and induction of glutathione transferases in the tumour cells. Biochem J. 257(1):215–20.

[37] Kuzmich S, Vanderveer LA, Walsh ES, LaCreta FP, Tew KD (1992) Increased levels of glutathione S-transferase pi transcript as a mechanism of resistance to ethacrynic acid. Biochem J. 281(Pt 1):219–24.

[38] Morrow CS, Smitherman PK, Townsend AJ (1998) Combined expression of multidrug resistance protein (MRP) and glutathione S-transferase P1-1 (GSTP1-1) in MCF7 cells and high level resistance to the cytotoxicities of ethacrynic acid but not oxazaphosphorines or cisplatin. Biochem Pharmacol. 56(8):1013–21.

[39] Nies AT, König J, Pfannschmidt M, Klar E, Hofmann WJ, Keppler D (2001) Expression of the multidrug resistance proteins MRP2 and MRP3 in human hepatocellular carcinoma. Int J Cancer. 94(4):492–9.

[40] Ji B, Ito K, Sekine S, Tajima A, Horie T (2004) Ethacrynic-acid-induced glutathione depletion and oxidative stress in normal and Mrp2-deficient rat liver. Free Radic Biol Med. 37(11):1718–29.

[41] Favre C, Carnovale CE, Monti JA, Carrillo MC (2001) Inhibition by interferon alpha-2b of rat liver regeneration: effect on ornithine decarboxylase and total protein synthesis. Biochem Pharmacol. 61(12):1587–93.

[42] Rosemberg S (1997) Principles of cancer management: biologic therapy. In: De Vita V, Hellman S RS, editors. Cancer: Principles and Practice of Oncology. Philadelphia: Lippincott-Raven. pp. 349–73.

[43] Ikeda K, Saitoh S, Suzuki Y, Kobayashi M, Tsubota A, Fukuda M, Koida I, Arase Y, Chayama K, Murashima N, Kumada H (1998) Interferon decreases hepatocellular carcinogenesis in patients with cirrhosis caused by the hepatitis B virus: a pilot study. Cancer. 82(5):827–35.

[44] Yoshida H, Shiratori Y, Moriyama M, Arakawa Y, Ide T, Sata M, Inoue O, Yano M, Tanaka M, Fujiyama S, Nishiguchi S, Kuroki T, Imazeki F, Yokosuka O, Kinoyama S, Yamada G, Omata M (1999) Interferon therapy reduces the risk for hepatocellular carcinoma: national surveillance program of cirrhotic and noncirrhotic patients with chronic hepatitis C in Japan. IHIT Study Group. Inhibition of Hepatocarcinogenesis by Interferon Therapy. Ann Intern Med. 131(3):174–81.

[45] Lai CL, Lau JY, Wu PC, Ngan H, Chung HT, Mitchell SJ, Corbett TJ, Chow AW, Lin HJ (1993) Recombinant interferon-alpha in inoperable hepatocellular carcinoma: a randomized controlled trial. Hepatology. 17(3):389–94.

[46] Llovet JM, Sala M, Castells L, Suarez Y, Vilana R, Bianchi L, Ayuso C, Vargas V, Rodés J, Bruix J (2000) Randomized controlled trial of interferon treatment for advanced hepatocellular carcinoma. Hepatology. 31(1):54–8.

[47] Dunk AA, Ikeda T, Pignatelli M, Thomas HC (1986) Human lymphoblastoid interferon. In vitro and in vivo studies in hepatocellular carcinoma. J Hepatol. 2(3):419–29.

[48] Yano H, Iemura A, Haramaki M, Ogasawara S, Takayama A, Akiba J, Kojiro M (1999) Interferon alfa receptor expression and growth inhibition by interferon alfa in human liver cancer cell lines. Hepatology. 29(6):1708–17.

[49] Murphy D, Detjen KM, Welzel M, Wiedenmann B, Rosewicz S (2001) Interferon-alpha delays S-phase progression in human hepatocellular carcinoma cells via inhibition of specific cyclin-dependent kinases. Hepatology. 33(2):346–56.

[50] Gutterman JU (1994) Cytokine therapeutics: lessons from interferon alpha. Proc Natl Acad Sci U S A. 91(4):1198–205.

[51] Creasey AA, Bartholomew JC, Merigan TC (1980) Role of G0-G1 arrest in the inhibition of tumor cell growth by interferon. Proc Natl Acad Sci U S A. 77(3):1471–5.

[52] Roos G, Leanderson T, Lundgren E (1984) Interferon-induced cell cycle changes in human hematopoietic cell lines and fresh leukemic cells. Cancer Res. 44(6):2358–62.

[53] Soini Y, Pääkkö P, Lehto VP (1998) Histopathological evaluation of apoptosis in cancer. Am J Pathol. 153(4):1041–53.

[54] Schulte-Hermann R, Grasl-Kraupp B, Bursch W (1995) Apoptosis and hepatocarcinogenesis. In: Jirtle J, editor. Liver Regeneration and Carcinogenesis. San Diego: Academic. pp. 141–78.

[55] Miyashita T, Harigai M, Hanada M, Reed JC (1994) Identification of a p53-dependent negative response element in the bcl-2 gene. Cancer Res. 54(12):3131–5.

[56] Miyashita T, Reed JC (1995) Tumor suppressor p53 is a direct transcriptional activator of the human bax gene. Cell. 80(2):293–9.

[57] Sangfelt O, Erickson S, Castro J, Heiden T, Einhorn S, Grandér D (1997) Induction of apoptosis and inhibition of cell growth are independent responses to interferon-alpha in hematopoietic cell lines. Cell Growth Differ. 8(3):343–52.

[58] Rodríguez-Villanueva J, McDonnell TJ (1995) Induction of apoptotic cell death in non-melanoma skin cancer by interferon-alpha. Int J Cancer. 61(1):110–4.

[59] Sedlak TW, Oltvai ZN, Yang E, Wang K, Boise LH, Thompson CB, Korsmeyer SJ (1995) Multiple Bcl-2 family members demonstrate selective dimerizations with Bax. Proc Natl Acad Sci U S A.92(17):7834–8.

[60] Yang J, Liu X, Bhalla K, Kim CN, Ibrado AM, Cai J, Peng TI, Jones DP, Wang X (1997) Prevention of apoptosis by Bcl-2: release of cytochrome c from mitochondria blocked. Science. 275(5303):1129–32.

[61] Nechushtan A, Smith CL, Hsu YT, Youle RJ (1999) Conformation of the Bax C-terminus regulates subcellular location and cell death. EMBO J. 18(9):2330–41.

[62] Shima Y, Nakao K, Nakashima T, Kawakami A, Nakata K, Hamasaki K, Kato Y, Eguchi K, Ishii N (1999) Activation of caspase-8 in transforming growth factor-beta-induced apoptosis of human hepatoma cells. Hepatology. 30(5):1215–22.

[63] Yamamoto M, Fukuda K, Miura N, Suzuki R, Kido T, Komatsu Y (1998) Inhibition by dexamethasone of transforming growth factor beta1-induced apoptosis in rat hepatoma cells: a possible association with Bcl-xL induction. Hepatology. 27(4):959–66.

[64] Buenemann CL, Willy C, Buchmann A, Schmiechen A, Schwarz M (2001) Transforming growth factor-beta1-induced Smad signaling, cell-cycle arrest and apoptosis in hepatoma cells. Carcinogenesis. 22(3):447–52.

[65] Gressner AM, Lahme B, Mannherz HG, Polzar B (1997) TGF-beta-mediated hepatocellular apoptosis by rat and human hepatoma cells and primary rat hepatocytes. J Hepatol. 26(5):1079–92.

[66] Rossmanith W, Schulte-Hermann R (2001) Biology of transforming growth factor beta in hepatocarcinogenesis. Microsc Res Tech. 52(4):430–6.

[67] Kanzler S, Meyer E, Lohse AW, Schirmacher P, Henninger J, Galle PR, Blessing M (2001) Hepatocellular expression of a dominant-negative mutant TGF-beta type II receptor accelerates chemically induced hepatocarcinogenesis. Oncogene. 20(36):5015–24.

[68] Matsuzaki K, Date M, Furukawa F, Tahashi Y, Matsushita M, Sakitani K, Yamashiki N, Seki T, Saito H, Nishizawa M, Fujisawa J, Inoue K (2000) Autocrine stimulatory mechanism by transforming growth factor beta in human hepatocellular carcinoma. Cancer Res. 60(5):1394–402.

[69] Santoni-Rugiu E, Jensen MR, Factor VM, Thorgeirsson SS (1999) Acceleration of c-myc-induced hepatocarcinogenesis by Co-expression of transforming growth factor (TGF)-

alpha in transgenic mice is associated with TGF-beta1 signaling disruption. Am J Pathol. 154(6):1693–700.

[70] Grasl-Kraupp B, Rossmanith W, Ruttkay-Nedecky B, Müllauer L, Kammerer B, Bursch W, Schulte-Hermann R (1998) Levels of transforming growth factor beta and transforming growth factor beta receptors in rat liver during growth, regression by apoptosis and neoplasia. Hepatology. 28(3):717–26.

[71] Huang YL, Chou CK (1998) Bcl-2 blocks apoptotic signal of transforming growth factor-beta in human hepatoma cells. J Biomed Sci. 5(3):185–91.

[72] de Luján Alvarez M, Ronco MT, Ochoa JE, Monti J a, Carnovale CE, Pisani GB, Lugano MC, Carrillo MC (2004) Interferon alpha-induced apoptosis on rat preneoplastic liver is mediated by hepatocytic transforming growth factor beta(1). Hepatology. 40(2):394–402.

[73] Hori Y, Takeyama Y, Ueda T, Shinkai M, Takase K, Kuroda Y (2000) Macrophage-derived transforming growth factor-beta1 induces hepatocellular injury via apoptosis in rat severe acute pancreatitis. Surgery. 127(6):641–9.

[74] Kamimura S, Tsukamoto H (1995) Cytokine gene expression by Kupffer cells in experimental alcoholic liver disease. Hepatology. 22(4 Pt 1):1304–9.

[75] Gao C, Gressner G, Zoremba M, Gressner AM (1996) Transforming growth factor beta (TGF-beta) expression in isolated and cultured rat hepatocytes. J Cell Physiol. 167(3):394–405.

[76] Matsuzaki K, Date M, Furukawa F, Tahashi Y, Matsushita M, Sugano Y, Yamashiki N, Nakagawa T, Seki T, Nishizawa M, Fujisawa J, Inoue K (2000) Regulatory mechanisms for transforming growth factor beta as an autocrine inhibitor in human hepatocellular carcinoma: implications for roles of smads in its growth. Hepatology. 32(2):218–27.

[77] Bissell DM, Wang SS, Jarnagin WR, Roll FJ (1995) Cell-specific expression of transforming growth factor-beta in rat liver. Evidence for autocrine regulation of hepatocyte proliferation. J Clin Invest. 96(1):447–55.

[78] Babior BM (2000) The NADPH oxidase of endothelial cells. IUBMB Life. 50(4-5):267–9.

[79] Quiroga AD, Alvarez MDL, Parody JP, Ronco MT, Francés DE, Pisani GB, Carnovale CE, Carrillo MC (2007) Involvement of reactive oxygen species on the apoptotic mechanism induced by IFN-alpha2b in rat preneoplastic liver. Biochem Pharmacol.73(11):1776–85.

[80] Herrera B, Murillo MM, Alvarez-Barrientos A, Beltrán J, Fernández M, Fabregat I (2004) Source of early reactive oxygen species in the apoptosis induced by transforming growth factor-beta in fetal rat hepatocytes. Free Radical Biol Med. 36(1):16–26.

[81] Park CH (1988) Vitamin C in leukemia and preleukemia cell growth. Prog Clin Biol Res. 259:321–30.

[82] Prasad KN, Kumar R (1996) Effect of individual and multiple antioxidant vitamins on growth and morphology of human nontumorigenic and tumorigenic parotid acinar cells in culture. Nutr Cancer. 26(1):11–9.

[83] Seifried HE, McDonald SS, Anderson DE, Greenwald P, Milner JA (2003) The antioxidant conundrum in cancer. Cancer Res. 63(15):4295–8.

[84] Uddin S, Majchrzak B, Woodson J, Arunkumar P, Alsayed Y, Pine R, Young PR, Fish EN, Platanias LC (1999) Activation of the p38 mitogen-activated protein kinase by type I interferons. J Biol Chem. 274(42):30127–31.

[85] Sarkar D, Su Z-Z, Lebedeva IV, Sauane M, Gopalkrishnan RV, Valerie K, Dent P, Fisher PB (2002) mda-7 (IL-24) Mediates selective apoptosis in human melanoma cells by inducing the coordinated overexpression of the GADD family of genes by means of p38 MAPK. Proc Natl Acad Sci U S A. 99(15):10054–9.

[86] Porras A, Zuluaga S, Black E, Valladares A, Alvarez AM, Ambrosino C, Benito M, Nebreda AR (2005) P38 alpha mitogen-activated protein kinase sensitizes cells to apoptosis induced by different stimuli. Mol Biol Cell. 15(2):922–33.

[87] Liu B, Fang M, Lu Y, Mills GB, Fan Z (2001) Involvement of JNK-mediated pathway in EGF-mediated protection against paclitaxel-induced apoptosis in SiHa human cervical cancer cells. Br J Cancer. 85(2):303–11.

[88] Park JM, Greten FR, Li Z-W, Karin M (2002) Macrophage apoptosis by anthrax lethal factor through p38 MAP kinase inhibition. Science. 297(5589):2048–51.

[89] Juretic N, Santibáñez JF, Hurtado C, Martínez J (2001) ERK 1,2 and p38 pathways are involved in the proliferative stimuli mediated by urokinase in osteoblastic SaOS-2 cell line. J Cell Biochem. 83(1):92–8.

[90] Yosimichi G, Nakanishi T, Nishida T, Hattori T, Takano-Yamamoto T, Takigawa M (2001) CTGF/Hcs24 induces chondrocyte differentiation through a p38 mitogen-activated protein kinase (p38MAPK), and proliferation through a p44/42 MAPK/extracellular-signal regulated kinase (ERK). Eur J Biochem. 268(23):6058–65.

[91] Razandi M, Pedram A, Levin ER (2000) Estrogen signals to the preservation of endothelial cell form and function. J Biol Chem. 275(49):38540–6.

[92] Jiménez B, Volpert OV, Crawford SE, Febbraio M, Silverstein RL, Bouck N (2000) Signals leading to apoptosis-dependent inhibition of neovascularization by thrombospondin-1. Nat Med. 6(1):41–8.

[93] Nakagami H, Morishita R, Yamamoto K, Yoshimura SI, Taniyama Y, Aoki M, Matsubara H, Kim S, Kaneda Y, Ogihara T (2001) Phosphorylation of p38 mitogen-activated protein kinase downstream of bax-caspase-3 pathway leads to cell death induced by high D-glucose in human endothelial cells. Diabetes. 50(6):1472–81.

[94] Grethe S, Ares MPS, Andersson T, Pörn-Ares MI (2004) p38 MAPK mediates TNF-induced apoptosis in endothelial cells via phosphorylation and downregulation of Bcl-x(L). Exp Cell Res. 298(2):632–42.

[95] Verma A, Deb DK, Sassano A, Uddin S, Varga J, Wickrema A, Platanias LC (2002) Activation of the p38 mitogen-activated protein kinase mediates the suppressive effects of type I interferons and transforming growth factor-beta on normal hematopoiesis. J Biol Chem. 277(10):7726–35.

[96] Herrera B, Fernández M, Roncero C, Ventura JJ, Porras A, Valladares A, Benito M, Fabregat I (2001) Activation of p38MAPK by TGF-beta in fetal rat hepatocytes requires radical oxygen production, but is dispensable for cell death. FEBS Lett. 499(3):225–9.

[97] Bulavin DV, Demidov ON, Saito S, Kauraniemi P, Phillips C, Amundson SA, Ambrosino C, Sauter G, Nebreda AR, Anderson CW, Kallioniemi A, Fornace AJ Jr, Appella E (2002) Amplification of PPM1D in human tumors abrogates p53 tumor-suppressor activity. Nat Genet. 31(2):210–5.

[98] Bulavin DV, Phillips C, Nannenga B, Timofeev O, Donehower LA, Anderson CW, Appella E, Fornace AJ Jr (2004) Inactivation of the Wip1 phosphatase inhibits mammary

tumorigenesis through p38 MAPK-mediated activation of the p16(Ink4a)-p19(Arf) pathway. Nat Genet. 36(4):343–50.

[99] Honmo S, Ozaki A, Yamamoto M, Hashimoto N, Miyakoshi M, Tanaka H, Yoshie M, Tamakawa S, Tokusashi Y, Yaginuma Y, Kasai S, Ogawa K (2007) Low p38 MAPK and JNK activation in cultured hepatocytes of DRH rats; a strain highly resistant to hepatocarcinogenesis. Mol Carcinog. 46(9):758–65.

[100] Mayer IA, Verma A, Grumbach IM, Uddin S, Lekmine F, Ravandi F, Majchrzak B, Fujita S, Fish EN, Platanias LC (2001) The p38 MAPK pathway mediates the growth inhibitory effects of interferon-alpha in BCR-ABL-expressing cells. J Biol Chem. 276(30):28570–7.

[101] Quiroga AD, de Lujan Alvarez M, Parody JP, Ronco MT, Carnovale CE, Carrillo MC (2009) Interferon-alpha2b (IFN-alpha2b)-induced apoptosis is mediated by p38 MAPK in hepatocytes from rat preneoplastic liver via activation of NADPH oxidase. Growth Factors. 27(4):214–27.

[102] Ulloa L, Doody J, Massagué J (1999) Inhibition of transforming growth factor-beta/SMAD signalling by the interferon-gamma/STAT pathway. Nature. 397(6721):710–3.

[103] Ghosh AK, Yuan W, Mori Y, Chen Sj, Varga J (2001) Antagonistic regulation of type I collagen gene expression by interferon-gamma and transforming growth factor-beta. Integration at the level of p300/CBP transcriptional coactivators. J Biol Chem. 276(14):11041–8.

[104] Higashi K, Inagaki Y, Fujimori K, Nakao A, Kaneko H, Nakatsuka I (2003) Interferon-gamma interferes with transforming growth factor-beta signaling through direct interaction of YB-1 with Smad3. J Biol Chem. 278(44):43470–9.

[105] Ishida Y, Kondo T, Takayasu T, Iwakura Y, Mukaida N (2004) The essential involvement of cross-talk between IFN-gamma and TGF-beta in the skin wound-healing process. J Immunol. 172(3):1848–55.

[106] Jeong W-I, Park O, Radaeva S, Gao B (2006) STAT1 inhibits liver fibrosis in mice by inhibiting stellate cell proliferation and stimulating NK cell cytotoxicity. Hepatology. 44(6):1441–51.

[107] Weng H, Mertens PR, Gressner AM, Dooley S (2007) IFN-gamma abrogates profibrogenic TGF-beta signaling in liver by targeting expression of inhibitory and receptor Smads. J Hepatol. 46(2):295–303.

[108] Yamamoto T, Matsuda T, Muraguchi A, Miyazono K, Kawabata M (2001) Cross-talk between IL-6 and TGF-beta signaling in hepatoma cells. FEBS Lett. 492(3):247–53.

[109] Foser S, Redwanz I, Ebeling M, Heizmann CW, Certa U (2006) Interferon-alpha and transforming growth factor-beta co-induce growth inhibition of human tumor cells. Cell Mol Life Sci. 63(19-20):2387–96.

[110] Roberts AB, Wakefield LM (2003) The two faces of transforming growth factor beta in carcinogenesis. Proc Natl Acad Sci U S A. 100(15):8621–3.

[111] Lei X, Yang J, Nichols RW, Sun L-Z (2007) Abrogation of TGFbeta signaling induces apoptosis through the modulation of MAP kinase pathways in breast cancer cells. Exp Cell Res. 313(8):1687–95.

[112] Park DY, Lee CH, Sol MY, Suh KS, Yoon SY, Kim JW (2003) Expression and localization of the transforming growth factor-beta type I receptor and Smads in preneoplastic lesions during chemical hepatocarcinogenesis in rats. J Korean Med Sci. 18(4):510–9.

[113] Alvarez MDL, Quiroga AD, Parody JP, Ronco MT, Francés DE, Carnovale CE, Carrillo MC (2009) Cross-talk between IFN-alpha and TGF-beta1 signaling pathways in preneoplastic rat liver. Growth Factors. 27(1):1–11.

[114] Kim SJ, Im YH, Markowitz SD, Bang YJ (2000) Molecular mechanisms of inactivation of TGF-beta receptors during carcinogenesis. Cytokine Growth Factor Rev. 11(1-2):159–68.

[115] Ghosh AK, Varga J (2007) The transcriptional coactivator and acetyltransferase p300 in fibroblast biology and fibrosis. J Cell Physiol. 213(3):663–71.

[116] Inagaki Y, Nemoto T, Kushida M, Sheng Y, Higashi K, Ikeda K, Kawada N, Shirasaki F, Takehara K, Sugiyama K, Fujii M, Yamauchi H, Nakao A, de Combrugghe B, Watanabe T, Okazaki I (2003) Interferon alfa down-regulates collagen gene transcription and suppresses experimental hepatic fibrosis in mice. Hepatology. 38(4):890–9.

[117] Alvarez MDL, Quiroga AD, Ronco MT, Parody JP, Ochoa JE, Monti JA, Carnovale CE, Carrillo MC (2006) Time-dependent onset of Interferon-alpha2b-induced apoptosis in isolated hepatocytes from preneoplastic rat livers. Cytokine. 36(5-6):245–53.

[118] Nambotin SB, Wands JR, Kim M (2011) Points of therapeutic intervention along the Wnt signaling pathway in hepatocellular carcinoma. Anticancer Agents Med Chem. 11(6):549–59.

[119] Parody JP, Alvarez ML, Quiroga AD, Ceballos MP, Frances DE, Pisani GB, Pellegrino JM, Carnovale CE, Carrillo MC (2010) Attenuation of the Wnt/beta-catenin/TCF pathway by in vivo interferon-alpha2b (IFN-alpha2b) treatment in preneoplastic rat livers. Growth Factors. 28(3):166–77.

[120] Gautschi O, Ratschiller D, Gugger M, Betticher DC, Heighway J (2007) Cyclin D1 in non-small cell lung cancer: a key driver of malignant transformation. Lung Cancer. 55(1):1–14.

[121] Kornmann M, Danenberg KD, Arber N, Beger HG, Danenberg PV, Korc M (1999) Inhibition of cyclin D1 expression in human pancreatic cancer cells is associated with increased chemosensitivity and decreased expression of multiple chemoresistance genes. Cancer Res. 59(14):3505–11.

[122] Wilson CL, Heppner KJ, Labosky P a, Hogan BL, Matrisian LM (1997) Intestinal tumorigenesis is suppressed in mice lacking the metalloproteinase matrilysin. Proc Natl Acad Sci U S A.94(4):1402–7.

[123] Powell WC, Knox JD, Navre M, Grogan TM, Kittelson J, Nagle RB, Bowden GT (1993) Expression of the metalloproteinase matrilysin in DU-145 cells increases their invasive potential in severe combined immunodeficient mice. Cancer Res. 53(2):417–22.

[124] Yamamoto H, Itoh F, Hinoda Y, Imai K (1995) Suppression of matrilysin inhibits colon cancer cell invasion in vitro. Int J Cancer. 61(2):218–22.

[125] Yan D, Wiesmann M, Rohan M, Chan V, Jefferson AB, Guo L, Sakamoto D, Caothien RH, Fuller JH, Reinhard C, Garcia PD, Randazzo FM, Escobedo J, Fantl WJ, Williams

LT (2001) Elevated expression of axin2 and hnkd mRNA provides evidence that Wnt/beta -catenin signaling is activated in human colon tumors. Proc Natl Acad Sci U S A. 98(26):14973–8.

[126] Jho E-hoon, Zhang T, Domon C, Joo C-K, Freund J-N, Costantini F (2002) Wnt/beta-catenin/Tcf signaling induces the transcription of Axin2, a negative regulator of the signaling pathway. Mol Cell Biol. 22(4):1172–83.

[127] Behrens J (1998) Functional Interaction of an Axin Homolog, Conductin, with -Catenin, APC, and GSK3. Science. 280(5363):596–9.

[128] Hart MJ, de los Santos R, Albert IN, Rubinfeld B, Polakis P (1998) Downregulation of beta-catenin by human Axin and its association with the APC tumor suppressor, beta-catenin and GSK3 beta. Curr Biol. 8(10):573–81.

[129] Fujimura N, Vacik T, Machon O, Vlcek C, Scalabrin S, Speth M, Diep D, Krauss S, Kozmik Z (2007) Wnt-mediated down-regulation of Sp1 target genes by a transcriptional repressor Sp5. J Biol Chem. 282(2):1225–37.

[130] de La Coste A, Romagnolo B, Billuart P, Renard CA, Buendia MA, Soubrane O, Fabre M, Chelly J, Beldjord C, Kahn A, Perret C (1998) Somatic mutations of the beta-catenin gene are frequent in mouse and human hepatocellular carcinomas. Proc Natl Acad Sci U S A. 95(15):8847–51.

[131] Miyoshi Y, Iwao K, Nagasawa Y, Aihara T, Sasaki Y, Imaoka S, Murata M, Shimano T, Nakamura Y (1998) Activation of the beta-catenin gene in primary hepatocellular carcinomas by somatic alterations involving exon 3. Cancer Res. 58(12):2524–7.

[132] Devereux TR, Anna CH, Foley JF, White CM, Sills RC, Barrett JC (1999) Mutation of beta-catenin is an early event in chemically induced mouse hepatocellular carcinogenesis. Oncogene. 18(33):4726–33.

[133] Calvisi DF, Factor VM, Loi R, Thorgeirsson SS (2001) Activation of beta-catenin during hepatocarcinogenesis in transgenic mouse models: relationship to phenotype and tumor grade. Cancer Res. 61(5):2085–91.

[134] Merle P, Kim M, Herrmann M, Gupte A, Lefrançois L, Califano S, Trépo C, Tanaka S, Vitvitski L, de la Monte S, Wands JR (2005) Oncogenic role of the frizzled-7/beta-catenin pathway in hepatocellular carcinoma. J Hepatol. 43(5):854–62.

[135] Merle P, de la Monte S, Kim M, Herrmann M, Tanaka S, Von Dem Bussche A, Kew MC, Trepo C, Wands JR (2004) Functional consequences of frizzled-7 receptor overexpression in human hepatocellular carcinoma. Gastroenterology. 127(4):1110–22.

[136] Willert J, Epping M, Pollack JR, Brown PO, Nusse R (2002) A transcriptional response to Wnt protein in human embryonic carcinoma cells. BMC Dev Biol. 2:8.

[137] Essers MAG, de Vries-Smits LMM, Barker N, Polderman PE, Burgering BMT, Korswagen HC (2005) Functional interaction between beta-catenin and FOXO in oxidative stress signaling. Science. 308(5725):1181–4.

[138] Hoogeboom D, Essers MAG, Polderman PE, Voets E, Smits LMM, Burgering BMT (2008) Interaction of FOXO with beta-catenin inhibits beta-catenin/T cell factor activity. J Biol Chem. 283(14):9224–30.

[139] Cavallo RA, Cox RT, Moline MM, Roose J, Polevoy GA, Clevers H, Peifer M, Bejsovec A (1998) Drosophila Tcf and Groucho interact to repress Wingless signalling activity. Nature. 395(6702):604–8.

[140] Levanon D, Goldstein RE, Bernstein Y, Tang H, Goldenberg D, Stifani S, Paroush Z, Groner Y (1998) Transcriptional repression by AML1 and LEF-1 is mediated by the TLE/Groucho corepressors. Proc Natl Acad Sci U S A. 95(20):11590–5.

[141] Bellan C, De Falco G, Tosi GM, Lazzi S, Ferrari F, Morbini G, Bartolomei S, Toti P, Mangiavacchi P, Cevenini G, Trimarchi C, Cinti C, Giordano A, Leoncini L, Tosi P, Cottier H (2002) Missing expression of pRb2/p130 in human retinoblastomas is associated with reduced apoptosis and lesser differentiation. Invest Ophthalmol Vis Sci. 43(12):3602–8.

[142] Ceballos MP, Parody JP, Alvarez MDL, Ingaramo PI, Carnovale CE, Carrillo MC (2011) Interferon-α2b and transforming growth factor-β1 treatments on HCC cell lines: Are Wnt/β-catenin pathway and Smads signaling connected in hepatocellular carcinoma? Biochem Pharmacol. 82(11):1682–91.

[143] Damdinsuren B, Nagano H, Kondo M, Natsag J, Hanada H, Nakamura M, Wada H, Kato H, Marubashi S, Miyamoto A, Takeda Y, Umeshita K, Dono K, Monden M (2006) TGF-beta1-induced cell growth arrest and partial differentiation is related to the suppression of Id1 in human hepatoma cells. Oncol Rep. 15(2):401–8.

[144] Ho J, Cocolakis E, Dumas VM, Posner BI, Laporte SA, Lebrun J-J (2005) The G protein-coupled receptor kinase-2 is a TGFbeta-inducible antagonist of TGFbeta signal transduction. EMBO J. 24(18):3247–58.

[145] Murata M, Nabeshima S, Kikuchi K, Yamaji K, Furusyo N, Hayashi J (2006) A comparison of the antitumor effects of interferon-alpha and beta on human hepatocellular carcinoma cell lines. Cytokine. 33(3):121–8.

[146] Labbé E, Letamendia A, Attisano L (2000) Association of Smads with lymphoid enhancer binding factor 1/T cell-specific factor mediates cooperative signaling by the transforming growth factor-beta and wnt pathways. Proc Natl Acad Sci U S A. 97(15):8358–63.

[147] Letamendia A, Labbé E, Attisano L (2001) Transcriptional regulation by Smads: crosstalk between the TGF-beta and Wnt pathways. J Bone Joint Surg Am. 83-A Suppl(Pt 1):S31–9.

[148] Lim SK, Hoffmann FM (2006) Smad4 cooperates with lymphoid enhancer-binding factor 1/T cell-specific factor to increase c-myc expression in the absence of TGF-beta signaling. Proc Natl Acad Sci U S A. 103(49):18580–5.

[149] Edlund S, Lee SY, Grimsby S, Zhang S, Aspenström P, Heldin C-H, Landström M (2005) Interaction between Smad7 and beta-catenin: importance for transforming growth factor beta-induced apoptosis. Mol Cell Biol. 25(4):1475–88.

Chapter 4

Tissue Occurrence of Carbonyl Products of Lipid Peroxidation and Their Role in Inflammatory Disease

Maria Armida Rossi

Additional information is available at the end of the chapter

1. Introduction

The lipid peroxidation is a diffuse process which regards the polyunsaturated fatty acids of lipids, when they are exposed to oxygen-derived free radicals.

The process occurs when oils or foods, vegetables or meats, or other materials are exposed to air and causes their alteration at least in part through the peroxidative decomposition of the fatty acids contained in their lipids.

The lipid peroxidation does not need the action of enzymes and brings to the progressive decomposition of the unsaturated fatty acids till to the formation of carbonylic end products, aldehydes and ketones.

Oxygen-derived free radicals can be produced by the effect of sun rays on O_2, but an important source of them is the cellular metabolism, too.

The interest and the importance of the lipid peroxidation arise from the fact that the polyunsaturated fatty acids are contained in the phospholipids present in all cellular membranes; their structure and function can be strongly modified by this process.

The cellular effects of the lipid peroxidation change according to its degree. A high lipoperoxidative rate can produce serious damages to the cells and their death; on the contrary, a low degree of it allows cell survival and may modulate tissue metabolism.

2. Steps of the lipoperoxidative process

The lipid peroxidation has many steps,as shown in the Figure 1. The process is started by the attach of free radicals to poly-unsaturated fatty acids of lipids. Free radicals are chemical species which have a single, unpaired electron in an outer orbit. Their molecular configuration is unstable and so they react with the adjacent molecules to acquire a more stable configuration. The polyunsaturated fatty acids contained in the phospholipids of cell membranes are a good target for their reaction; in the attach of free radicals to the unsaturated fatty acids of lipids a methylen group near a double bond can give the electron required by the free radical to form the electon pair. So the unsaturated fatty acid has become a free radical and reacts with another molecule, starting the propagation phase which characterizes the lipoperoxidative process. In our cells the molecular oxygen is always present and can react with the lipid radical to form a lipoperoxide. This molecule has un unstable configuration too. The formed lipoperoxides react with adjacent membrane molecules, either other lipids or proteins. The reaction of a lipoperoxide with a protein molecule changes it in a reactive free radical; the so activated protein can interact with another protein to give a protein complex or interacts with lipids to form lipofuscin molecules. The presence of lipofuscin is frequent in the tissues of old people; this fact was well known by the anatomists already in the past century; the mechanism of their formation has been clarified with the discovery of the lipoperoxidative process. Beside the reaction with other molecules, the lipoperoxides can break to give more stable end products, aldehydes and ketones. These carbonylic end products of lipid peroxidation are formed above all in the microsomes where the rate of the lipid peroxidation is strong, but they can diffuse and react with various molecular targets both within the cell and outside it.

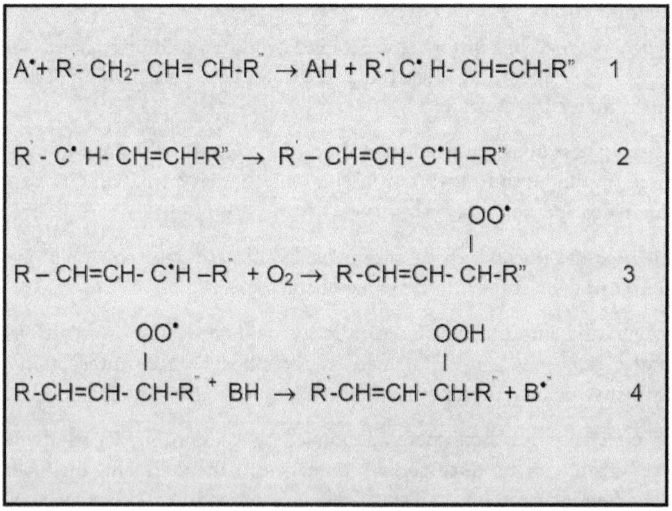

Figure 1. First steps of the lipid peroxidative process. 1) The free radical A$^\bullet$ abstracts an electron from a near molecule, which becomes a free radical; the target molecule is often a polyunsaturated fatty acid. 2) Its molecular configuration is unstable, so a shift of the double bond occurs. 3) This still unstable free radical binds O$_2$ and becomes a peroxide. 4) The peroxide captures an electron from the molecule B and forms an hydro-peroxide. Now the molecule B is a free radical. The further fate of the hydroperoxide is its fragmentation in small carbonylic compounds (not shown iin the Figure)

The first reports of the actual occurrence of the lipid peroxidation in tissues include the researches separately carried on by Comporti M et al.[1] and by Recknagel RO and Ghoshal AK. [2] to explain the liver damage induced by the rat treatment with CCl₄. Both these works used methods of investigation quite modern for those years and brought important findings to understand the structure and the functions of the different cell compartments: nucleus, mithochondria, microsomes, lysosomes. The further experimental studies on the steps and the effects of the lipid peroxidation have been deeply facilitated by Benedetti al. [3] who were able to develop a method to synthetize its carbonylic end-products, above all the aldehyde 4-hydroxy-2,3-trans-nonenal (HNE), whose chemical structure is shown in the Figure 2. HNE was shown to be produced in good amounts when the lipid peroxidation was stimulated; furthermore several experimental researches found that this aldehyde was the more cytotoxic end product of the lipoperoxidative process [4]. The first experimental works on the effects used millimolar concentrations of the aldehyde which were rather high; later the researchers found that it could display several biological effects at concentrations micromolar or less, which can be easily found in tissues even in normal conditions.

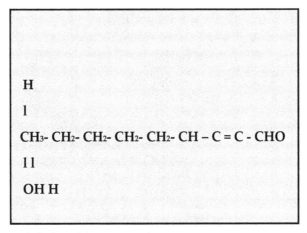

Figure 2. Structure of 4-hydroxy-2,3-trans-nonenal.

3. Lipoperoxidative effects on cell compartments

The damage to the different cell structures induced by the rat treatment with CCl₄ are similar to the alterations found in different pathological processes and are followed by similar changes in the tissue metabolism. In fact all the cell structures, mithocondria, microsomes, lysosomes, nuclei, are delimitated by membranes where the lipid peroxidation can take place and cause damage, bringing to changes in their functions. a) Effects of the lipid peroxidation on microsomes. The action of toxic compounds on cells leads to a quite rapid swelling of rough and smooth endoplasmic reticulum; the ribosomes dissociate from the rough endoplasmic reticulum and the protein synthesis decreases within less than 30 minutes [5,6]. The inhibition of microsomal glucose-6-phosphatase activity is seen very early in the action of toxic compounds; afterwords the activity of several other enzymes (hexokinase, lactate dehydrogenase, alpha and beta polymerases, 5'nucleotidase) has been found to decrease under the effect of the end products of the lipid peroxidation. However the effects of the inhibition of the protein synthesis can be seen only after different hours because the cells have a reserve of preformed proteins which can be used. The microsomes are the site of the drug metabolizing enzyme system (d.m.e.s.) which metabolizes different compounds, either endogenous components, such as different hormones, or various xenobiotics. The result of the changes induced by the d.m.e.s. on a compound can be different: the compound can be inactivated, it can change its functions or it can even acquire a toxic action. . . . CCl_4 has solvent properties in high amounts, but much smaller quantities can induce biological toxic effects through its homolitical cleavage catalysed by the d.m.e.s. In fact CCl_4 fission generates free radicals able to trigger the lipid peroxidative process, starting from microsomal membranes. In the rats intoxicated with CCl_4 an early effect is the decrease of the hepatic content of cytochrome P_{450}, which is part of the d.m.e.s., the enzyme system which metabolizes the haloalkane, generating the free radicals responsible of many of its dangerous effects. The lipid peroxidation increases strongly in the liver of rats treated with this haloalkane; it is started by the free radicals generated by CCl_4 fission in the microsomes. The decrease of the cytochrome P_{450} and the damage to the liver endoplasmic reticulum lead to an apparent and quite interesting "paradox", shown by Ugazio et al. [7], i.e. the pre-treatment with a sublethal dose of CCl_4 protects the rats from the subsequent administration of a higher, potentially lethal dose. In fact the pretrearment impairs the hepatic microsome metabolic ability and so the subsequent haloalkane dose is less metabolized and it is unable to cause a serious liver damage and the animal death. A single, non lethal dose, of CCl_4 induces fatty liver in rats; if the treatmemt is unrepeated this degenerative process can be reversible and the hepatic tissue returns to a normal anatomic aspect and to its usual physiological functions. The demonstration that the toxicity of small doses of this haloalkane was not due to its solvent properties but was the consequence of its cleavage by the d.m.e.s underlined the importance of the interaction between the various xenobiotics, foods or drugs or air inquinants, and the living organism, human or animal. A different behaviour of the cell metabolism due to genetic factors or to different other causes, such as diseases or the assumption of various substances, can higly modify the response to xenobiotics and their effects on the health. An important step in the studies on CCl_4 toxicity was the finding that the pre-treatment of rats with antioxidants (DPPD, GSH, propyl gallate) could prevent them both from liver damage and cell death, suggesting the role of an oxidative mechanism in the development of its toxic action [8,9]. b) Lipoperoxidative effects on mitochondria. The effects of the lipid peroxidation induced by toxic compounds can be seen also in mitochondria which show a swelling and a change in ATP synthesis [9,10]. The studies with the electron microscope revealed the damage to mitochondrial components. In the first phases of the mitochondrial swelling the production of ATP can increase for an easier entry of the substrates in the organelle through the more permeable membrane, then it decreases and stops. In many intoxication both the decrease of the synthesis of ATP and the damage of the plasma membrane contribute to an increase of the Ca^{2+} influx in cells, which comes before the cell necrosis. [11]. Ca^{2+} concentration is strictly checked in cells; the cytosolic free calcium is maintained at concentrations < 0.1 μM , which are much lower than the extracellular ones. This control is very important because an increase in Ca^{2+} activates several enzymes (ATPases, phospholipases, proteases, endonucleases) which can damage the same cell structures. c) Effects on lysosomes. The lysosomes are damaged by the attack of free radicals and by the onset of the lipid peroxidation in their membrane (12,13); so their lytic enzymes can be released in the cytoplasm. In the injured cells the intracellular pH tend to be acid; so the released lysosomal enzymes can be activated and destroy important cell components. The damage of the lysosomal membrane can lead to the enzymatic digestion of proteins, RNA, DNA and the cell dies by necrosis. The occurrence of the lipid peroxidation in lysosomes can also lead to the inactivation of their lytic enzymes if the lipoperoxidative rate is very high; Krohne et al. [14] have shown that the lipid peroxidaction end products, HNE and malonaldehyde (MDA), inactivated lysosomal cysteine proteases by covalent binding to their active center. d) Lipid peroxidation-induced changes in the nucleus. The cell nucleus has a membrane, like the other cell organelles; if the lipid peroxidation occurs in the nuclear membrane, it can cause serious damage. The nuclear importance is due to the presence of the DNA molecule; a damage to the DNA can lead to alterations in the codified proteins. If these changes involve important sites of the molecule, the protein can be no more functional. Some alterations in the DNA molecule are lethal, others lead to vital, but modified cells. Some changes in the DNA molecule can bring to the generation of transformed cells which show different changes in their morphology, metabolism and behaviour toward the near cells. The reaction of the different products of the lipid peroxidation with DNA has been extensively studied [15,16]; it can lead to the formation of adducts to DNA bases, which have profound mutagenic potential. The alterations of DNA molecule are believed to be important in the pathogenesis of cancer; a special attention has been given to the oncogenes and antioncogenes, which play an important role in regulating cell division.

4. Effects of lipid peroxidation in inflammation

The lipid peroxidation plays an important role in inflammation; in this process its presence is constant and its degree can reach high values.

Inflammation is the local response to any tissue damage. It is characterized by two main events: 1. changes in the blood flow in the microcirculation of the injured site. 2. recruitment of leukocytes, neutrophils and monocyte-macrophages; these cells phagocyte and destroy the agents of the tissue injury: bacteria, virus, parasites, dead cells, tissue debris. The leukocytes which gain the damaged tissue are activated by cytokines (IL-1, IL-6, TNF, MCP-1), which trigger the respiratory or phagocytic burst in them. This process is characterized by a strong increase of the consumption of oxygen, which is used to produce the superoxide anion ($O_2^{\bullet-}$); its synthesis is catalysed by the NADPH (nicotinamide adenine dinucleotide phosphate) oxidase [17]. The NADPH oxidase is formed by a complex of proteins which are located both in the plasma membrane and in the cytoplasm in the resting neutrophil. When the neutrophil is activated by different stimuli (the phagocytosis itself, various cytokines), the components of the NADPH oxidase assemble on the membrane of the phagosome and the enzymatic complex can reduce oxygen to superoxide anion as shown in the following reaction:

$$NADPH + O_2 \rightarrow NADP^+ + O_2^\bullet$$

Chemotactic compounds

- Microrganisms and microbic compounds
- Complement components (C3a,C5a)
- Leukotriene B_4
- Lipid peroxidation products

    4-hydroxyexenal (HEE)
    4-hydroxyoctenal (HOE)
    4-hydroxynonenal (HNE)

- Cytokines

    Interleukin-8 (specific for neutrophils)
    Monocyte chemotactic protein-1
    (specific for monocyte-macrophages)

Figure 3. List of the principal chemotactic compounds.

The superoxide anion is a free radical and so it is highly reactive.Two molecules of superoxide anion can react together and form the hydrogen peroxide (H_2O_2); this molecule has a low bactericidal power and it is also used as a disinfectant in pharmacology. In the tissues the hydrogen peroxide is used in a reaction catalysed by the myeloperoxidase (MPO) to form hypochlorite (OCl$^\bullet$). The microbial power of the hypochlorite is very strong; furthermore it can oxidase protein and lipids and so it can trigger the lipid peroxidation. The hydrogen peroxide can also be converted to the hydroxyl radical (OH$^\bullet$), a free radical with a very short lifetime; in fact it reacts with the nearest molecule to acquire a more stable configuration. Both the anion superoxide and the hydroxyl radical are able to induce the lipid peroxidation and this fact explains its steady presence in inflammation. The occurrence of the lipoperoxidative process may lead to a worsening of the tissue damage, but it also contributes to the recruitment of leukocytes, both neutrophils, and monocyte-macrophages since some lipid peroxidation end products display a chemotactic power, as shown in the Figure 2. The migration of leukocytes from blood to the inflamed tissue requires several passages [18]. Both the leukocytes and the endothelial cells need the presence of adhesion molecules on their surface to allow the leukocyte binding to the microcirculation of the damaged tissue. At first the binding is not firm and allows the leukocyte rolling on the endothelial surface; afterwards it becomes very strong and this firm adhesion is followed by the leukocyte passage outside the blood vessels to gain the site of the inflammation.

The chemotactic compounds or chemotaxins display different actions on the leukocytes. The term "chemotaxis" refers to the ability of a molecule to stimulate the oriented migration of a cell in the presence of a chemical gradient of the chemotactic compound or chemotaxin; the leukocytes have specific receptors for the different chemotaxins and move toward the site where the chemotaxins have the highest concentration. Beside this property, the chemotactic compounds display many other functions on the leukocytes: they induce the phagocyte burst, activate the adhesion molecules which are expressed on the plasmamembrane of the neutrophils, promote the synthesis of different cytokines, expecially by the macrophages.

a. Chemotactic activity of the products of the lipid peroxidation. The lipid peroxidation end product HNE has been shown to display a chemotactic power toward the polymorphonuclear leukocytes. At first this property was found by Curzio et al. (19) on rat neutrophils; the chemotactic concentrations of this aldehyde ranged from 10 μM to 0.1 μM. These doses are rather low and are devoid of any cytotoxic property. HNE chemotactic activity was initially demonstrated "in vitro" by the use of a Boyden chamber. This chamber has two compartments separated by a filter made of a mixture of cellulose esters with a pore size of 3 μ; the cells are placed in the upper chamber, the solution containing the substances to be tested in the lower one. The so mounted chamber is incubated at 37ºC for 75 min; then the chamber is removed and opened; the filter is removed, fixed in ethanol and stained with haematoxylin. The cell migration can be evaluated under the light microscope by the leading front technique. The first demonstration of HNE chemotactic power was obtained "in vitro", but afterwards it was confirmed by Schaur et al. [20] who carried on "in vivo" experimental researches. They induced an aseptic inflammation in the subcutaneous tissue of a rat leg by injecting in it some polydextrane Sephadex G-200; in control rats they inoculated Sephadex alone, while in the experimental group of rats they inoculated Sephadex together with a solution of preformed HNE. When they examinaed the histological samples obtained from the two groups of rats, they found the migration of neutrophils in both of them, but their number was much more higher around the Sephadex plus HNE. In their experimental researches the authors excluded the presence of any cytotoxic effects by the aldehyde concentrations able to stimulate the oriented migration of the neutrophils. Beside HNE, other 4-hydroxy-alkenals have been shown to display a chemotactic activity toward rat neutrophils: 4-hydroxy-2,3-hexenal(HEE) and 4-hydroxy-2,3-octenal(HOE). HOE was the most active of the lipoperoxidative end products; it could stimulate the oriented migration of neutrophils even at very low concentrations [21] between 10^{-11} and 10^{-8} M. Most chemotactic compounds can activate a phosphoinositide specific phospholipase C (PL-C) [22]; their stimulation of PL-C activity is mediated by a regulatory G protein and leads to the production of diacylglycerol and inositol-1,4,5-tris-phosphate (Ins-P$_3$). The diacylglycerol activates the protein kinase C and the Ins-P$_3$ promotes the mobilization of Ca^{++} from intracellular stores. The well known chemotaxin N-formylmethionyl-leucyl-phenylalanine (fMLP) increases the PL-C activity of neutrophils and its action is prevented by the cell pretreatment with pertussis toxin, which ADP ribosylates the alpha subunit of some G proteins. The chemotactic 4-hydroxy-alkenals formed by the lipid peroxidation have been found to activate the PL-C [23] of rat neutrophils and a good correspondence could be found between the concentrations able to increase the PL-C activity and those which regulated the cell migration. The pretreatment of neutrophils with pertussis toxin prevented the activation of PL-C by HOE, too; this finding suggested that its mechanism of action was like that of other well known chemotaxins. This discovery of the stimulation of an enzyme activity by very low doses of 4-hydroxyalkenals represented a clean change in the evaluation of the lipoperoxidative process and of the functions of its end-products. The first experimental studies on the biological effects of the lipid peroxidation supplied a lot of proofs about the inhibition of several enzymes in tissues where the lipid peroxidation rate was stimulated [5] or about the decrease of their activity in tissue homogenates or in subcellular fractions incubated in the presence of high concentrations of HNE [24].

b.   Activation of the exocytosis by 4-hydroxynonenal. HNE was found to induce the exocytosis in DMSO-differentiated HL-60 cells. [25] This human promyelocitic cell line was chosen because it could be induced to differentiate toward the granulocytic cell line and therefore it represented a good in vitro model to study the mechanism of action of a chemotactic compound, like 4-hydroxynonenal. The exocytosis was valued by measuring the secretion of ß-glucuronidase, an enzyme of the azure granules, by the cells incubated in the presence of different HNE concentrations. The exocytosis was triggered by HNE doses between 10^{-8} and 10^{-6} M, which are wholly devoid of any cytotoxic power. The lack of any effect on the cell viability was checked by measuring the release of lactate dehydrogenase (LDH) in the cells incubated at 37°C for 1 hour in the presence of different HNE concentrations; the presence of HNE between 0.01 and 1.0 µM failed to induce any increase of the enzyme loss by the cells in the incubation period [25].

c.   Stimulation of IL-8 release by 4-hydroxynonenal. I recently found HNE ability to change the release of the chemokine interleukin-8 (IL-8) in DMSO–differentiated HL-60 cells [26]; the aldeyde failed to modify the intracellular concentration of IL-8, but after 30 min of incubation it began to enhance the chemokine release. The increase of IL-8 level in the cell suspensions incubated in the presence of HNE was quite slow and became remarkable only after 1 h. This fact suggested that the effects shown by the aldehyde both on the chemotaxis and on the exocytosis were not mediated through the release of IL-8.

d.   4-hydroxynonenal induced synthesis of cyclooxygenase-2. The vascular reactions of inflammation are regulated by many chemicals mediators; among them the prostaglandins influence several cell functions. The prostaglandins play an important role in inflammation; above all the PGE$_2$ and the PGD$_2$ induce vasodilation and increase the permeability of post-capillary venules. These prostaglandins are produced from arachidonic acid by two cyclooxigenases, COX-1 and COX-2; the COX-1 is constitutively, while the COX-2 is inducible. The COX-2 js present in leukocytes and mastzellen and is induced by different mediators of inflammation. HNE has been shown to induce the synthesis of COX-2 [27] too; this finding underlines the importance of the lipid peroxidation role in inflammation.

5. Positive and negative actions of inflammation

Inflammation has many positive effects and it is considered a defensive response of the organism, but it is followed by negative aspects which may contribute to increase the tissue damage, as shown in Figure 3.

The leukocytes which reach a damaged tissue can remove the injurious agents. They can phagocyte and kill the microrganisms of an infectious disease; they also phagocyte the dead cells or the cell debris which are left in any damaged tissue.

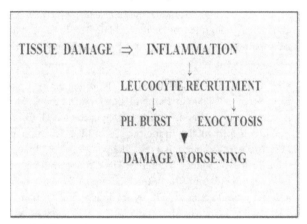

Figure 4. Main events in inflammation.

The Figure 3 underlines that the positive functions of the leukocytes are also followed by unpleasant effects, which can be caused both by the phagocytic burst and by the exocytosis. The phagocytic burst leads to the generation of free radicals and reactive oxygen species (ROS) which can diffuse outside the phagocytic cells and amplify the effects of the initial injurious agent. Moreover the induction of the lipid peroxidation by the ROS and the free radicals can worsen the tissue damage. . The exocytosis is a kind of physiological, controlled secretion of lysosomal enzymes by neutrophils and is activated by several chemotactic agent; their azurofil granules fuse with the plasma membrane and release their content in the extracellular space. In this way different lytic enzymes can diffuse in the inflammed tissue; the blood stasis which is always present in the late phases of the inflammatory process favours the lowering of pH in the damaged tissue; this fact leads to the activation of the released lysosomal enzymes and the tissue itself can be damaged.

6. Action of lipid peroxidation on atherosclerosis

Medical progress has brought good successes against many diseases in the past century. Antibiotics can win many infectious agents; the progress in surgery can correct cardiac malformations; the consequences of a vessel obstruction can be obviated by the insertion of a by pass.

In our times, when the life is becoming longer, the atherosclerosis represents a serious problem which can compromise the life of many people. The complications of the atherosclerosis are becoming the main causes of death.

Its pathological lesion is the atheroma, which is localized in the arteries of big and medium calibre: aorta, carotids, coronaries.

The atheroma is characterized by an accumulation of cholesterol and cholesterol esters both inside and outside the cells. It contains lipid-loaded cells, called foam cells. They are thought to derive from monocytes or smooth muscular cells, which have migrated in the arterial intima and have been engulfed by oxidized LDL. The lipid peroxidation plays an important role in the pathogenesis of this diffuse process through its intervent in LDL oxidation.

The modification of LDL by oxidation leads to its unregulated uptake by intimal macrophages to form foam cells[28]. In the oxidation of LDL the lipid peroxidation is stimulated and it contributes to modify their apolipoprotein. HNE, the major lipid peroxidation end product is formed also in the process of LDL oxidation and is present in the oxidized-LDL. Esterbauer et al. found that the aldehyde alone could modify the LDL. He incubated native LDL in the presence of different HNE concentrations and observed its covalent binding to the apolipoprotein B with the blockage of the epsilon-amino groups on lysine residues. Both the modification of LDL by oxidation and its modification by HNE binding were associated with an increased degradation by macrophages and a lipid loading of them.

The migration of macrophages toward the arterial intima is stimulated by chemotactic compounds, like the migration of leukocytes to a site of inflammation. The oxidized LDL have been shown to stimulate the synthesis of the monocyte chemotactic protein-1 (MCP-1) by macrophages [29]. This cytokine has a chemotactic power specific for the monocyte-macrophages; however the macrophage recruitment in the atheroma could be also favoured by the 4-hydroxyalkenals which have been found in the oxidized LDL: 4-hydrohexenal (HEE), 4-hydroxyoctenal (HOE) and 4-hydroxynonenal (HNE). These aldehydes are lipid peroxidation end products which display both a cytotoxic and a chemotactic power. They are likely to be produced by the oxidation of the LDL which have reached the intima of arteries and can contribute to the recruitment of monocyte-macrophages[30]; a direct cytotoxic effect on the foam cells of the atheroma was considered unlikely by Muller because it required higher levels of the aldehydes.

7. The lipid peroxidation role in ischemia-reperfusion.

The onset of the lipid peroxidation in a tissue requires the presence of molecular oxygen; however the ischemia can induce changes of the cell metabolism which may increase the tissue damage if the blood supply returns. This unexpected fact happens in the ischemia-reperfusion[31]. During ischemia the lack of oxygen causes the catabolism of ATP with an increased production of ipoxantine. which is an oxidable substrate for the xanthine dehydrogenase. Moreover in the ischemic tissue there is the conversion of the native xanthine dehydrogenase to a superoxide producing-oxidase; this conversion is thought to be produced by a calcium triggered protease. In the reperfusion the O_2 which reaches the tissue is transformed by the xanthine oxidase in superoxide anion; this free radical contribute to extend the tissue damage induced by the ischemia.

The tissue necrosis triggers an inflammatory process and the leukocytes which can reach the tissue in the reperfusion can worsen the tissue damage through the production of ROS and the release of lytic enzymes.

8. Conclusions

Tne lipid peroxidation can be regarded as a common process which happens in our cells. In fact low levels of the lipid peroxidation end products have been shown in tissues even in normal conditions [20]. The rate of the lipid peroxidation can be stimulated by the ROS and the free radicals which can arise also from the normal metabolism of cells.

The lipid peroxidation rate can be increased by some xenobiotics; the experimental works about the action of the haloalkane CCl_4 have been the source of the first explanations of its effects [1-3].

In any inflammation the leukocytes, above all the neutrophils and the macrophages, produce the superoxide anion and other free radicals, which can increase the lipoperoxidative rate[18]. A high degree of the lipid peroxidation is followed by some unavoidable damages to the tissue; some lesions can be reversible and can be repaired by the normal reparative process or by the aid of a pharmachological support; however a low alteration of the tissue integrity and function can follow any inflammatory event .

The aging is viewed by some authors [32] as the sum of the repeated tissue damages which occur in our life and the lipid peroxidation can take a part in them.

Another aspect of the lipid peroxidation regards its possible modulation of the normal metabolism; low concentrations of HNE, the major lipid peroxidation product, can modulate the activity of some enzymes, like the phosphoinositide-dependent phospholipase C [23] and several other enzymes [24].

Author details

Maria Armida Rossi

Department of Experimental Medicine and Oncology, University of Turin, Turin, Italy

Acknowledgement

The Author is grateful to his teacher, the Prof. Mario Umberto Dianzani (University of Turin, Turin, Italy) and to the late Prof. Hermann Esterbauer (University of Gratz, Gratz, Austria).

9. References

[1]   Comporti M, Saccocci C, Dianzani MU. Effect of CCl₄ in vitro and in vivo on lipid peroxidation of rat liver homogenates and subcellular fractions. Enzymologia 1965;29(3)185-204

[2]   Recknagel RO, Ghoshal AK. Lipoperoxidation of rat liver microsomal lipids induced by carbon tetrachloride. Nature 1966;210(5041) 1162-63

[3]   Benedetti A, Comporti M, Esterbauer H. Identification of 4-hydroxynonenal as a cytotoxic product originating from the production of liver microsomal lipids. Biochim Biophys Acta 1980; 620(2), 281-96

[4]   Esterbauer H, Schaur RJ, Zollner H. Chemistry and bio-chemistry of 4-hydroxy-nonenal, malonaldehyde and related aldehydes. Free Radic Biol Med. 1991;11(9) 81-128

[5]   Glende EA, Hruszkawycz AM, Recknagel RO. Critical role of lipid peroxidation in carbon tetrachloride-induced loss of aminopyrine demethylase, cytochrome P-450 and glucose-6-phosphatase. Biochem Pharmacol. 1976;25(19) 2163-70

[6]   Bidlack WR, Tappel AL. Damage to microsomal membrane by lipid peroxidation. Lipids 1973; 8(4)177-82

[7]   Ugazio G, Torrielli MV, Burdino E, Sawyer BC, Slater TF. Long range effects of products of carbon tetrachloride-stimulated lipid peroxidation. Biochem Soc Trans. 1976; 4(2) 353-6

[8]   Dianzani MU, Ugazio G. Lipoperoxidation after carbon tetrachloride poisoning in rats previously treated with antioxidants. Chem Biol Interact. 1973; 6(2) 67-79

[9]   Christophersen BO. The inhibitory effect of reduced gluthathione on the lipid peroxidation of the microsomal fraction and mitochondria. Biochem J. 1968; 106(2) 515-22

[10] Hyams DE, Issebalbacher KJ. Prevention of fatty liver by administration of adenosine triphosphate. Nature 1964; 204,1196-7

[11] Dianzani MU, Viti I. The content and distribution of cyto-chrome c in the fatty liver of rats. Biochem J. 1955; 59(1) 141-5

[12] Yoshikawa T, Yokoe N, Takamura S, Kato A, HottaT, Matsumura N, Ikezaki M, Hosokawa K, Kondo M. Lipid peroxidation and lysosomal enzymes in D-galactosamine hepatitis and its protection by vitamin E. Gastroenterol Jpn. 1979; 14(1) 31-9

[13] Mak IT. Misra HP,,Weglicki WB. Temporal relationship of free radical-induced lipid peroxidation and loss of latent enzyme activity in highly enriched hepatic lysosomes. J Biol Chem. 1983; 258(22) 13733-7

[14] Krohne TU, Kaemmer E, Hotz EG, Kopitz J. Lipid peroxidation products reduce lysosomal protease activities in human retinal pigment. J Exp Eye Res. 2010; 90(2) 261-6

[15] Lim P, Wesenschelln GE, Holland V, Lee DH, Pfeifer GP, Rodriguez H, Termini J. Peroxyl radical mediated oxidative DNA base damage. Biochemistry 2004; 43(49) 15339-48

[16] Winczura A, Zdzalik D, Tudek B. Damage of DNA and proteins by major lipid peroxidation products in genoma stability. Free Rad Res. 2012;46(4):442-59

[17] Babior BM. NADPH oxidase. Curr Opin Immunol. 2004;(16)42-7

[18] Muller WA. Leukocyte-endothelial-cell interactions in leukocyte transmigration and the inflammatory response. Trends Immunol. 2003;24(6) 327-34

[19] Curzio M., Torrielli MV., Giroud JP., Esterbauer H., Dianzani MV. Neutrophil chemotactic responses to aldehydes. Res Commun Chem Pathol Pharmacol. 1982;36(3)463-476

[20] Schaur RJ, Dussing G, Kink E,Schauenstein E, Poscch W, Kukovetz E, Egger G. The lipid peroxidation product 4-hydroxynonenal is formed by– and is able to attract rat neutrophils in vivo. Free Rad Res.1994; 20(4) 365-373

[21] Curzio M.,Esterbauer H.Dianzani MU. Chemotactic power of 4-hydroxy-octenal. IRCS Med Sci. 1983;11,521

[22] Sandborg RR, Smolen JE Biology of disease. Early biochemical events in leukocyte activation. Lab Invest 1988,59(3) 300-320

[23] Rossi MA, Di Mauro C,Esterbauer H, Fidale F, Dianzani MU. Activation of phosphoinositide-specific phospholipase C by the chemotactic aldehydes 4-hydroxy-2,3-trans-nonenal and 4-hydroxy-2,3-trans-octenal. Cell Biochem Funct. 1994;12(4) 275-280

[24] Dianzani MU. Biochemical effects of saturated and unsaturated aldehydes. In: McBrien DCH and Slater TF (eds) Free Radicals, Lipid Peroxidation and Cancer. Academic Press, London; 1981. p.129-158,

[25] Maggiora M, Dianzani MU, Rossi MA Effect of 4-hydroxy-nonenal. A lipid peroxidation product, on exocytosis in HL-60 cells. Cell Biochem Funct. 2002;20(4)303-307

[26] Maggiora M., Rossi M.A. Changes in IL-8 release and intracellular content in DMSO-differentiated HL-60 cells after treatment with 4-hydroxynonenal. Cell Biochem. Funct. 2008; 26(5) 566-570

[27] Uchida K, Kumagai T. 4-hydroxynonenal as a COX-2 inducer. Mol Aspects Med. 2003, 24(4-5) 213-8

[28] Hoff HF,O'Neil J, Chisolm GM 3rd, Cole TB, Quehenberger O, Esterbauer H, Jürgens G. Modification of low density lipoprotein with 4-hydroxynonenal induces uptake by macrophages. Arteriosclerosis. 1989; 9(4) 538-4

[29] Wang GP, Deng ZD, Ni J, Qu ZL. Oxidized low density lipoprotein and very low density lipoprotein enhance expression of monocyte chemoattractant protein-1 in rabbit peritoneal exudate macrophages. Atherosclerosis 1997; 133(1) 31-6

[30] Muller K, Hardwick SJ, Marchant CE, Law NS, Waeg G, Esterbauer H. Carpenter KL, Mitchinson MJ. Cytotoxic and chemotactic potencies of several aldehydic components of oxidized low density lipoproteins for human monocyte-macrophage. FEBS Lett. 1996,388(2-3)165-168

[31] McCord JM, Roy RS, Schaffer SW. Free radicals and myocardial ischemia. The role of xanthina oxidase. Adv Myocardiol. 1985;5:183-9

[32] Balaban RS, Nemoto S, Finkel T. Mitochondria,oxidants and aging. Cell 2005,120(4) 483-95

Chapter 5

Lipid Peroxidation and Antioxidants in Arterial Hypertension

Teresa Sousa, Joana Afonso, Félix Carvalho and António Albino-Teixeira

Additional information is available at the end of the chapter

1. Introduction

This chapter aims at giving a critical overview of the major oxidant and antioxidant changes in arterial hypertension, summarizing the experimental and clinical evidence about the involvement of oxidative stress in the pathophysiology of hypertension, either as a cause or a consequence of this disease. This review also provides a description of the biomarkers commonly used to evaluate lipid peroxidation and antioxidant defenses in experimental and human hypertension. Finally, we review the strategies (antioxidants, antihypertensive drugs) known to prevent or ameliorate oxidative damage, both in animal models of hypertension and hypertensive patients.

2. Pathophysiological role of oxidative stress in arterial hypertension

2.1. ROS sources and oxidative pathways involved in the pathogenesis of hypertension

In aerobic organisms, the beneficial effects of oxygen come with the price of reactive oxygen species (ROS) formation. These highly bioactive and short-lived molecules can interact with lipids, proteins and nucleic acids, causing severe molecular damage. However, living organisms have evolved specific mechanisms to adapt to the coexistence of ROS. In physiological conditions, there is a delicate balance between oxidants and antioxidants that not only protects our cells from the detrimental effects of reactive oxygen species (ROS), but also allows the existence of redox signaling processes that regulate cellular and organ functions. However, the disruption of redox homeostasis, leading to persistent high levels of ROS, is potentially pathological [1, 2]. Besides ROS, another group of molecules collectively designated as reactive nitrogen species (RNS) also exerts important functions in diverse physiological and pathological redox signaling processes. The excess of RNS is often termed nitrosative stress [3, 4].

ROS can be classified into two main categories: free radicals [e.g. superoxide (O_2^-), hydroxyl (HO·), peroxyl (ROO·)], which are highly reactive species due to the presence of one or more unpaired electrons, and non-radical oxidants [e.g. singlet oxygen ($^1O_2$) hydrogen peroxide (H_2O_2), hypochlorous acid (HOCl)] that have generally more specific reactivity and higher stability [3, 5, 6]. RNS include nitric oxide (·NO) and nitrogen dioxide radicals (·NO$_2$ and also non radicals such as nitrous acid (HNO$_2$), peroxynitrite (ONOO$^-$), peroxynitrous acid (ONOOH) and alkyl peroxynitrites (ROONO) [3]. Among biological ROS and RNS, O_2^-, H_2O_2, ·NO and ONOO$^-$ appear to be especially relevant in neuronal, renal and vascular control of blood pressure [3, 7 ,8] (Table 1). Major sources of ROS (and also RNS) within these systems include, but are not limited to, NADPH oxidases, xanthine oxidase, mitochondrial respiratory chain enzymes, ·NO synthases and myeloperoxidase [3, 8, 9].

	Free radicals	Non radical oxidants
ROS	O_2^- HO· ROO·	H_2O_2 HClO
RNS	·NO	ONOO-

Table 1. Reactive oxidant species involved in cardiovascular and renal physiology or pathophysiology

NADPH oxidases (Nox) are enzyme complexes that catalyze the reduction of molecular oxygen using NADPH as an electron donor. Generally, the product of the electron transfer reaction is O_2^- but H_2O_2 is also rapidly formed from dismutation of Nox-derived O_2^- due to the presence of superoxide dismutase (SOD) in the cells or by spontaneous reaction. Nox-derived ROS have been shown to play a role in host defense and also in diverse signaling processes [10]. The Nox family comprises seven members (Nox1-5 and Duox1-2) with distinct tissue distribution and functions [10, 11]. So far, only Nox1, Nox2 and Nox4 have been shown to play relevant roles in hypertension pathophysiology [5, 8, 10]. These isoforms are localized in major sites of blood pressure control. For example, Nox1, Nox2 and Nox4 are expressed in the central nervous system where they appear to regulate sympathetic nerve activity [8]. Nox2 and Nox4 participate in the regulation of renal functions and contribute to end-organ damage associated with hypertension [8, 12]. In the vasculature, Nox1 controls smooth muscle cell growth and migration, Nox2 contributes to endothelial dysfunction and Nox4 controls vascular smooth muscle cell differentiation and improves endothelial-dependent vasodilatation [8, 13, 14].

Xanthine oxidoreductase has two inter-convertible forms, xanthine dehydrogenase (XDH) and xanthine oxidase (XO), that participate in purines metabolism catalyzing the conversion of hypoxanthine to xanthine and xanthine to uric acid [15, 16]. XDH preferentially uses NAD$^+$ as an electron acceptor while the oxidase reduces molecular oxygen in a reaction that generates O_2^- and H_2O_2 [15, 16]. The XO form predominates in oxidative stress conditions and may contribute to endothelium dysfunction due to its localization in the luminal surface of vascular endothelium [16, 17]. Besides the production of ROS by XO, both XDH and XO generate uric acid which possesses antioxidant properties, such as scavenging of ONOO$^-$ and HO·, prevention of oxidative inactivation of endothelium enzymes and stabilization of Vitamin C

[18-22]. On the other hand, uric acid may also have prooxidant and proinflammatory effects [23, 24]. Indeed, high systemic levels of uric acid are associated with increased cardiovascular disease and poor outcome but it is not clear whether these effects reflect deleterious actions of uric acid or the oxidative damage caused by XO-derived ROS [23, 25].

Mitochondrial respiratory chain enzymes are primary intracellular sources of ROS. More than 90% of the total oxygen consumed by aerobic organisms is utilized by mitochondrial oxidases which produce ATP in a process coupled to the reduction of cellular oxygen to water [26]. About 1-4% of the oxygen used in these reactions is converted to O_2^- and H_2O_2 which can be largely detrimental to mitochondrial functions if not adequately detoxified [26-28]. ROS levels in the mitochondria are regulated by the respiratory rate and manganese SOD [29]. Hypertensive animals have increased mitochondrial ROS production in the vessels, kidney and CNS [30-32].

·NO synthases (NOS) constitute a family of enzyme isoforms (neuronal NOS, nNOS; inducible NOS, iNOS; endothelial NOS, eNOS) that produce ·NO in a reaction that converts L-arginine to L-citrulline [28]. However, in conditions of limited bioavailability of the cofactor tetrahydrobiopterin, or the substrate L-arginine, these enzymes become unstable and reduce molecular oxygen to O_2^- instead of ·NO production (uncoupled NOS) [28, 29]. NOS uncoupling is more often described for eNOS and is triggered by oxidative/nitrosative stress [28, 33]. Numerous experimental studies have shown that arterial hypertension is associated with eNOS dysregulation and endothelial dysfunction [34, 35].

Myeloperoxidase (MPO) is a heme protein secreted by activated neutrophiles and monocytes in inflammatory conditions and produces several oxidizing molecules that can affect lipids and proteins [28, 36]. MPO uses H_2O_2 to produce ROS such as HOCl, chloramines, tyrosyl radicals and nitrogen dioxides [36, 37]. Although MPO-derived ROS have a primary role in microbial killing, they also cause tissue damage in the heart, vessels, kidney and brain and appear to contribute to endothelial dysfunction [37, 40]. Figure 1 illustrates the major sources of ROS and/or RNS generation.

Of all the putative oxidative pathways involved in the pathogenesis of hypertension, the impairment of endothelial-dependent vasorelaxation by O_2^- is by far the most studied [41-44]. In conditions of increased O_2^- bioavailability, this ROS rapidly inactivates endothelial-derived ·NO leading to endothelial dysfunction [41]. In addition, O_2^- may also modulate vascular tone by increasing intracellular Ca^{2+} concentration in vascular smooth muscle cells and endothelial cells [45]. The imbalance between O_2^- and ·NO also affects the renal function, leading to enhanced sodium reabsorption and increased $ONOO^-$ formation, which contributes to tissue damage [12, 46]. In the CNS, elevated O_2^- generation also appears to contribute to hypertension by reducing the cardiovascular depressor actions of ·NO in the rostral ventrolateral medulla [47]. In recent years H_2O_2 has also emerged as a pivotal molecule in the pathophysiology of arterial hypertension [48-50]. Of note, H_2O_2 seems to be even more harmful than O_2^- due to its higher life span and diffusibility within and between cells [7, 51]. Furthermore, the conversion of O_2^- to H_2O_2 appears to be favored in cardiovascular diseases since the expression and activity of SOD is enhanced by inflammatory cytokines in hypertension or in response to the pressor peptide, angiotensin II [7]. Several prohypertensive effects have been described for H_2O_2, such as increased vasoconstriction, vascular hypertrophy and hyperplasia, decreased diuresis and natriuresis and also increased spinal sympathetic outflow [7, 50, 52-58]. Increasing evidence has also shown that H_2O_2 amplifies oxidative stress by stimulating ROS generation by NADPH oxidases, XO and eNOS [7, 51]. In addition, H_2O_2 also appears to enhance the activation of the intrarenal renin-angiotensin system, a major regulator of blood pressure and renal function [49]. Altogether, these effects propagate H_2O_2 generation and prolong the redox pathologic signaling involved in blood pressure dysregulation. The oxidative mechanisms contributing to hypertension are summarized in Table 2.

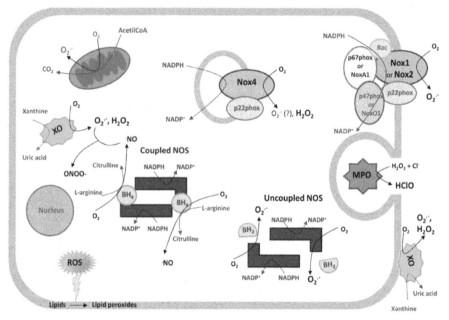

Figure 1. Sources of ROS and/or RNS generation - In normal cells, 1–2% of electrons carried by the mitochondrial electron transport chain leak from this pathway and pass directly to oxygen generating superoxide radical (O_2^-) which can be a source of other ROS. O_2^- can also be formed by xanthine oxidase (XO) which catalyzes the oxidation of hypoxanthine and xanthine. All NOX enzymes utilize NADPH as an electron donor and catalyze transfer of electrons to molecular oxygen to generate O_2^- and/or H_2O_2. Nitric Oxide synthases (NOS) generate ·NO and L-citrulline from arginine and O_2. Under pathologic conditions of oxidative stress, or when tetrahydrobiopterin (BH4) or L-arginine are deficient, NOS enzymes become structurally unstable (uncoupled NOS) resulting in production of O_2^- rather than ·NO. Activated monocytes also secrete a heme enzyme, myeloperoxidase(MPO), that uses H_2O_2 as a substrate to generate products that can oxidize lipids and proteins. One of these oxidants is hypochlorous acid (HOCl) which plays a critical role in host defenses against invading bacteria, viruses, and tumor cells but may also injure normal tissue. Within cell membranes, ROS can trigger lipid peroxidation, a self-propagating chain-reaction that can result in significant tissue damage.

Affected organ	Oxidative stress consequences	Major ROS and RNS involved
Vasculature	Impaired endothelium-dependent vasodilation	O_2^{-}, ·NO, ONOO^{-}
	Increased vasoconstriction	O_2^{-}, H_2O_2
	Increased hypertrophy and hyperplasia	O_2^{-}, H_2O_2
Kidney	Decreased blood flow	O_2^{-}, H_2O_2
	Increased salt reabsorption	O_2^{-}, H_2O_2
	Tissue damage	HO·, HClO, ONOO^{-}
Brain/Spinal cord	Increased sympathetic efferent activity	O_2^{-}, H_2O_2

Table 2. Putative oxidative pathways leading to arterial hypertension

2.2. Evidence for redox changes in experimental and human hypertension

In the last two decades several studies have consistently observed increased oxidative stress in experimental and human arterial hypertension. Studies in diverse experimental models of hypertension have demonstrated raised prooxidant activity and ROS levels, altered antioxidant defenses and increased ROS-mediated damage, both at peripheral and central sites of cardiovascular regulation [8, 33, 59]. In human hypertensive patients there is also evidence of redox dysfunction. O_2^{-} release from peripheral polymorphonuclear leucocytes is higher in hypertensive patients than in normotensive subjects [60]. Plasma H_2O_2 production is also raised in hypertensive patients. Furthermore, among still normotensive subjects, those with a family history of hypertension have a higher H_2O_2 production [61, 62]. An elevation of several oxidative stress byproducts, such as malondialdehyde, 8-isoprostanes, 8-oxo-2'-deoxyguanosine, oxidized low density lipoproteins, carbonyl groups and nitrotyrosine has also been observed in plasma or serum, urine or blood cells of hypertensive patients [63-66]. Furthermore, both enzymatic and non-enzymatic antioxidant defenses appear to be significantly reduced in human hypertension [65, 67]. Alterations of redox biomarkers in human hypertension are summarized in Table 3.

	Biomarker	Evaluated in:	Alteration in hypertensive patients	References
ROS/RNS	O_2^{-}	Peripheral PMN	↑	[60]
	H_2O_2	Plasma	↑	[61, 62]
		Lymphocytes	↑	[68]
	NOx	Plasma	↓	[69]
		Urine	↓	[70]

	Biomarker	Evaluated in:	Alteration in hypertensive patients	References
Prooxidant enzymes	NADPH oxidase activity	Mononuclear cells	↑	[71]
	p22phox (Nox subunit) mRNA and protein expression	Mononuclear cells	↑	[71]
Oxidative or nitrosative stress byproducts	Malondialdehyde (MDA)/Thiobarbituric acid reactive substances (TBARS)	Plasma	↑	[67, 72]
		Erythrocytes	↑	[64, 73, 74]
		Mononuclear cells and whole blood	↑	[65]
	F2-Isoprostane (or 8-isoprostane or 8-epi-PGF2α)	Plasma	↑	[63, 66, 74]
		Urine	↑	[63, 74, 75]
	8-Oxo-2'-deoxyguanosine		↑	[65]
			↑	[76, 77]
	Carbonyl groups	Serum	↑	[64]
	Oxidized low density lipoproteins	Plasma	↑	[63, 78]
	3-Nitrotyrosine	Plasma	↑	[66, 79]
Redox status	GSSG/GSH	Mononuclear cells and whole blood	↑	[65]
	GSH/GSSG	Erythrocytes	↓	[74]
Antioxidants	GSH	Mononuclear cells and whole blood	↓	[65]
		Erytrocytes	↓	[64]
	Uric acid	Plasma	↑	[79]
		Serum	↑	[80]
	Vitamin C (ascorbic acid)	Plasma	↓	[67]
		Serum	↓	[81]
	Vitamin E (α-Tocopherol,	Erytrhocytes	↓	[67]
	Total antioxidant status (TAS)	Plasma	↓	[63, 82]
	Ferric reducing activity of plasma (FRAP)	Plasma	↓	[74, 83]

Biomarker	Evaluated in:	Alteration in hypertensive patients	References
SOD activity	Erythrocytes	↓	[64, 74]
	Whole blood and mononuclear cells	↓	[65]
Catalase activity	Erythrocytes	↑ or ↓	[64, 74]
	Whole blood and mononuclear cells	↓	[65]
Glutathione peroxidase activity	Erythrocytes	↓	[64, 74]
	Whole blood and mononuclear cells	↓	[65]
Glutathione-S-transferase activity	Erythrocytes	↑	[64]

NOx- nitrites and nitrates; PMN – Polymorphonuclear leucocytes; GSH – reduced glutathione; GSSG- oxidized glutathione;

Table 3. Altered oxidative/nitrosative stress biomarkers in human arterial hypertension

2.3. Oxidative stress as a cause for arterial hypertension

Whether oxidant imbalance is a cause or a consequence of high blood pressure remains a debatable question. The hypothesis that oxidative stress contributes to arterial hypertension is supported by several lines of evidence: (1) the induction of oxidative stress by the administration of lead or the glutathione synthesis inhibitor, buthionine sulfoximine, or the SOD inhibitor, sodium diethyldithiocarbamate, increases blood pressure in rats [48, 84]; (2) the infusion of H_2O_2 into the renal medulla leads to hypertension [48]; the treatment of hypertensive animals with antioxidants or inhibitors of ROS production prevents or attenuates hypertension [50, 85-87]; (3) the manipulation of genes related to ROS generation or elimination can alter blood pressure [88, 89]; (4) the *in vitro* exposure of cells and tissues to exogenous oxidants reproduces events involved in the pathophysiology of hypertension [43]; (5) systemic and tissue redox dysfunction appears to precede the blood pressure elevation [90].

2.4. Oxidative stress as a consequence of arterial hypertension

Arterial hypertension is associated with oscillatory shear stress and vascular stretch caused by increased vascular pressure. These mechanical forces are known to induce oxidative stress and vascular damage [91]. Furthermore, there is evidence that lowering blood pressure *per se* causes reduction of oxidative stress and improvement in endothelial function [92]. Several antihypertensive drugs with distinct mechanisms of action have been shown to decrease oxidant biomarkers in experimental and human hypertension [93-95]. However, there is limited evidence supporting the use of antioxidants to lower blood pressure in human hypertensive patients [5, 92]. Nevertheless, the failure of these studies does not exclude a role for oxidative stress in human essential hypertension but instead suggests that the antioxidant supplementation approach was not the appropriate therapeutic strategy [96].

3. Biomarkers of redox status in arterial hypertension

The evaluation of redox status may provide valuable information about the pathogenesis and progression of arterial hypertension and related cardiovascular and renal diseases. However, the short lifetime of ROS turns their assessment in animal models and humans a significant challenge, leading to a growing interest in the development and validation of oxidative stress biomarkers. Traditional approaches to evaluate oxidant status have frequently relied on indirect measurements of ROS bioavailability (e.g. evaluation of prooxidant and antioxidant activity, oxidized products from ROS and the GSH/GSSG ratio) as indicators of normal biological processes, pathogenic processes, or pharmacologic responses to therapeutic intervention [9, 96-99].

A biomarker of oxidative stress is classically defined as a biological molecule whose chemical structure has been modified by ROS and that can be used to reliably assess oxidative stress status in animal models and humans [100]. The ideal biomarker of oxidative stress depends on its ability to contribute to an early indication of disease severity and/or its progression, as well as to evaluate therapy efficacy. The measurement of redox status biomarkers may also help to clarify the pathophysiologic mechanisms mediating oxidative injury and may allow the prediction of disease. Ideally, biomarkers of oxidative damage for human studies would be evaluated in specimens that can be collected relatively easily, such as blood or urine. However, to serve these purposes, an ideal biomarker of oxidative damage should fulfill several conditions, such as: a) being a stable product, not susceptible to artifactual induction, oxidation, or loss during sample handling, processing, analysis, and storage; b) having a well-established relationship with the generation of ROS and/or progression of disease; c) allowing direct assessment in a target tissue or being able to generate a valid substitute that quantitatively reflects the oxidative modification of the target tissue; d) being present at concentrations high enough to be a significant detectable product; e) showing high specificity for the reactive species in question and free of erroneous factors from dietary intake; f) being noninvasive; g) being measurable by a specific, sensitive, reproducible and inexpensive assay; h) being measurable across populations; i) being present in concentrations that do not vary widely in the same persons under the same conditions at different times [97].

3.1. Systemic and tissue antioxidant defenses

ROS are involved in many biological processes including cell growth, differentiation, apoptosis, immunity and defense against micro-organisms [1, 101, 102]. Low or moderate concentrations of ROS are beneficial for living organisms. However, high concentrations of ROS can cause direct damage of macromolecules such as DNA, proteins, carbohydrates, and lipids, or disrupt redox signaling and control pathways, leading to a myriad of human diseases [103]. ROS bioavailability is determined by the balance between their production by prooxidant enzymes and their clearance by various antioxidant compounds and enzymes [1]. As defined by Halliwell and Gutteridge, an antioxidant is any substance that, at low concentration, is able to significantly delay or inhibit the oxidation of an oxidizable substrate [104]. Biological antioxidant defenses have evolved to match the diversity of prooxidants and several enzymatic and non-enzymatic molecules exist in cells and body fluids to control ROS levels within the physiological range [105]. The coordinated action of antioxidants results in the interception and deactivation of the damaging species. For example, the radical chain events initiated by free radicals can be terminated by the interaction of radicals with different non-enzymatic antioxidants [e.g. GSH, ascorbic acid, uric acid, α-tocopherol, etc] or prevented by specialized enzymatic defenses such as SOD, catalase and glutathione peroxidase (GPx) [105, 106]. The reduction of antioxidants bioavailability disrupts redox homeostasis leaving organisms more vulnerable to oxidative damage. Therefore, antioxidants may be useful biomarkers for risk stratification and disease prognostication.

3.2. Enzymatic antioxidants defenses

All eukaryotic cells possess powerful antioxidant enzymes which are responsible for neutralizing ROS.The first line of defense against ROS is achieved by SOD which is active in catalyzing the detoxification of O_2^-. This radical can be readily converted into H_2O_2 by SOD enzymes present in the cytosol and organelles (Cu,Zn-SOD or SOD-1), mitochondria (Mn-SOD or SOD-2) and extracellular fluids (EC-SOD or SOD-3) [36, 107, 108]. H_2O_2 generated in this reaction can be further decomposed to water and oxygen. This is achieved primarily by catalase in the peroxisomes and also by GPx enzymes in the cytosol and mitochondria [107, 108]. GPx are selenium-containing enzymes whose activity is dependent on GSH availability [108]. Besides neutralizing H_2O_2, GPx also degrades lipid hydroperoxides to lipid alcohols [36]. These reactions lead to the oxidation of GSH to GSSG. Catalase and GPx are differentially required for the clearance of high-levels or low-levels of H_2O_2, respectively [36]. Figure 2 illustrates major antioxidant enzymatic pathways.

In addition to these key antioxidant enzymatic defenses, there are other specialized enzymes with direct and/or indirect antioxidant functions. Glutathione reductase (GR) is responsible for the replenishment of GSH from GSSG disulphide. Glutathione-S-transferase catalyzes the conjugation of GSH with reactive electrophiles and is also involved in the detoxification of some carbonyl-, peroxide- and epoxide-containing metabolites produced within the cell in oxidative stress conditions [109]. Peroxiredoxins are selenium-independent enzymes that decompose H_2O_2, organic hydroperoxides and peroxynitrite [110]. Thioredoxin (Trx) and glutaredoxin (Grx) systems include several enzymes that regulate the thiol-disulphide state of proteins and influence their structure and function [110]. Trx isoforms reduce disulphide bonds in proteins, especially in peroxiredoxins and Trx reductase regenerates the oxidized Trx. Grx protects proteins SH-groups from irreversible oxidation by catalyzing S-glutathionylation and restores functionally active thiols through catalysis of deglutathionylation [110]. Grx enzymes are functionally coupled to GR which reduces the GSSG produced in the deglutathionylation reaction [110].

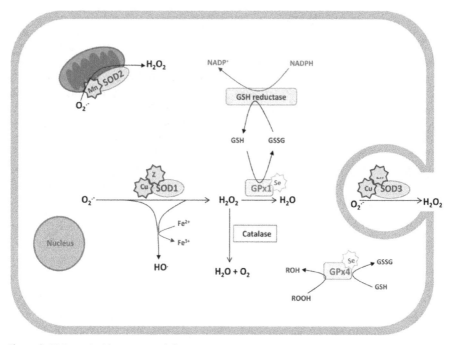

Figure 2. Major antioxidant enzyme defenses

Hypertensive patients have reduced activity and decreased content of antioxidant enzymes, including SOD, GPx, and catalase [43]. However, several studies have also described an adaptive increase in antioxidant enzyme activities in some experimental models of hypertension [50, 111, 112]. The uncoordinated activity of antioxidant enzymes may aggravate oxidative stress. For example, the increased dismutation of O_2^- by SOD significantly increases H_2O_2 concentration, and may lead to deleterious consequences for the tissue in the absence of compensation of catalase and GPx activities [113]. Examples of altered antioxidant defenses in human and experimental hypertension are shown in Table 3 and Table 4, respectively.

Biomarker	Evaluated in:	Alteration	Hypertension Model	Reference
SOD	Brain	↓ expression and activity of Mn-SOD	Spontaneously hypertensive rats (SHR)	[114]
		↓ Cu, Zn-SOD activity	Stroke prone spontaneously hypertensive rats (SHRSP)	[115]
		↓ expression and activity of SOD1 and SOD2	SHR	[116]

Biomarker	Evaluated in:	Alteration	Hypertension Model	Reference
	Kidney	↓ expression of EC-SOD	SHR	[117]
		↑ SOD activity	Angiotensin II (Ang II) induced hypertension	[49]
		↓ expression of SOD1 and SOD3	SHR	[118]
	Arteries	↑ SOD activity	Hypertension induced by renin-angiotensin system (RAS) activation	[50]
		↓ SOD activity	SHR	[119, 120]
		↑ expression and activity of Cu, Zn-SOD and Mn-SOD	SHR	[121]
Catalase	Brain	↓ Catalase expression and activity	SHR	[116]
		↓ Catalase activity	Renovascular hypertensive rat; SHR	[122, 123]
	Kidney	↑ Catalase activity	SHR; Ang II-induced hypertension;	[49, 122]
		↓ Catalase expression	SHR	[124]
		↑ Catalase expression	SHR	[125]
	Arteries	↑ Catalase activity	Hypertension induced by RAS activation	[50]
		↓ Catalase activity	SHR	[120]
GPx	Brain	↓ GPx activity	SHR	[122]
	Kidney	↑ GPx activity	Ang II-induced hypertension	[49]
		↓ GPx expression	SHR	[124, 125]
		↓ GPx activity	SHR	[122]
	Arteries	↑ GPx activity	Hypertension induced by RAS activation	[50]
		↓ GPx activity	SHR	[120]
		↓ GPx expression	Salt- sensitive hypertension (Ovariectomized female rats)	[126]

Table 4. Alterations in major antioxidant enzyme defenses in the brain, kidney and arteries in experimental models of hypertension

3.3. Non-enzymatic antioxidants defenses

Non-enzymatic antioxidants such as GSH, ascorbic acid (vitamin C) and α-tocopherol (vitamin E) play an excellent role in protecting the cells from oxidative damage [4]. GSH has a potent electron-donating capacity that renders GSH both a potent antioxidant *per se* and a conventional cofactor for enzymatic reactions that require readily available electron pairs. In physiological conditions, GSH is present inside the cells mainly in its reduced form and less than 10 percent of total GSH exists in the oxidized form, GSSG [127]. Therefore, intracellular GSH status can be used as a sensitive marker of the cell health and resistance to toxicity. Furthermore, it has been demonstrated that GSH depletion can lead to cell apoptosis [128]. The measurements of GSH and GSSG have been considered useful indicators of the status of oxidative stress [4, 129]. Vitamins E and C are among the major dietary antioxidants. The vitamins have received considerable attention in clinical trials of primary and secondary prevention of cardiovascular diseases (CVD) and cancer. Vitamin E is found in lipoproteins, cell membranes and extracellular fluids. It terminates lipid peroxidation processes and converts O_2^- and $HO\cdot$ to less reactive forms [130]. Vitamin C, a water soluble antioxidant, is found in high concentrations in the adrenal and pituitary glands, liver, brain, spleen and pancreas. It is hydrophilic and can directly scavenge ROS and lipid hydroperoxides. Vitamin C can also restore oxidized vitamin E and can spare selenium [131]. Carotenoids, such as β-carotene are lipid soluble antioxidants that function as efficient scavengers of $^1O_2$ but may also quench $ROO\cdot$ radicals [108]. Uric acid is a highly abundant aqueous antioxidant, considered to be the main contributor for the antioxidant capacity in the plasma [96, 132]. It has the ability to quenche $HO\cdot$ and $ONOO^-$ and may prevent lipid peroxidation [21, 132]. The scavenging of $ONOO^-$ by uric acid is significantly increased in the presence of Vitamin C and cysteine which regenerate the urate radical formed in these reactions. Uric acid also acts as a chelator of iron in extracellular fluids [16]. However, once inside the cells, uric acid appears to exert prooxidant effects. It is not clear whether the correlation between the raised plasma levels of uric acid and cardiovascular risk are due to increased ROS generation by XO or to the prooxidative effects of uric acid itself. Some authors speculate that the increased concentrations of urate might be an adaptive mechanism that confers protection from oxidative damage [132]. It is likely that uric acid effects have different consequences depending on the surrounding microenvironment [21]. Bilirrubin, the end-product of heme catabolism, also appears to function as a chain-breaking antioxidant [133]. Low circulating bilirrubin levels are considered a risk factor for cardiovascular diseases [134]. Plasma albumin, the predominant plasma protein, is also an antioxidant due to its sulfhydryl groups and is able to scavenge MPO-derived chlorinated reactive species and $ROO\cdot$ radicals [108, 135].

The combined antioxidant activities of aqueous- and lipid-antioxidants, including GSH, vitamins, uric acid, bilirrubin, albumin, etc, can be evaluated in the plasma and serum by several assays that measure the ability of the antioxidants present in the sample to inhibit the oxidation of the cation radical ABTS$^+$ [2,2'-azino-bis(3-ethylbenzothiazoline-6-sulphonic acid] (Total Antioxidant Status assay), to reduce a ferric-tripyridyltriazine complex (Ferric Reducing Ability of Plasma, FRAP assay) or to trap free radicals (Oxygen Radical Absorbance Capacity, ORAC assay; Total radical Trapping Parameter, TRAP) [50, 136-138].

The measurement of the overall antioxidant capacity may be more representative of the *in vivo* balance between oxidizing and antioxidant compounds than the evaluation of individual antioxidants [139]. Nevertheless, these assays have also some limitations. First, they correlate poorly with each other as the various antioxidants react differently in each assay. Second, in biological fluids, uric acid appears to account for more than 50% of the total antioxidant activity measured in most assays [108]. However, the putative protective effect of uric acid is debatable [140, 141].

Under conditions of high ROS levels it is expected a decrease of non-enzymatic antioxidants defenses in plasma, since the need for neutralization ROS species implies a higher consumption of endogenous antioxidants. For example, decreased levels of antioxidant vitamins C and E have been demonstrated in newly diagnosed untreated hypertensive patients compared with normotensive control subjects [142-144].

3.4. Systemic, urinary and tissue markers of lipid peroxidation

Measuring oxidative stress in biological systems is complex and requires accurate quantification of ROS or damaged biomolecules. One method to quantify oxidative stress is to measure lipid peroxidation. Lipids that contain unsaturated fatty acids with more than one double bond are particularly susceptible to the action of free radicals. The peroxidation of lipids disrupts biological membranes and is thereby highly deleterious to its structure and function [145]. A large number of by-products are formed during this process and can be measured by different assays. Common biomarkers of lipid peroxidation damage include hydroperoxides, which are primary products generated in the initial stages of lipid peroxidation, and secondary products formed at later lipid peroxidation stages, such as malondialdehyde (MDA) or F2-isoprostanes (Table 5) [146, 147] .The lag time required for the exponential generation of lipid peroxidation products can also be used to evaluate the susceptibility of lipid molecules to free radical damage. Therefore, lipids with higher resistance to oxidative stress exhibit longer lag times than those which are easily attacked by free radicals [147].

3.5. F2-isoprostanes

F2-isoprostanes are prostaglandin F2α isomers primarily produced by free radical-catalyzed peroxidation of the polyunsaturated fatty acid (PUFA), arachidonic acid [97]. Although there is also evidence of F2-isoprostane formation by the action of cyclooxygenase, it is currently assumed that systemic and urinary F2-isoprostanes are mostly derived from free radical-induced lipid peroxidation, independently of cyclooxygenase enzymatic activity. Therefore, F2-isoprostanes have been regarded as reliable biomarkers of oxidative stress. Furthermore, F2-isoprostanes have been shown to exert potent vasoconstrictor effects on animal and human vessels, suggesting a pathogenic role in cardiovascular diseases and have been extensively used as markers of lipid peroxidation in human diseases [74, 75,148]. Their high stability and presence in measurable concentrations in many biological tissues and fluids, under physiological and pathological conditions, has also allowed the establishment of reference intervals and the comparison or monitoring of disease states

[97,149,150]. Urine specimens are particularly suited for F2-isoprostanes measurements. First, the ex vivo formation of F2 isoprostanes is minimized in these samples due to the low urinary lipid content, avoiding the need for time-sensitive sample processing [97, 149, 151, 152]. Second, they provide a noninvasive route for systemic oxidative stress evaluation. Although they can also be locally produced in the kidney, many studies have demonstrated that urinary F2-isoprostanes are mainly derived from free F2-isoprostanes filtered from the circulation [97, 149, 151, 152]. Only hydrolyzed isoprostanes are excreted into the urine whereas blood plasma samples contain both free and esterified isoprostanes. Since plasma samples have considerable amounts of arachidonic acid, the addition of preservatives, such as butylated hydroxytoluene (BHT) and indomethacin, and the storage at -80°C, are recommended to avoid degradation and/or ex vivo formation of F2-isoprostanes [97.

3.6. TBARS

The free radical attack to PUFAs in cellular membranes leads to the disruption of cell structure and function. MDA, one of the end products of these oxidative reactions, can be detected in several biological fluids and tissues and is therefore used as a biomarker of lipid peroxidation and oxidative stress [153]. High MDA levels indicate a high rate of lipid peroxidation [154]. The reaction of MDA with 2-thiobarbituric acid (TBA) is frequently used to estimate oxidative stress [155]. MDA reacts with TBA under conditions of high temperature and acidity generating 2-thiobarbituric acid reactive substances (TBARS) that can be measured either spectrophotometrically or spectrofluorometrically. However, these products can also be formed by sample autooxidation under assay conditions or by cross-reactivity with non-MDA substrates such as bile pigments, proteins, carbohydrates and other aldehydes. Therefore, TBARS measurements often originate doubts due to their limited specificity as markers of lipid peroxidation [156]. Nevertheless, undesirable autooxidation and non-MDA substrates reactivity can be minimized by adding BHT during sample preparation. Plasma TBARS measurements have been reported to correlate with some clinical features of cardiovascular disease, preeclampsia, ischemia/reperfusion, chronic kidney disease and cerebrovascular disorders [157-160]. Since the TBARS assay may overestimate MDA, other methods can be used to evaluate lipid peroxidation products, such as the lipid hydroperoxide (LPO) test. The principle of the LPO test is that in the presence of hemoglobin, lipid hydroperoxides are reduced to hydroxyl derivates with the equimolar production of a methylene blue product, which can be quantified spectrophotometrically [161].

3.7. HNE

The aldehyde 4-hydroxy-2-nonenal (4-HNE) is one of the most cytotoxic products of free radical attack on ω6-PUFA, namely arachidonic and linoleic acids, being able to react with diverse biological molecules such as proteins, peptides, phospholipids and nucleic acids. It also acts as an important mediator of oxidant-induced signaling, cellular proliferation and apoptosis [97, 162]. 4-HNE can be detected in plasma and several biological tissues under physiological conditions but its generation is significantly raised in pathological states associated with oxidative stress [97, 162-164]. Renovascular hypertensive rats showed increased 4-HNE deposition in the intima of injured mesenteric arteries, suggesting the presence of free radical injury and cytotoxicity induced by 4-HNE [163]. A wide diversity of effects have been demonstrated for 4-HNE depending on its concentration. Concentrations below 0.1 microM are within the physiological range and appear to induce chemotaxis and stimulation of guanylate cyclase and phospholipase C [165]. 4-HNE concentrations between 1-20 microM inhibit DNA and protein synthesis and stimulate phospholipase A2. Higher concentrations (100 microM and above) are cytotoxic and genotoxic leading to cell death [165]. Thus, 4-HNE represents a broad indicator of lipid peroxidation.

3.8. Early stage of lipid peroxidation products

Lipid hydroperoxides are the primary products of lipid peroxidation and can further react to form secondary products including aldehydes such as MDA and 4-HNE [166,167]. Therefore, lipid hydroperoxides may be used to evaluate initial stage or acute lipid peroxidation while MDA and 4-HNE appear to be more representative of chronic oxidative stress. Recent reports described that 13-hydroperoxyoctadecadienoic acid (13-HPODE), a precursor to 3-hydroxyoctadecadienoic acid (13-HODE) is able to react with proteins forming adducts by covalently binding to specific amino acid residues. The Hexanoyl-Lysine (HEL) adduct results from the oxidative modification of ω6-PUFAs such as linoleic acid, the predominant PUFA in the human diet, and arachidonic acid [168]. HEL may be another useful biomarker for detecting and quantifying the earlier stages of lipid peroxidation. Monoclonal antibodies and ELISA kits have been developped, and HEL can be detected in oxidatively modified LDL, in human atherosclerotic lesions, human urine and serum. It has been also reported that HEL is formed in rat muscle during exercise and that its formation is inhibited by antioxidants such as flavonoids [169].

The lipid peroxidation biomarkers most commonly evaluated in hypertensive patients or experimental hypertension are shown in Table 3 and Table 5, respectively.

Lipid peroxidation biomarker	Measured in:	Alteration	Experimental model of hypertension	References
MDA	Plasma	↑	SHR	[170]
	Aorta	↑	Salt-loaded SHR	[94]
TBARS	Plasma	↑	Hypertension induced by RAS activation	[50]
	Plasma	↑	Ang II-induced hypertension	[171]
	Plasma, Heart	↑	Mineralocorticoid-induced hypertension	[172]
	Urine	↑	Salt-sensitive hypertension	[173]
	Aorta, Left Ventricle	↑	Ang II-induced hypertension	[174]

Lipid peroxidation biomarker	Measured in:	Alteration	Experimental model of hypertension	References
F2-**Isoprostanes**	Plasma	↑	Salt-sensitive hypertension	[175]
	Plasma	↑	Glucocorticoid-induced hypertension	[87]
	Urine	↑	SHR	[170, 176]
4-HNE	Mesenteric arteries	↑	Renovascular hypertension	[163]
	Aorta	↑	SHRSP	[177]
4-HNE adducts	Blood	↑	SHR	[178]

Table 5. Lipid peroxidation biomarkers in experimental hypertension

3.9. Other prooxidant biomarkers

Besides antioxidants and lipid peroxidation parameters, there are other important indexes of oxidant status. These include the expression and activity of prooxidant enzymes, ROS concentration, byproducts formed by ROS/RNS interaction with DNA (8-hydroxy-2-deoxyguanosine) or proteins (3-nitrotyrosine, carbonyl groups) and redox-sensitive transcription factors such as nuclear factor kappa B (NF-KB). Major sources of cellular ROS include Nox enzymes, mitochondrial electron transport enzymes, uncoupled NOS, XO and MPO. Table 6 summarizes several prooxidant biomarkers evaluated in experimental models of hypertension.

Biomarkers of prooxidant status	Evaluated in:	Alteration	Hypertension model	Reference
Mitochondrial electron transport chain enzymes/ mitochondrial ROS production	Brain	Oxidative impairment of mitochondrial enzymes	SHR	[30]
	Kidney	↑ mitochondrial ROS production	SHR, Mineralocorticoid hypertension	[179, 180]
NADPH oxidase family enzymes (or NOXs)	Brain	↑ NADPH oxidase activity	Salt-loaded SHRSP	[181]
	Kidney	↑ Nox4 expression, ↑ NADPH oxidase activity	Ang II–induced hypertension	[49]
	Arteries	↑ **expression of NAD(P)H oxidase subunits (p67(phox) and gp91(phox)**	Ang II-induced hypertension	[182]
		↑ Nox1 and Nox4 expression	SHRSP	[183]

Biomarkers of prooxidant status	Evaluated in:	Alteration	Hypertension model	Reference
		↑ NADPH oxidase activity	Ang II-induced hypertension; Hypertension induced by RAS activation	[50, 184-186]l
eNOS	Arteries	Uncoupled eNOS/↑ eNOS-derived ROS	SHR	[34]
	Arteries	↑ eNOS expression/↓ eNOS activity	SHR	[35]
XO	Arteries	↑XO activity	SHR	[187]
	Arteries	↑XO expression	SHR	[188]
H_2O_2	Kidney, Blood /urine	↑ renal production/↑ production in plasma /↑ urinary excretion	Ang II-induced hypertension	[49]
MPO	Arteries	↑ MPO activity	SHRSP	[189]
	Kidney, Heart, Brain	↑ MPO activity	Renovascular hypertension	[190]
GSH/GSSG	Kidney	↓ ratio	SHR	[191]
	Plasma, Heart, kidney	↓ ratio	Salt-sensitive hypertension	[192]
3-nitrotyrosine	Kidney	↑ expression	SHR	[193]
Protein carbonyl groups	Arteries, Heart, Kidney	↑ expression	SHR	[194]
	Kidney	↑ expression	SHR	[195]
8-Hydroxy-2-deoxyguanosine (8-OH-dG)	Arteries, Heart, Kidney,	↑ expression	SHR	[194].
NF-κB	Kidney	↑ activation	SHR; Ang II-induced hypertension	[49, 196]
	Arteries	↑ activation	Ang II-induced hypertension	[197]

Table 6. Other Prooxidant status biomarkers in experimental hypertension

4. Prophylactic and therapeutic strategies to reduce oxidative damage in arterial hypertension

A plethora of studies has demonstrated that hypertension is associated with an imbalance between oxidants and antioxidants that leads to altered cell signaling and oxidative damage.

Therefore, extensive research has been conducted in order to identify the ROS involved in blood pressure dysregulation, as well as the major prooxidant enzymes and antioxidant defenses that contribute to the loss of redox homeostasis in cardiovascular and renal systems. Furthermore, studies on experimental models of hypertension recognized several important neurohumoral stimuli responsible for ROS overproduction and also the main targets for ROS-induced dysfunction [8, 43]. Therapeutic interventions to reduce oxidative stress in hypertension have mostly relied on the administration of drugs that increase antioxidant capacity or inhibit ROS generation. In addition, other strategies aimed at reducing the activation of neurohumoral pathways that stimulate ROS production (upstream mediators) or at blocking/repairing the downstream targets affected by ROS have also been tested [196, 198, 199].

4.1. Targeting oxidative stress in experimental hypertension

The pharmacological modulation of ROS bioavailability in animal models of hypertension has been useful to demonstrate a causative role for oxidative stress in the pathophysiology of hypertension [43, 50]. However, the blood pressure lowering efficacy of these strategies appears to differ when comparing distinct experimental models [48, 50, 85, 200, 201]. This is probably because the development of each animal model was based on a particular etiological factor presumably responsible for human hypertension, such as high salt intake, overactivation of the renin-angiotensin system, genetic factors or renal disease. Since these factors may stimulate different redox pathways, the effectiveness of an antioxidant in one model does not necessarily translate to other models or to human essential hypertension which is known to have a multifactorial nature. Another important observation is that treatments with antioxidants or ROS inhibitors are generally more effective in preventing rather than reversing the hypertension [49, 50, 87, 202]. Indeed, there are several studies demonstrating that ROS activate feed-forward mechanisms that amplify the cardiovascular and renal dysfunction [8, 43, 49, 51]. Once triggered, these pathways may be sufficient to sustain the deleterious effects of oxidative stress even after ROS blockade or elimination [49]. *In vivo* drug treatments targeting oxidative stress in experimental models of hypertension are reviewed below and their effects on blood pressure are summarized in Table 7.

4.1.1. Drugs inhibiting ROS production

Apocynin is a methoxy-substituted catechol (4-hydroxy-3-methoxy-acetophenone), originally extracted from the roots of the tradicional medicinal herb *Picrorhiza kurroa* which has anti-inflammatory properties [203]. Several experimental studies have used apocynin for its ability to inhibit Nox enzymes. The mechanism of inhibition involves the blockade of translocation of cytosolic protein subunits to the membrane which is crucial for the activation of Nox1 and Nox2 [204]. Thus, the effect of apocynin is restricted to inducible Nox enzymes that require cytosolic activators and it does not seem to affect constitutively active Nox isoforms and their putative physiological actions [204]. However, to be an effective Nox inhibitor, apocynin has to undergo a peroxidase-mediated oxidation to be converted into the metabolically active diapocynin [205-207]. The activation of apocynin occurs in the presence of MPO and H_2O_2 [205, 207]. This fact suggests that apocynin may function only in conditions of high inflammatory and prooxidant activity. Apocynin has also been shown to have direct antioxidant properties, being able to scavenge H_2O_2 derived products [205, 207]. However, it can also function as a prooxidant in resting cells [203]. Nevertheless, it is possible that when administered in conditions of enhanced oxidative stress, the protective effect prevails.

Gp91ds-tat is a chimeric peptide that specifically inhibits NADPH oxidase by preventing the assembly of its subunits. It is constituted by a segment of gp91phox (*gp91ds*) important for the interaction of this membrane subunit with the cytosolic subunit, p47phox, and by a *tat* peptide from the HIV virus, which allows the uptake of the peptide into the cell [208, 209]. However, since it is a peptide it may have poor oral bioavailability and may induce sensitization reactions. Furthermore, the *tat* segment may have side effects on cellular signaling and activity [204, 208]. Thus, it is not suitable for long treatments or to clinical use in the treatment of human cardiovascular diseases. Although it was designed to block Nox2, it may also inhibit Nox1 given the substantial degree of homology between the two isoforms [204]. As for apocynin, Nox4 is not likely to be affected by gp91ds-tat since it is constitutively active and does not require the activation of cytosolic subunits [204].

Allopurinol and its metabolite **oxypurinol** are hypoxanthine and xanthine analogs, respectively, that inhibit XO activity [16]. At low concentrations, allopurinol is a competitive inhibitor of XO, while at higher concentrations it behaves as a non-competitive inhibitor [16]. XO rapidly metabolizes allopurinol into oxypurinol, a noncompetitive inhibitor of the enzyme which has a much higher half-life and is therefore responsible for most of the pharmacological effects of allopurinol [16]. In addition, both allopurinol and oxypurinol have intrinsic antioxidant properties, being able to scavenge ROS such as O_2^-, HO· and HClO [210-212]. However, these effects appear to require higher doses than those required for XO inhibition [210]. Allopurinol is approved for the treatment of human patients with gout or hyperuricemia, but it has also potential therapeutic application in cardiovascular diseases. Most common adverse effects are nauseas, diarrheas, hypersensitivity reactions and skin rash [16].

4.1.2. Antioxidants

Tempol (4-hydroxy-2,2,6,6-tetramethylpiperidine-1-oxyl) is a membrane-permeable nitroxide that catalyzes the conversion of O_2^- to H_2O_2 thus functioning as a SOD mimetic [213, 214]. Tempol protects the lipids or proteins from oxidative damage and interacts with other antioxidants to promote the reduction of oxidized lipids [214]. The main antihypertensive effect of this drug is related to the reduction of the O_2^- interaction with ·NO which improves vasodilation [213, 214]. It also promotes natriuresis by enhancing the vasodilation of renal medullary vessels in a ·NO independent manner [214]. Indeed, tempol has been shown to have sympatholytic actions, being able to inhibit afferent, peripheral and central activation of the sympathetic nervous system [214]. These actions are responsible for the rapid fall of blood pressure and heart rate after acute intravenous administration of tempol [214]. Nevertheless, some studies reported that the formation of H_2O_2 by tempol can counteract its vasodilator, natriuretic and antihypertensive effects in models of hypertension where H_2O_2 plays a more prominent role than O_2^- [48, 50, 53]. The co-administration of catalase in these conditions restores the protective action of tempol [48, 50].

N-acetylcysteine (NAC) is a thiol containing compound. It is the acetylated derivative of the aminoacid L-cysteine and a precursor for reduced glutathione (GSH) [215, 216]. It appears to have direct antioxidant action since its free thiol can interact with the electrophilic groups of ROS [215]. However, this effect does not seem likely to occur in vivo because NAC has poor oral bioavailability being rapidly metabolized into GSH, among other metabolites [216]. Thus, the main protective action of NAC is probably related to its role as a GSH precursor, which then detoxifies reactive species either by enzymatic or non-enzymatic reactions [216]. In humans, NAC is approved as a mucolytic agent because it destroys the disulphide bridges of mucoproteins [215]. It is also used as an antidote for acetaminophen poisoning which dramatically depletes hepatic GSH content causing severe damage [217]. NAC may also have potential therapeutical applications in the treatment of heart diseases [218].

Polyethylene glycol-catalase is the conjugated form of the enzyme catalase with polyethylene glycol (PEG) which enhances the stability in aqueous solution, reduces immunogenicity and decreases sensitivity to proteolysis, thus increasing the circulatory half-life of catalase [219]. PEG also enhances the catalase association with cells [219]. The antioxidant effect of PEG-catalase results from the enzymatic degradation of H_2O_2 to water.

Ebselen (2-phenyl-1,2-benzisoselenazol-3[2H]-one) is a lipid-soluble seleno-organic compound that mimics glutathione peroxidase activity, being able to react with H_2O_2 and organic hydroperoxides including membrane-bound phospholipid and cholesterylester hydroperoxides [220]. It appears to reduce lipid peroxidation initiated by hydroperoxides but not free radicals initiators [221]. In addition, ebselen reacts rapidly with $ONOO^-$. The ebselen selenoxid product yielded in this reaction is regenerated to ebselen by GSH, which allows its reutilization as a defense against $ONOO^-$ [222, 223]. Ebselen also directly inhibits inflammation-related enzymes such as 5-lipoxygenase, NO synthases, protein kinase C, NADPH oxidase and H^+/K^+-ATPase by reacting with the SH group, leading to the formation of a selenosulphide complex [221]. Some authors have also proposed that the antioxidant and anti-inflammatory actions of ebselen are mediated through interactions with the thioredoxin (Trx) system [220]. Reduced Trx is important for growth and redox regulation by thiol redox control [220]. Ebselen was found to be an excellent substrate for mammalian TrxR and a highly efficient oxidant of reduced Trx. It also seems to function as a Trx peroxidase or peroxiredoxin mimic, thus contributing to the elimination of H_2O_2 and lipid hydroperoxides [220]. Ebselen has been used in clinical trials for the treatment of patients with acute ischemic stroke or delayed neurological deficit after aneurismal subarachnoid hemorrhage [224, 225].

Vitamin C (ascorbic acid) is a water soluble antioxidant found in the body as an ascorbate anion. It acts as a free radical scavenger [226]. Although this effect requires higher concentrations than those achieved in the plasma by oral administration, ascorbate appears to concentrate in tissues in much higher levels than those found in the plasma and can act effectively as a ROS scavenger [227]. In addition, it reduces membrane lipid peroxidation and regenerates Vitamin E [226]. Recent reports also suggest that Vitamin C can suppress NADPH oxidase activity [227].

Vitamin E is a generic term for a group of compounds including tocopherols and tocotrienols. The isoform α-tocopherol appears to be the most abundant in vivo [227]. Vitamin E terminates the propagation of the free radical chain reaction in lipid membranes and inhibits LDL oxidation [226, 227]. Vitamin E can also have non antioxidant actions primarily through the regulation of enzymes involved in signal transduction. Enzymes inhibited by vitamin E include protein kinases C and B, protein tyrosine kinase, lipoxygenases, mitogen activated protein kinases, phospholipase A2 and cyclooxygenase-2. In contrast, vitamin E has stimulatory effects on protein tyrosine phosphatase and diacylglycerol kinase [228]. Both vitamins C and E have been shown to stimulate the activation of NOS activity and increase NO synthesis in endothelial cells and thus may contribute to improved endothelial-dependent vasodilation in hypertension [229]. However, although Vitamins C and E are generally considered to be non-toxic, they can undergo oxidation and generate pro-oxidant molecules [226]. Nevertheless, it appears that this is more likely to occur with Vitamin E, especially in the absence of sufficient Vitamin C to regenerate the α-tocopherol radical [227, 230, 231].

Alpha-lipoic acid (1,2-dithiolane-3-pentanoic acid or thioctic acid) has a wide range of effects on cell functions, acting as an antioxidant, a metal chelator and a signaling mediator [232]. Both lipoic acid (LA) and its reduced form dihydrolipoic acid (DHLA), may scavenge HO· and HClO, although neither species is able to neutralize H_2O_2 [232]. DHLA also regenerates Vitamins C and E and does not become a free radical after reacting with these species. Furthermore, LA and DHLA chelate transition metals, thus reducing the metal-catalyzed free radical damage [232]. LA also contributes to improve antioxidant defenses by increasing the intracellular levels of Vitamin C and GSH. Many of LA protective actions have been attributed to its interference in cell signaling processes [232]. For example, LA effect on GSH appears to be mediated by nuclear factor erythroid 2- related factor 2 (Nrf2), an important transcription factor regulating gene transcription through the Antioxidant Response Element. LA was also shown to interact with several kinases and protein phosphatases [232]. Its interaction with components of the insulin signaling cascade also appears to improve glucose disposal in animal models of diabetes and human diabetic patients [232]. In addition, LA improves endothelial NO synthesis and endothelial-dependent vasodilation and prevents deleterious modifications of thiol groups in Ca^{2+} channels [232]. It has also important anti-inflammatory effects by inhibiting the activation of NF-KB, a transcription factor that regulates the expression of proinflammatory genes [232].

Pyrrolidine dithiocarbamate (PDTC), a low-molecular weight thiol compound, has the ability to scavenge oxygen radicals and to chelate metals [233, 234]. It may also act as a prooxidant and a thiol group modulator [233]. PDTC has been shown to interfere with the activation of several transcription factors, being a potent inhibitor of NF-KB [233, 234]. PDTC can also activate other signaling pathways, such as the extracellular signal-regulated kinase (ERK), c-Jun N-terminal kinase (JNK) and the transcription factor Heat Shock Factor (HSF) [233, 235].

5, 6, 7, 8-Tetrahydrobiopterin (BH₄) is a key cofactor of NOS [236, 237]. It is involved in the formation and stabilization of eNOS and iNOS [236, 238]. In the absence of BH₄, NOS can become uncoupled and starts producing O_2^- instead of NO [33, 237]. Furthermore, BH₄ also possesses direct antioxidant activity, being able to scavenge O_2^- and HO· [239]. The protective effects of BH₄ on the development of hypertension appear to be due an increase in eNOS activity, a reduction in O_2^- production and a decrease in iNOS expression [199].

Drug	Antihypertensive effect	Lack of antihypertensive effect
Apocynin	Prevented/attenuated mineralocorticoid-induced hypertension [86, 240] Prevented/reversed glucocorticoid-induced hypertension [241] Prevented/reversed adrenocorticotropic hormone-induced hypertension [242] Prevented the development of Ang II-induced hypertension in mice [186] Prevented the development of renovascular hypertension [243] Prevented the development of hypertension induced by RAS activation [50] Reduced blood pressure in borderline and spontaneous hypertension [244] Attenuated salt-sensitive hypertension [245] Normalized blood pressure in a model of hypertension induced by disruption of dopamine D₂ receptor [246]	Failed to prevent the hypertension induced by chronic infusion of endothelin-1 [200] Failed to prevent hypertension in transgenic mice overexpressing renin or angiotensinogen [247, 248] Failed to prevent Ang II-induced hypertension in rats [249, 250]
Gp91ds-tat	Attenuated the blood pressure rise induced by Ang II in mice [209]	Failed to attenuate salt-sensitive hypertension [251]
Allopurinol	Attenuated salt-sensitive hypertension [252] Prevented glucocorticoid-induced hypertension [253]	Failed to prevent or attenuate mineralocorticoid-induced hypertension [254] Failed to prevent glucocorticoid-induced hypertension [255] Failed to prevent or attenuate adrenocorticotropic-induced hypertension [242] Failed to prevent the development of hypertension induced by the blockade of nitric oxide synthesis [256] Failed to prevent the progression of hypertension in young SHR [257]
Oxypurinol	Reduced blood pressure in SHR [258]	
Tempol	Attenuated hypertension in SHR [201] Prevented the progression of hypertension in salt-loaded SHRSP [259] Attenuated mineralocorticoid-induced hypertension [260]	Failed to prevent Ang II-induced hypertension [264] Failed to attenuate hypertension induced by inhibition of superoxide dismutase [48]

Drug	Antihypertensive effect	Lack of antihypertensive effect
	Prevented/attenuated glucocorticoid-induced hypertension [87] Attenuated salt-sensitive hypertension [261] Prevented the development of renovascular hypertension [243] Attenuated high-volume hypertension [262] Attenuated hypertension induced by NO inhibition [112] Partially prevented/reversed adrenocorticotropic hormone-induced hypertension [263]	Failed to prevent hypertension induced by RAS activation [50]
NAC	Attenuated hypertension in young SHR [265] Prevented the development of glucocorticoid-induced hypertension [202] Prevented the development of adrenocorticotropic hormone-induced hypertension [266] Markedly reduced salt-sensitive hypertension [267] Prevented/ attenuated hypertension induced by nitric oxide synthesis inhibition [268]	Failed to reduce blood pressure in adult SHR [265] Failed to attenuate glucorticoid-induced hypertension [202] Failed to reverse adrenocorticotropic-induced hypertension [266] Failed to prevent the development of hypertension induced by the blockade of nitric oxide synthesis [256]
PEG-catalase	Prevented the development of hypertension induced by RAS activation [50] Transiently decreased blood pressure in Ang II-hypertensive rats [49] Reduced blood pressure in high-volume hypertension in mice [269]	Lacked a sustained antihypertensive effect in Ang II-induced hypertension [49]
Ebselen	Attenuated the blood pressure rise induced by Ang II in mice overexpressing p22phox in vascular smooth muscle and in littermate control mice [270]	Failed to prevent the development of hypertension induced by the blockade of nitric oxide synthesis [256]
Vitamin C	Prevented the progression of hypertension induced by salt administration in SHRSP and in SHR [229, 271] Attenuated salt-induced hypertension [272, 273]	Failed to prevent adrenocorticotropic hormone-induced hypertension [274]
Vitamin E	Prevented the progression of hypertension induced by salt administration in SHRSP [229] Attenuated hypertension in young SHRSP [275] Attenuated salt-induced hypertension [273]	Failed to prevent adrenocorticotropic hormone-induced hypertension [274]
Lipoic acid	Reduced blood pressure in SHR [276] Prevented fructose-induced hypertension [277] Prevented/attenuated salt-induced hypertension [278] Prevented mineralocorticoid-induced	

Drug	Antihypertensive effect	Lack of antihypertensive effect
	hypertension [279]	
PDTC	Perinatal administration ameliorated hypertension in SHR offsprings [280] Prevented /Reduced hypertension in SHR [32, 196] Attenuated mineralocorticoid-induced hypertension [260]	
BH₄	Suppressed the development of hypertension in SHR [199] Reduced hypertension in SHR [281]	Failed to attenuate hypertension in castrated SHR [281] Failed to prevent the development of adrenocorticotropic hormone in rat [282]

Table 7. Effect of chronic treatment with antioxidants or inhibitors of ROS production on blood pressure

4.2. Antioxidant approaches in human hypertension

Although there is considerable evidence of oxidative stress involvement in the pathophysiology of hypertension, the attempts to demonstrate benefits from antioxidant therapy in human cardiovascular diseases have been very disappointing [5, 96, 283]. Most of the large trials regarding the effects of diet supplementation with Vitamin C, Vitamin E and β-carotene failed to show significant improvements in blood pressure and other cardiovascular endpoints [5, 283]. Furthermore, some of them also led to the conclusion that antioxidant treatment with Vitamin E or β-carotene may even be harmful [283-285]. In contrast, smaller clinical trials have provided some evidence of antioxidant treatment advantages. For example, some studies showed that systemic Vitamin C levels inversely correlates with blood pressure and that Vitamin C supplementation effectively attenuates hypertension [142, 286]. Vitamin E and lipoic acid have also been shown to improve vascular function, though there is not consistent evidence of a blood pressure lowering effect of these agents in human patients [5, 287, 288]. Nevertheless, it has been demonstrated that a high consumption of dietary fruits and vegetables increases plasma antioxidant capacity and reduces blood pressure [289, 290]. Thus, it appears that a diet rich in fruits and vegetables is a better strategy than antioxidant supplementation to improve antioxidant status and cardiovascular health [5]. Overall, the clinical trials with antioxidant supplements have been very unsatisfactory and are in disagreement with the findings obtained in experimental hypertension studies. There are some possible justifications for the disappointing outcomes of these trials. First, the type of the drug used as well as the dose and duration of the therapy might not be adequate [5, 291, 292]. Most trials followed an antioxidant strategy based in the administration of ROS scavengers such as Vitamins C and E. However, these drugs do not neutralize H_2O_2 which has been shown to play a relevant role in the pathophysiology of hypertension and other cardiovascular diseases [5, 7, 48-50]. Furthermore, it is known that human blood and tissues have plenty antioxidants and that several stimuli induce an adaptive increase of enzymatic antioxidant defenses which can mask the benefits of exogenously administered antioxidants [293]. In addition, the antioxidant doses used in most of the experimental studies have been much higher than those tested in human patients [291]. So, there is the possibility that in humans the antioxidants did not achieve effective concentrations to neutralize ROS. Furthermore, it is not known if orally administered antioxidants can reach the precise sites of increased ROS production as oxidative stress is heterogeneously distributed throughout the organs, tissues and cellular compartments [5, 96, 291]. Indeed, the unspecific scavenging of ROS may even interfere with many important physiological functions in a deleterious manner [29, 96]. Another important limitation of most antioxidants tested is that they can exert themselves prooxidant effects in the absence of a coordinated antioxidant response [5, 96]. For example, Vitamin E needs to be regenerated by Vitamin C otherwise it may cause oxidative damage [231].

There are also drawbacks in clinical trials design. In large trials of antioxidant supplementation, patients have not been recruited accordingly to their redox status [5, 294]. It is unlikely that a beneficial effect of antioxidant therapy would be observed in patients without previous evidence of increased oxidative stress. Another important consideration is that these clinical trials often have heterogeneous populations in terms of the etiology of cardiovascular disease [295]. Indeed, most studies have indiscriminately enrolled any patient at cardiovascular risk [294]. This is in obvious contrast to the homogeneous populations analyzed in experimental studies. Furthermore, some of the patients may be at an advanced stage of disease exhibiting irreversible damage insusceptible to antioxidant interventions [5, 29]. It should also be highlighted that many patients enrolled in these studies were already being treated with drugs such as aspirin, lipid-lowering agents and some antihypertensive drugs which can themselves interfere with oxidant status and mask the effects of additional therapy with antioxidants [5, 92, 296, 297].

So far, most interventions aimed at reducing oxidative stress in human hypertension have relied on antioxidant supplementation. However, it is possible that a strategy based on the inhibition of ROS production is more effective than the antioxidant interventions [5, 96]. The disruption of cardiovascular redox status is most likely triggered by an increase in prooxidant activity rather than a reduction in antioxidant defenses. Indeed, many neurohumoral or ambiental prohypertensive stimuli (angiotensin II, aldosterone, high-salt intake) are known to upregulate the expression and activity of prooxidant enzymes [5, 8, 43]. Nevertheless, there are already some studies that investigated the cardiovascular effects of prooxidant enzyme inhibition. Patients treated with allopurinol showed improvements in vascular function [298, 299]. However, a blood pressure lowering effect of this XO inhibitor has been shown only in newly diagnosed hypertensive adolescents and in hyperuricemic patients with normal renal function [300, 301]. Furthermore, the combination of allopurinol with antihypertensive drugs did not provide additional benefits on blood pressure [299]. This is probably because XO is not a major contributor to the development of hypertension, even though its activity may be increased in pathophysiological conditions [5]. Indeed, compelling evidence indicates that NADPH oxidases are the main contributors to ROS overproduction in cardiovascular and renal diseases [5, 8, 302]. Moreover, Nox-derived ROS are known to amplify redox dysfunction by inducing the activation of other prooxidant enzymes, such as XO, mitochondrial enzymes and NOS synthases [7, 51]. Since many antihypertensive drugs block upstream activators of Nox enzymes, it is not surprising that the inhibition of XO by allopurinol does not improve blood pressure control in patients already treated with antihypertensive drugs. To date, no Nox inhibitors have been tested in clinical trials although some specific Nox inhibitors have already been developed and patented [96, 296]. Future strategies to demonstrate the benefits of oxidative stress reduction in cardiovascular diseases should include the testing of specific Nox inhibitors in human patients. Moreover, the development of reliable oxidative stress biomarkers for risk stratification and monitoring of therapy is also highly desirable [96, 296]. Table 8 summarizes the possible reasons for the failure of antioxidants in clinical trials.

Limitations related to the drug treatment	Limitations related to the clinical trials design
Inadequate dose or duration of therapy Lack of effect on non-radical oxidants such as H_2O_2 Lack of effect on prooxidant activity Inaccessibility of ROS scavengers to intracellular sites of increased ROS production Some antioxidants may themselves become prooxidants in the absence of a coordinated antioxidant response Unspecific scavenging of ROS may disrupt physiological functions	Lack of previous evidence of increased redox dysfunction in patients analyzed Heterogeneous populations in terms of the etiology of cardiovascular disease Some patients may be at an advanced stage of disease exhibiting irreversible damage Patients treated simultaneously with drugs that interfere with oxidant status (aspirin, lipid lowering agents, antihypertensive drugs) Lack of validated oxidative stress biomarkers for risk stratification and monitoring of therapy

Table 8. Possible reasons for the failure of clinical trials with antioxidants in cardiovascular diseases

4.3. Antihypertensive treatments with direct and indirect antioxidant effects

It is known that first-line antihypertensive drugs such as angiotensin II receptor blockers (ARB) and angiotensin converting enzyme inhibitors (ACEi) can reduce oxidative stress due to their inhibitory effect on angiotensin II, which is a major stimulus for the activation or upregulation of Nox enzymes [5, 296]. ROS such as O_2^- and H_2O_2 are widely recognized as important downstream mediators of Ang II physiological and pathological effects [303]. Nevertheless, some of these antihypertensive drugs also possess antioxidant effects independently of RAS inhibition. For example, captoptil, a thiol-containing ACEi, is a ROS scavenger and a metal chelator [304]. The ARBs candesartan and olmesartan also exhibit antioxidant effects independent of AT_1 receptor blockade or blood pressure control [305-307]. In addition, other agents belonging to the beta-blocker or calcium channel blocker drug classes have also been shown to exert antioxidant effects unrelated to their blood pressure lowering action. The beta-blockers carvedilol and nebivolol appear to possess ROS scavenging properties as well as inhibitory effects on ROS production, such as the inhibition of Nox activation [308, 309]. In addition, nebivolol also increases NO release from the endothelium, thus attenuating oxidative stress effects on endothelium-dependent vasodilation [309, 310]. The calcium channel blocker lacidipine has also been demonstrated to have a potent antioxidant activity and to reduce the intracellular production of ROS induced by oxidized LDL [311, 312]. Therefore, even though convincing evidence is lacking regarding a clinical therapeutic effect of antioxidants, there is extensive data showing that currently approved antihypertensive treatments have the ability to modify oxidative stress status.

5. Conclusions

Extensive experimental evidence has shown that unbalanced ROS and/or RNS production can disturb several physiological functions, leading to the genesis and progression of arterial hypertension. Many studies have observed marked alterations in direct and indirect oxidative stress biomarkers, such as lipid peroxidation products, prooxidant enzymes and antioxidant defenses. However, most clinical trials with antioxidants have failed to demonstrate a protective effect on blood pressure and cardiovascular function. This does not necessarily exclude a role for oxidative stress in human cardiovascular diseases but instead suggests that other approaches should be addopted to recover redox homeostasis. The inhibition of Nox enzymes appears to be a promising strategy as these enzymes are major sources of ROS overproduction at cardiovascular and renal sites of blood pressure control. Indeed, several drugs already in use for the treatment of hypertension (e.g. ARBs, ACEi, the β-blocker nebivolol) or dyslipidemia (statins) are known to reduce the activation of Nox enzymes. In addition, there is an urgent need to implement universally validated approaches to evaluate oxidative status in human patients. These should cover a broader range of redox biomarkers and would add valuable information for risk stratification and therapeutic monitoring in human patients.

Author details

Teresa Sousa, Joana Afonso and António Albino-Teixeira
Department of Pharmacology and Therapeutics, Faculty of Medicine, University of Porto, Portugal

Félix Carvalho
REQUIMTE, Laboratory of Toxicology, Department of Biological Sciences, Faculty of Pharmacy, University of Porto, Portugal

6. References

[1]   Droge, W. (2002) Free radicals in the physiological control of cell function. Physiol Rev 82: 47-95.

[2]   Jones, D.P. (2008) Radical-free biology of oxidative stress. Am J Physiol Cell Physiol 295: C849-868.

[3]   Turko, I.V., Murad, F. (2002) Protein nitration in cardiovascular diseases. Pharmacol Rev 54: 619-634.

[4] Valko, M., Leibfritz, D., Moncol, J., Cronin, M.T., Mazur, M., Telser, J. (2007) Free radicals and antioxidants in normal physiological functions and human disease. Int J Biochem Cell Biol 39: 44-84.

[5] Paravicini, T.M., Touyz, R.M. (2008) NADPH oxidases, reactive oxygen species, and hypertension: clinical implications and therapeutic possibilities. Diabetes Care 31 Suppl 2: S170-180.

[6] Pham-Huy, L., He, H., Pham-Huy, C. (2008) Free Radicals, Antioxidants in Disease and Health. Int J Biomed Sci 4: 89-96.

[7] Ardanaz, N., Pagano, P.J. (2006) Hydrogen peroxide as a paracrine vascular mediator: regulation and signaling leading to dysfunction. Exp Biol Med (Maywood) 231: 237-251.

[8] Datla, S.R., Griendling, K.K. (2010) Reactive oxygen species, NADPH oxidases, and hypertension. Hypertension 56: 325-330.

[9] Griendling, K.K., FitzGerald, G.A. (2003) Oxidative stress and cardiovascular injury: Part I: basic mechanisms and in vivo monitoring of ROS. Circulation 108: 1912-1916.

[10] Bedard, K., Krause, K.-H. (2007) The NOX family of ROS-generating NADPH oxidases: Physiology and pathophysiology. Physiol Rev 87: 245-313.

[11] Geiszt, M., Leto, T.L. (2004) The Nox family of NAD(P)H oxidases: Host defense and beyond. J Biol Chem 279: 51715-51718.

[12] Nistala, R., Whaley-Connell, A., Sowers, J.R. (2008) Redox control of renal function and hypertension. Antiox Redox Signal 10: 2047-2089.

[13] Brandes, R.P., Takac, I., Schroeder, K. (2011) No Superoxide-No Stress? Nox4, the Good NADPH Oxidase! Arterioscler Thromb Vasc Biol 31: 1255-1257.

[14] Brown, D.I., Griendling, K.K. (2009) Nox proteins in signal transduction. Free Rad Biol Med 47: 1239-1253.

[15] Berry, C.E., Hare, J.M. (2004) Xanthine oxicloreductase and cardiovascular disease: molecular mechanisms and pathophysiological implications. J Physiol (London) 555: 589-606.

[16] Pacher, P., Nivorozhkin, A., Szabo, C. (2006) Therapeutic effects of xanthine oxidase inhibitors: renaissance half a century after the discovery of allopurinol. Pharmacol Rev 58: 87-114.

[17] Houston, M., Estevez, A., Chumley, P., Aslan, M., Marklund, S., Parks, D.A., Freeman, B.A. (1999) Binding of xanthine oxidase to vascular endothelium - Kinetic characterization and oxidative impairment of nitric oxide-dependent signaling. J Biol Chem 274: 4985-4994.

[18] Becker, B.F. (1993) Towards the physiological function of uric acid. Free Rad Biol Med 14: 615-631.

[19] Hink, H.U., Santanam, N., Dikalov, S., McCann, L., Nguyen, A.D., Parthasarathy, S., Harrison, D.G., Fukai, T. (2002) Peroxidase properties of extracellular superoxide dismutase - Role of uric acid in modulating in vivo activity. Arterioscler Thromb Vasc Biol 22: 1402-1408.

[20] Hooper, D.C., Spitsin, S., Kean, R.B., Champion, J.M., Dickson, G.M., Chaudhry, I., Koprowski, H. (1998) Uric acid, a natural scavenger of peroxynitrite, in experimental allergic encephalomyelitis and multiple sclerosis. Proc Natl Acad Sci USA 95: 675-680.

[21] Kuzkaya, N., Weissmann, N., Harrison, D.G., Dikalov, S. (2005) Interactions of peroxynitrite with uric acid in the presence of ascorbate and thiols: implications for uncoupling endothelial nitric oxide synthase. Biochem Pharmacol 70: 343-354.

[22] Sevanian, A., Davies, K.J., Hochstein, P. (1985) Conservation of vitamin C by uric acid in blood. J Free Rad Biol Med 1: 117-124.

[23] Johnson, R.J., Kang, D.H., Feig, D., Kivlighn, S., Kanellis, J., Watanabe, S., Tuttle, K.R., Rodriguez-Iturbe, B., Herrera-Acosta, J., Mazzali, M. (2003) Is there a pathogenetic role for uric acid in hypertension and cardiovascular and renal disease? Hypertension 41: 1183-1190.

[24] Waring, W.S., Webb, D.J., Maxwell, S.R.J. (2000) Uric acid as a risk factor for cardiovascular disease. QJM 93: 707-713.

[25] Kirschbaum, B. (2001) Renal regulation of plasma total antioxidant capacity. Med Hypotheses 56: 625-629.

[26] Storey, K.B. (1996) Oxidative stress: Animal adaptations in nature. Braz J Med Biol Res 29: 1715-1733.

[27] Addabbo, F., Montagnani, M., Goligorsky, M.S. (2009) Mitochondria and Reactive Oxygen Species. Hypertension 53: 885-892.

[28] Sugamura, K., Keaney, J.F., Jr. (2011) Reactive oxygen species in cardiovascular disease. Free Radic Biol Med 51: 978-992.

[29] Munzel, T., Gori, T., Bruno, R.M., Taddei, S. (2010) Is oxidative stress a therapeutic target in cardiovascular disease? Eur Heart J 31: 2741-2748.

[30] Chan, S.H., Wu, K.L., Chang, A.Y., Tai, M.H., Chan, J.Y. (2009) Oxidative impairment of mitochondrial electron transport chain complexes in rostral ventrolateral medulla contributes to neurogenic hypertension. Hypertension 53: 217-227.

[31] Dikalova, A.E., Bikineyeva, A.T., Budzyn, K., Nazarewicz, R.R., McCann, L., Lewis, W., Harrison, D.G., Dikalov, S.I. (2010) Therapeutic Targeting of Mitochondrial Superoxide in Hypertension. Circ Res 107: 106-U221.

[32] Elks, C.M., Mariappan, N., Haque, M., Guggilam, A., Majid, D.S., Francis, J. (2009) Chronic NF-{kappa}B blockade reduces cytosolic and mitochondrial oxidative stress and attenuates renal injury and hypertension in SHR. Am J Physiol Renal Physiol 296: F298-305.

[33] Briones, A.M., Touyz, R.M. (2010) Oxidative stress and hypertension: current concepts. Curr Hypertens Rep 12: 135-142.

[34] Li, H., Witte, K., August, M., Brausch, I., Godtel-Armbrust, U., Habermeier, A., Closs, E.I., Oelze, M., Munzel, T., Forstermann, U. (2006) Reversal of endothelial nitric oxide synthase uncoupling and up-regulation of endothelial nitric oxide synthase expression lowers blood pressure in hypertensive rats. J Am Coll Cardiol 47: 2536-2544.

[35] Vera, R., Sanchez, M., Galisteo, M., Concepcion Villar, I., Jimenez, R., Zarzuelo, A., Perez-Vizcaino, F., Duarte, J. (2007) Chronic administration of genistein improves endothelial dysfunction in spontaneously hypertensive rats: involvement of eNOS, caveolin and calmodulin expression and NADPH oxidase activity. Clin Sci 112: 183-191.

[36] Wassmann, S., Wassmann, K., Nickenig, G. (2004) Modulation of oxidant and antioxidant enzyme expression and function in vascular cells. Hypertension 44: 381-386.

[37] Malle, E., Buch, T., Grone, H.J. (2003) Myeloperoxidase in kidney disease. Kidney Int 64: 1956-1967.

[38] Green, P.S., Mendez, A.J., Jacob, J.S., Crowley, J.R., Growdon, W., Hyman, B.T., Heinecke, J.W. (2004) Neuronal expression of myeloperoxidase is increased in Alzheimer's disease. J Neurochem 90: 724-733.

[39] La Rocca, G., Di Stefano, A., Eleuteri, E., Anzalone, R., Magno, F., Corrao, S., Loria, T., Martorana, A., Di Gangi, C., Colombo, M., Sansone, F., Patane, F., Farina, F., Rinaldi, M., Cappello, F., Giannuzzi, P., Zummo, G. (2009) Oxidative stress induces myeloperoxidase expression in endocardial endothelial cells from patients with chronic heart failure. Basic Res Cardiol 104: 307-320.

[40] Xu, J., Xie, Z., Reece, R., Pimental, D., Zou, M.H. (2006) Uncoupling of endothelial nitric oxidase synthase by hypochlorous acid: role of NAD(P)H oxidase-derived superoxide and peroxynitrite. Arterioscler Thromb Vasc Biol 26: 2688-2695.

[41] Cai, H., Harrison, D.G. (2000) Endothelial dysfunction in cardiovascular diseases - The role of oxidant stress. Circ Res 87: 840-844.

[42] Didion, S.P., Ryan, M.J., Baumbach, G.L., Sigmund, C.D., Faraci, F.M. (2002) Superoxide contributes to vascular dysfunction in mice that express human renin and angiotensinogen. Am J Physiol Heart Circ Physiol 283: H1569-H1576.

[43] Lassegue, B., Griendling, K.K. (2004) Reactive oxygen species in hypertension; An update. Am J Hypertens 17: 852-860.

[44] Rathaus, M., Bernheim, J. (2002) Oxygen species in the microvascular environment: regulation of vascular tone and the development of hypertension. Nephrol Dial Transplant 17: 216-221.

[45] Touyz, R.M. (2003) Reactive oxygen species in vascular biology: role in arterial hypertension. Exp Rev Cardiov Ther 1: 91-106.

[46] Cowley, A.W., Jr. (2008) Renal Medullary Oxidative Stress, Pressure-Natriuresis, and Hypertension. Hypertension 52: 777-786.

[47] Tai, M.H., Wang, L.L., Wu, K.L.H., Chan, J.Y. (2005) Increased superoxide anion in rostral ventrolateral medulla contributes to hypertension in spontaneously hypertensive rats via interactions with nitric oxide. Free Rad Biol Med 38: 450-462.

[48] Makino, A., Skelton, M.M., Zou, A.P., Cowley, A.W., Jr. (2003) Increased renal medullary H2O2 leads to hypertension. Hypertension 42: 25-30.

[49] Sousa, T., Oliveira, S., Afonso, J., Morato, M., Patinha, D., Fraga, S., Carvalho, F., Albino-Teixeira, A. (2012) Role of H(2) o(2) in Hypertension, Renin-Angiotensin System Activation and Renal Medullary Disfunction Caused by Angiotensin II. Br J Pharmacol: in press.

[50] Sousa, T., Pinho, D., Morato, M., Marques-Lopes, J., Fernandes, E., Afonso, J., Oliveira, S., Carvalho, F., Albino-Teixeira, A. (2008) Role of superoxide and hydrogen peroxide in hypertension induced by an antagonist of adenosine receptors. Eur J Pharmacol 588: 267-276.

[51] Cai, H. (2005) NAD(P)H oxidase-dependent self-propagation of hydrogen peroxide and vascular disease. Circ Res 96: 818-822.

[52] Asghar, M., Banday, A.A., Fardoun, R.Z., Lokhandwala, M.F. (2006) Hydrogen peroxide causes uncoupling of dopamine D1-like receptors from G proteins via a

mechanism involving protein kinase C and G-protein-coupled receptor kinase 2. Free Rad Biol Med 40: 13-20.

[53] Chen, Y.F., Cowley, A.W., Jr., Zou, A.P. (2003) Increased H(2)O(2) counteracts the vasodilator and natriuretic effects of superoxide dismutation by tempol in renal medulla. Am J Physiol Regul Integr Comp Physiol 285: R827-833.

[54] Gao, Y.J., Hirota, S., Zhang, D.W., Janssen, L.J., Lee, R. (2003) Mechanisms of hydrogen-peroxide-induced biphasic response in rat mesenteric artery. Br J Pharmacol 138: 1085-1092.

[55] Gao, Y.J., Lee, R. (2001) Hydrogen peroxide induces a greater contraction in mesenteric arteries of spontaneously hypertensive rats through thromboxane A(2) production. Br J Pharmacol 134: 1639-1646.

[56] Lin, H.H., Chen, C.H., Hsieh, W.K., Chiu, T.H., Lai, C.C. (2003) Hydrogen peroxide increases the activity of rat sympathetic preganglionic neurons in vivo and in vitro. Neuroscience 121: 641-647.

[57] Rodriguez-Martinez, M.A., Garcia-Cohen, E.C., Baena, A.B., Gonzalez, R., Salaices, M., Marin, J. (1998) Contractile responses elicited by hydrogen peroxide in aorta from normotensive and hypertensive rats. Endothelial modulation and mechanism involved. Br J Pharmacol 125: 1329-1335.

[58] Thakali, K., Davenport, L., Fink, G.D., Watts, S.W. (2006) Pleiotropic effects of hydrogen peroxide in arteries and veins from normotensive and hypertensive rats. Hypertension 47: 482-487.

[59] Didion, S., Chrissobolis, S., Faraci, F.M. (2008) Oxidative Stress in Hypertension. In: Miwa, Beckman, Muller, editors. Aging Medicine: Oxidative Stress in Aging 3: 229-251.

[60] Kristal, B., Shurtz-Swirski, R., Chezar, J., Manaster, J., Levy, R., Shapiro, G., Weissman, I., Shasha, S.M., Sela, S. (1998) Participation of peripheral polymorphonuclear leukocytes in the oxidative stress and inflammation in patients with essential hypertension. Am J Hypertens 11: 921-928.

[61] Lacy, F., Kailasam, M.T., O'Connor, D.T., Schmid-Schonbein, G.W., Parmer, R.J. (2000) Plasma hydrogen peroxide production in human essential hypertension - Role of heredity, gender, and ethnicity. Hypertension 36: 878-884.

[62] Lacy, F., O'Connor, D.T., Schmid-Schonbein, G.W. (1998) Plasma hydrogen peroxide production in hypertensives and normotensive subjects at genetic risk of hypertension. J Hypertens 16: 291-303.

[63] Dhawan, V., Jain, S. (2004) Effect of garlic supplementation on oxidized low density lipoproteins and lipid peroxidation in patients of essential hypertension. Mol Cell Biochem 266: 109-115.

[64] Kedziora-Kornatowska, K., Czuczejko, J., Pawluk, H., Kornatowski, T., Motyl, J., Szadujkis-Szadurski, L., Szewczyk-Golec, K., Kedziora, J. (2004) The markers of oxidative stress and activity of the antioxidant system in the blood of elderly patients with essential arterial hypertension. Cell Mol Biol Lett 9: 635-641.

[65] Redon, J., Oliva, M.R., Tormos, C., Giner, V., Chaves, J., Iradi, A., Saez, G.T. (2003) Antioxidant activities and oxidative stress byproducts in human hypertension. Hypertension 41: 1096-1101.

[66] Zhou, L., Xiang, W., Potts, J., Floyd, M., Sharan, C., Yang, H., Ross, J., Nyanda, A.M., Guo, Z. (2006) Reduction in extracellular superoxide dismutase activity in African-American patients with hypertension. Free Rad Biol Med 41: 1384-1391.

[67] Wen, Y., Killalea, S., McGettigan, P., Feely, J. (1996) Lipid peroxidation and antioxidant vitamins C and E in hypertensive patients. Ir J Med Sci 165: 210-212.

[68] Labios, M., Martinez, M., Gabriel, F., Guiral, V., Ruiz-Aja, S., Beltran, B., Munoz, A. (2008) Effects of eprosartan on mitochondrial membrane potential and H_2O_2 levels in leucocytes in hypertension. J Hum Hypertens 22: 493-500.

[69] Lyamina, N.P., Dolotovskaya, P.V., Lyamina, S.V., Malyshev, I.Y., Manukhina, E.B. (2003) Nitric oxide production and intensity of free radical processes in young men with high normal and hypertensive blood pressure. Med Sci Monit 9: CR304-310.

[70] Sierra, M., Gonzalez, A., Gomez-Alamillo, C., Monreal, I., Huarte, E., Gil, A., Sanchez-Casajus, A., Diez, J. (1998) Decreased excretion of nitrate and nitrite in essential hypertensives with renal vasoconstriction. Kidney Int 54: S10-S13.

[71] San Jose, G., Moreno, M.U., Olivan, S., Beloqui, O., Fortuno, A., Diez, J., Zalba, G. (2004) Functional effect of the p22(phox) -930(A/G) polymorphism on p22(phox) expression and NADPH oxidase activity in hypertension. Hypertension 44: 163-169.

[72] Russo, C., Olivieri, O., Girelli, D., Faccini, G., Zenari, M.L., Lombardi, S., Corrocher, R. (1998) Anti-oxidant status and lipid peroxidation in patients with essential hypertension. J Hypertens 16: 1267-1271.

[73] Kedziora-Kornatowska, K., Kornatowski, T., Bartosz, G., Pawluk, H., Czuczejko, J., Kedziora, J., Szadujkis-Szadurski, L. (2006) Production of nitric oxide, lipid peroxidation and oxidase activity of ceruloplasmin in blood of elderly patients with primary hypertension. Effects of perindopril treatment. Aging Clin Exp Res 18: 1-6.

[74] Rodrigo, R., Prat, H., Passalacqua, W., Araya, J., Guichard, C., Bachler, J.P. (2007) Relationship between oxidative stress and essential hypertension. Hypertension Res 30: 1159-1167.

[75] Minuz, P., Patrignani, P., Gaino, S., Degan, M., Menapace, L., Tommasoli, R., Seta, F., Capone, M.L., Tacconelli, S., Palatresi, S., Bencini, C., Del Vecchio, C., Mansueto, G., Arosio, E., Santonastaso, C.L., Lechi, A., Morganti, A., Patrono, C. (2002) Increased oxidative stress and platelet activation in patients with hypertension and renovascular disease. Circulation 106: 2800-2805.

[76] Espinosa, O., Jimenez-Almazan, J., Chaves, F.J., Tormos, M.C., Clapes, S., Iradi, A., Salvador, A., Fandos, M., Redon, J., Saez, G.T. (2007) Urinary 8-oxo-7,8-dihydro-2 '-deoxyguanosine (8-oxo-dG), a reliable oxidative stress marker in hypertension. Free Rad Res 41: 546-554.

[77] Rosello-Lleti, E., Garcia de Burgos, F., Morillas, P., Cortes, R., Martinez-Dolz, L., Almenar, L., Grigorian, L., Orosa, P., Portoles, M., Bertomeu, V., Rivera, M. (2011) Impact of Cardiovascular Risk Factors and Inflammatory Status on Urinary 8-OHdG in Essential Hypertension. Am J Hypertens 25: 236-242.

[78] Maggi, E., Marchesi, E., Ravetta, V., Falaschi, F., Finardi, G., Bellomo, G. (1993) Low-density lipoprotein oxidation in essential hypertension. J Hypertens 11: 1103-1111.

[79] Santangelo, L., Cigliano, L., Montefusco, A., Spagnuolo, M.S., Nigro, G., Golino, P., Abrescia, P. (2003) Evaluation of the antioxidant response in the plasma of healthy or hypertensive subjects after short-term exercise. J Hum Hypertens 17: 791-798.

[80] Alderman, M.H., Cohen, H., Madhavan, S., Kivlighn, S. (1999) Serum uric acid and cardiovascular events in successfully treated hypertensive patients. Hypertension 34: 144-150.

[81] Tse, W.Y., Maxwell, S.R.J., Thomason, H., Blann, A., Thorpe, G.H.G., Waite, M., Holder, R. (1994) Antioxidant status in controlled and uncontrolled hypertension and its relationship to endothelial damage. J Hum Hypertens 8: 843-849.

[82] Subash, P., Premagurumurthy, K., Sarasabharathi, A., Cherian, K.M. (2010) Total antioxidant status and oxidative DNA damage in a South Indian population of essential hypertensives. J Hum Hypertens 24: 475-482.

[83] Kashyap, M.K., Yadav, V., Sherawat, B.S., Jain, S., Kumari, S., Khullar, M., Sharma, P.C., Nath, R. (2005) Different antioxidants status, total antioxidant power and free radicals in essential hypertension. Mol Cell Biochem 277: 89-99.

[84] Vaziri, N.D., Wang, X.Q., Oveisi, F., Rad, B. (2000) Induction of oxidative stress by glutathione depletion causes severe hypertension in normal rats. Hypertension 36: 142-146.

[85] Baumer, A.T., Kruger, C.A., Falkenberg, J., Freyhaus, H.T., Rosen, R., Fink, K., Rosenkranz, S. (2007) The NAD(P)H oxidase inhibitor apocynin improves endothelial NO/superoxide balance and lowers effectively blood pressure in spontaneously hypertensive rats: comparison to calcium channel blockade. Clin Exp Hypertens 29: 287-299.

[86] Beswick, R.A., Dorrance, A.M., Leite, R., Webb, R.C. (2001) NADH/NADPH oxidase and enhanced superoxide production in the mineralocorticoid hypertensive rat. Hypertension 38: 1107-1111.

[87] Zhang, Y., Croft, K.D., Mori, T.A., Schyvens, C.G., McKenzie, K.U., Whitworth, J.A. (2004) The antioxidant tempol prevents and partially reverses dexamethasone-induced hypertension in the rat. Am J Hypertens 17: 260-265.

[88] Dikalova, A., Clempus, R., Lassegue, B., Cheng, G.J., McCoy, J., Dikalov, S., Martin, A.S., Lyle, A., Weber, D.S., Weiss, D., Taylor, R., Schmidt, H., Owens, G.K., Lambeth, J.D., Griendling, K.K. (2005) Nox1 overexpression potentiates angiotensin II-induced hypertension and vascular smooth muscle hypertrophy in transgenic mice. Circulation 112: 2668-2676.

[89] Godin, N., Liu, F., Lau, G.J., Brezniceanu, M.-L., Chenier, I., Filep, J.G., Ingelfinger, J.R., Zhang, S.-L., Chan, J.S.D. (2010) Catalase overexpression prevents hypertension and tubular apoptosis in angiotensinogen transgenic mice. Kidney Int 77: 1086-1097.

[90] Wilcox, C.S. (2005) Oxidative stress and nitric oxide deficiency in the kidney: a critical link to hypertension? Am J Physiol-Regul Integr Comp Physiol 289: R913-R935.

[91] Paravicini, T.M., Touyz, R.M. (2006) Redox signaling in hypertension. Cardiov Res 71: 247-258.

[92] Grossman, E. (2008) Does increased oxidative stress cause hypertension? Diabetes Care 31 Suppl 2: S185-189.

[93] Baykal, Y., Yilmaz, M.I., Celik, T., Gok, F., Rehber, H., Akay, C., Kocar, I.H. (2003) Effects of antihypertensive agents, alpha receptor blockers, beta blockers, angiotensin-converting enzyme inhibitors, angiotensin receptor blockers and calcium channel blockers, on oxidative stress. J Hypertens 21: 1207-1211.

[94] de Cavanagh, E.M., Ferder, L.F., Ferder, M.D., Stella, I.Y., Toblli, J.E., Inserra, F. (2010) Vascular structure and oxidative stress in salt-loaded spontaneously hypertensive rats: effects of losartan and atenolol. Am J Hypertens 23: 1318-1325.

[95] Sugiura, T., Kondo, T., Kureishi-Bando, Y., Numaguchi, Y., Yoshida, O., Dohi, Y., Kimura, G., Ueda, R., Rabelink, T.J., Murohara, T. (2008) Nifedipine improves endothelial function role of endothelial progenitor cells. Hypertension 52: 491-498.

[96] Wingler, K., Hermans, J.J., Schiffers, P., Moens, A., Paul, M., Schmidt, H.H. (2011) NOX1, 2, 4, 5: counting out oxidative stress. Br J Pharmacol 164: 866-883.

[97] Dalle-Donne, I., Rossi, R., Colombo, R., Giustarini, D., Milzani, A. (2006) Biomarkers of oxidative damage in human disease. Clin Chem 52: 601-623.

[98] de Zwart, L.L., Meerman, J.H., Commandeur, J.N., Vermeulen, N.P. (1999) Biomarkers of free radical damage applications in experimental animals and in humans. Free Radic Biol Med 26: 202-226.

[99] Atkinson, A. (2001) Biomarkers and surrogate endpoints: preferred definitions and conceptual framework. Clin Pharmacol Ther 69: 89-95.

[100] Offord, E., van Poppel, G., Tyrrell, R. (2000) Markers of oxidative damage and antioxidant protection: current status and relevance to disease. Free Radic Res 33 Suppl: S5-19.

[101] Hancock, J.T., Desikan, R., Neill, S.J. (2001) Role of reactive oxygen species in cell signalling pathways. Biochem Soc Trans 29: 345-350.

[102] Sohal, R.S., Dubey, A. (1994) Mitochondrial oxidative damage, hydrogen peroxide release, and aging. Free Radic Biol Med 16: 621-626.

[103] Taniyama, Y., Griendling, K.K. (2003) Reactive oxygen species in the vasculature: molecular and cellular mechanisms. Hypertension 42: 1075-1081.

[104] Halliwell, B., Gutteridge, J.M. (1989) Free Rad Biol Med, 2nd edition, Clarendon Press, Oxford, UK.

[105] Sies, H. (1993) Strategies of antioxidant defense. Eur J Biochem 215: 213-219.

[106] Frei, B., Stocker, R., Ames, B.N. (1988) Antioxidant defenses and lipid peroxidation in human blood plasma. Proc Natl Acad Sci U S A 85: 9748-9752.

[107] Shull, S., Heintz, N.H., Periasamy, M., Manohar, M., Janssen, Y.M., Marsh, J.P., Mossman, B.T. (1991) Differential regulation of antioxidant enzymes in response to oxidants. J Biol Chem 266: 24398-24403.

[108] Young, I.S., Woodside, J.V. (2001) Antioxidants in health and disease. J Clin Pathol 54: 176-186.

[109] Hayes, J.D., Pulford, D.J. (1995) The glutathione S-transferase supergene family: regulation of GST and the contribution of the isoenzymes to cancer chemoprotection and drug resistance. Crit Rev Biochem Mol Biol 30: 445-600.

[110] Kalinina, E.V., Chernov, N.N., Saprin, A.N. (2008) Involvement of thio-, peroxi-, and glutaredoxins in cellular redox-dependent processes. Biochemistry (Mosc) 73: 1493-1510.

[111] Farmand, F., Ehdaie, A., Roberts, C.K., Sindhu, R.K. (2005) Lead-induced dysregulation of superoxide dismutases, catalase, glutathione peroxidase, and guanylate cyclase. Environ Res 98: 33-39.

[112] Sainz, J., Wangensteen, R., Rodriguez Gomez, I., Moreno, J.M., Chamorro, V., Osuna, A., Bueno, P., Vargas, F. (2005) Antioxidant enzymes and effects of tempol on the development of hypertension induced by nitric oxide inhibition. Am J Hypertens 18: 871-877.

[113] Gille, G., Sigler, K. (1995) Oxidative stress and living cells. Folia Microbiol (Praha) 40: 131-152.

[114] Chan, P., Liao, S.S., Hsu, C.T., Lee, Y.S., Tomlinson, B., Kuo, J.S., Cheng, J.T. (1999) Superoxide dismutase gene expression and activity in the brain of spontaneously hypertensive rats and normotensive rats. Chin Med J (Engl) 112: 1119-1124.

[115] Nozoe, M., Hirooka, Y., Koga, Y., Sagara, Y., Kishi, T., Engelhardt, J.F., Sunagawa, K. (2007) Inhibition of Rac1-derived reactive oxygen species in nucleus tractus solitarius decreases blood pressure and heart rate in stroke-prone spontaneously hypertensive rats. Hypertension 50: 62-68.

[116] Chan, S.H., Tai, M.H., Li, C.Y., Chan, J.Y. (2006) Reduction in molecular synthesis or enzyme activity of superoxide dismutases and catalase contributes to oxidative stress and neurogenic hypertension in spontaneously hypertensive rats. Free Radic Biol Med 40: 2028-2039.

[117] Adler, S., Huang, H., Loke, K.E., Xu, X., Tada, H., Laumas, A., Hintze, T.H. (2001) Endothelial nitric oxide synthase plays an essential role in regulation of renal oxygen consumption by NO. Am J Physiol Renal Physiol 280: F838-843.

[118] Simao, S., Gomes, P., Pinto, V., Silva, E., Amaral, J.S., Igreja, B., Afonso, J., Serrao, M.P., Pinho, M.J., Soares-da-Silva, P. (2011) Age-related changes in renal expression of oxidant and antioxidant enzymes and oxidative stress markers in male SHR and WKY rats. Exp Gerontol 46: 468-474.

[119] Potenza, M.A., Addabbo, F., Montagnani, M. (2009) Vascular actions of insulin with implications for endothelial dysfunction. Am J Physiol Endocrinol Metab 297: E568-577.

[120] Zheng, H., Yu, Y.S. (2012) Chronic hydrogen-rich saline treatment attenuates vascular dysfunction in spontaneous hypertensive rats. Biochem Pharmacol 83: 1269-1277.

[121] Ulker, S., McMaster, D., McKeown, P.P., Bayraktutan, U. (2003) Impaired activities of antioxidant enzymes elicit endothelial dysfunction in spontaneous hypertensive rats despite enhanced vascular nitric oxide generation. Cardiovasc Res 59: 488-500.

[122] Polizio, A.H., Pena, C. (2005) Effects of angiotensin II type 1 receptor blockade on the oxidative stress in spontaneously hypertensive rat tissues. Regul Pept 128: 1-5.

[123] Xu, Y., Gao, Q., Gan, X.B., Chen, L., Zhang, L., Zhu, G.Q., Gao, X.Y. (2011) Endogenous hydrogen peroxide in paraventricular nucleus mediates sympathetic activation and enhanced cardiac sympathetic afferent reflex in renovascular hypertensive rats. Exp Physiol 96: 1282-1292.

[124] Fortepiani, L.A., Reckelhoff, J.F. (2005) Increasing oxidative stress with molsidomine increases blood pressure in genetically hypertensive rats but not normotensive controls. Am J Physiol Regul Integr Comp Physiol 289: R763-770.

[125] Kumar, U., Chen, J., Sapoznikhov, V., Canteros, G., White, B.H., Sidhu, A. (2005) Overexpression of inducible nitric oxide synthase in the kidney of the spontaneously hypertensive rat. Clin Exp Hypertens 27: 17-31.

[126] Zhang, L., Fujii, S., Kosaka, H. (2007) Effect of oestrogen on reactive oxygen species production in the aortas of ovariectomized Dahl salt-sensitive rats. J Hypertens 25: 407-414.

[127] Kosower, N.S., Kosower, E.M. (1978) The glutathione status of cells. Int Rev Cytol 54: 109-160.

[128] Franco, R., DeHaven, W.I., Sifre, M.I., Bortner, C.D., Cidlowski, J.A. (2008) Glutathione depletion and disruption of intracellular ionic homeostasis regulate lymphoid cell apoptosis. J Biol Chem 283: 36071-36087.

[129] Schulz, J.B., Lindenau, J., Seyfried, J., Dichgans, J. (2000) Glutathione, oxidative stress and neurodegeneration. Eur J Biochem 267: 4904-4911.

[130] Singh, U., Devaraj, S., Jialal, I. (2005) Vitamin E, oxidative stress, and inflammation. Annu Rev Nutr 25: 151-174.

[131] Smith, A.R., Visioli, F., Hagen, T.M. (2002) Vitamin C matters: increased oxidative stress in cultured human aortic endothelial cells without supplemental ascorbic acid. FASEB J 16: 1102-1104.

[132] Waring, W.S. (2002) Uric acid: an important antioxidant in acute ischaemic stroke. QJM 95: 691-693.

[133] Stocker, R., Yamamoto, Y., McDonagh, A.F., Glazer, A.N., Ames, B.N. (1987) Bilirubin is an antioxidant of possible physiological importance. Science 235: 1043-1046.

[134] Lin, J.P., Vitek, L., Schwertner, H.A. (2010) Serum bilirubin and genes controlling bilirubin concentrations as biomarkers for cardiovascular disease. Clin Chem 56: 1535-1543.

[135] Hu, M.L., Louie, S., Cross, C.E., Motchnik, P., Halliwell, B. (1993) Antioxidant protection against hypochlorous acid in human plasma. J Lab Clin Med 121: 257-262.

[136] Benzie, I.F., Strain, J.J. (1996) The ferric reducing ability of plasma (FRAP) as a measure of "antioxidant power": the FRAP assay. Anal Biochem 239: 70-76.

[137] Wayner, D.D., Burton, G.W., Ingold, K.U., Locke, S. (1985) Quantitative measurement of the total, peroxyl radical-trapping antioxidant capability of human blood plasma by controlled peroxidation. The important contribution made by plasma proteins. FEBS Lett 187: 33-37.

[138] Yanes, L., Romero, D., Iliescu, R., Cucchiarelli, V.E., Fortepiani, L.A., Santacruz, F., Bell, W., Zhang, H., Reckelhoff, J.F. (2005) Systemic arterial pressure response to two weeks of Tempol therapy in SHR: involvement of NO, the RAS, and oxidative stress. Am J Physiol Regul Integr Comp Physiol 288: R903-908.

[139] Ghiselli, A., Serafini, M., Natella, F., Scaccini, C. (2000) Total antioxidant capacity as a tool to assess redox status: critical view and experimental data. Free Radic Biol Med 29: 1106-1114.

[140] Kanbay, M., Solak, Y., Dogan, E., Lanaspa, M.A., Covic, A. (2010) Uric acid in hypertension and renal disease: the chicken or the egg? Blood Purif 30: 288-295.

[141] Young, I.S. (2001) Measurement of total antioxidant capacity. J Clin Pathol 54: 339.

[142] Duffy, S.J., Gokce, N., Holbrook, M., Huang, A., Frei, B., Keaney, J.F., Jr., Vita, J.A. (1999) Treatment of hypertension with ascorbic acid. Lancet 354: 2048-2049.

[143] Newaz, M.A., Nawal, N.N., Rohaizan, C.H., Muslim, N., Gapor, A. (1999) alpha-Tocopherol increased nitric oxide synthase activity in blood vessels of spontaneously hypertensive rats. Am J Hypertens 12: 839-844.

[144] Ward, N.C., Hodgson, J.M., Puddey, I.B., Mori, T.A., Beilin, L.J., Croft, K.D. (2004) Oxidative stress in human hypertension: association with antihypertensive treatment, gender, nutrition, and lifestyle. Free Radic Biol Med 36: 226-232.

[145] Gutteridge, J.M., Halliwell, B. (1990) The measurement and mechanism of lipid peroxidation in biological systems. Trends Biochem Sci 15: 129-135.

[146] Halliwell, B., Whiteman, M. (2004) Measuring reactive species and oxidative damage in vivo and in cell culture: how should you do it and what do the results mean? Br J Pharmacol, 142: 231-55.

[147] Vincent, H.K., Taylor, A.G. (2006) Biomarkers and potential mechanisms of obesity-induced oxidant stress in humans. Int J Obes (Lond) 30: 400-418.

[148] Janicka, M., Kot-Wasik, A., Kot, J., Namiesnik, J. (2010) Isoprostanes-biomarkers of lipid peroxidation: their utility in evaluating oxidative stress and analysis. Int J Mol Sci 11: 4631-4659.

[149] Soffler, C., Campbell, V.L., Hassel, D.M.(2010) Measurement of urinary F2-isoprostanes as markers of in vivo lipid peroxidation: a comparison of enzyme immunoassays with gas chromatography-mass spectrometry in domestic animal species. J Vet Diagn Invest. 22: 200-9.

[150] Greco, A., Minghetti, L., Levi, G. (2000) Isoprostanes, novel markers of oxidative injury, help understanding the pathogenesis of neurodegenerative diseases. Neurochem Res 25: 1357-1364.

[151] Kadiiska, M.B., Gladen, B.C., Baird, D.D., Germolec, D., Graham, L.B., Parker, C.E., Nyska, A., Wachsman, J.T., Ames, B.N., Basu, S., Brot, N., Fitzgerald, G.A., Floyd, R.A., George, M., Heinecke, J.W., Hatch, G.E., Hensley, K., Lawson, J.A., Marnett, L.J., Morrow, J.D., Murray, D.M., Plastaras, J., Roberts, L.J. 2nd, Rokach, J., Shigenaga, M.K., Sohal, R.S., Sun, J., Tice, R.R., Van Thiel, D.H., Wellner, D., Walter, P.B., Tomer, K.B., Mason, R.P., Barrett, J.C.(2005) Biomarkers of oxidative stress study II: are oxidation products of lipids, proteins, and DNA markers of CCl4 poisoning? Free Radic Biol Med 38:698–710.

[152] Milne GL, Yin H, Brooks JD, et al.: 2007, Quantification of F2-isoprostanes in biological fluids and tissues as a measure of oxidant stress. Methods Enzymol 433:113–126.

[153] Todorova, I., Simeonova, G. Kyuchukova, D., Dinev, D., Gadjeva, V. (2005) Reference values of oxidative stress parameters (MDA, SOD, CAT) in dogs and cats. Comp Clin Path 13: 190–194.

[154] Huszar, G, Vigue, L. (1994) Correlation between the rate of lipid peroxidation and cellular maturity as measured by creatine kinase activity in human spermatozoa. J Androl 15: 71-7.

[155] Janero, D.R. (1990) Malondialdehyde and thiobarbituric acid-reactivity as diagnostic indices of lipid peroxidation and peroxidative tissue injury. Free Radic Biol Med 9: 515-540.

[156] Yeo, H.C., Helbock, H.J., Chyu, D.W., Ames, B.N. (1994) Assay of malondialdehyde in biological fluids by gas chromatography-mass spectrometry. Anal Biochem 220: 391-396.

[157] Khaira, A., Mahajan, S., Kumar, A., Saraya, A., Tiwari, S.C., Prakash, S., Gupta, A., Bhowmik, D., Agarwal, S.K. (2011) Endothelial function and oxidative stress in chronic kidney disease of varying severity and the effect of acute hemodialysis. Ren Fail 33: 411-417.

[158] Rumley, A.G., Woodward, M., Rumley, A., Rumley, J., Lowe, G.D. (2004) Plasma lipid peroxides: relationships to cardiovascular risk factors and prevalent cardiovascular disease. QJM 97: 809-816.

[159] Satoh, K. (1978) Serum lipid peroxide in cerebrovascular disorders determined by a new colorimetric method. Clin Chim Acta 90: 37-43.

[160] Yoneyama, Y., Sawa, R., Suzuki, S., Doi, D., Yoneyama, K., Otsubo, Y., Araki, T. (2002) Relationship between plasma malondialdehyde levels and adenosine deaminase activities in preeclampsia. Clin Chim Acta 322: 169-173.

[161] Tateishi, T., Yoshimine, N, Kuzuya, F. (1987) Serum lipid peroxide assayed by a new colorimetric method. Exp Gerontol 22 (2): 103-111.

[162] Uchida, K. (2003) 4-Hydroxy-2-nonenal: a product and mediator of oxidative stress. Prog Lipid Res 42: 318-343.

[163] Suzuki, H., Nakazato, K., Asayama, K., Masawa, N., Takamata, M., Sakata, N. (2002) The role of oxidative stress on pathogenesis of hypertensive arterial lesions in rat mesenteric arteries. Acta Histochem Cytochem 35: 287-293.

[164] Blasig, I.E., Grune, T., Schonheit, K., Rohde, E., Jakstadt, M., Haseloff, R.F., Siems, W.G. (1995) 4-Hydroxynonenal, a novel indicator of lipid peroxidation for reperfusion injury of the myocardium. Am J Physiol 269: H14-22.

[165] Esterbauer, H., Schaur, R., Zollner, H. (1991) Chemistry and biochemistry of 4-hydroxynonenal, malonaldehyde and related aldehydes. Free Rad Biol Med 11: 81-128.

[166] Michel, F., Bonnefont-Rousselot, D., Mas, E., Drai, J., Therond, P. (2008) [Biomarkers of lipid peroxidation: analytical aspects]. Ann Biol Clin (Paris) 66: 605-620.

[167] Abuja, P.M., Albertini, R. (2001) Methods for monitoring oxidative stress, lipid peroxidation and oxidation resistance of lipoproteins. Clin Chim Acta 306: 1-17.

[168] Kato, Y., Yoshida, A., Naito, M., Kawai, Y., Tsuji, K., Kitamura, M., Kitamoto, N., Osawa, T. (2004) Identification and quantification of N(epsilon)-(Hexanoyl))lysine in human urine by liquid chromatography/tandem mass spectrometry. Free Radic Biol Med 37: 1864-1874.

[169] Kato, Y., Miyake, Y., Yamamoto, K., Shimomura, Y., Ochi, H., Mori, Y., Osawa, T. (2000) Preparation of a monoclonal antibody to N(epsilon)-(Hexanonyl)lysine: application to the evaluation of protective effects of flavonoid supplementation against exercise-induced oxidative stress in rat skeletal muscle. Biochem Biophys Res Commun 274: 389-393.

[170] Duarte, J., Perez-Palencia, R., Vargas, F., Ocete, M.A., Perez-Vizcaino, F., Zarzuelo, A., Tamargo, J. (2001) Antihypertensive effects of the flavonoid quercetin in spontaneously hypertensive rats. Br J Pharmacol 133: 117-124.

[171] Elmarakby, A.A., Imig, J.D. (2010) Obesity is the major contributor to vascular dysfunction and inflammation in high-fat diet hypertensive rats. Clin Sci (Lond) 118: 291-301.

[172] Galisteo, M., Garcia-Saura, M.F., Jimenez, R., Villar, I.C., Zarzuelo, A., Vargas, F., Duarte, J. (2004) Effects of chronic quercetin treatment on antioxidant defence system and oxidative status of deoxycorticosterone acetate-salt-hypertensive rats. Mol Cell Biochem 259: 91-99.

[173] De Miguel, C., Guo, C., Lund, H., Feng, D., Mattson, D.L. (2010) Infiltrating T lymphocytes in the kidney increase oxidative stress and participate in the development of hypertension and renal disease. Am J Physiol Renal Physiol 300: F734-742.

[174] Zhang, G.X., Kimura, S., Nishiyama, A., Shokoji, T., Rahman, M., Abe, Y. (2004) ROS during the acute phase of Ang II hypertension participates in cardiovascular MAPK activation but not vasoconstriction. Hypertension 43: 117-124.

[175] Meng, S., Roberts, L.J., 2nd, Cason, G.W., Curry, T.S., Manning, R.D., Jr. (2002) Superoxide dismutase and oxidative stress in Dahl salt-sensitive and -resistant rats. Am J Physiol Regul Integr Comp Physiol 283: R732-738.

[176] Paliege, A., Pasumarthy, A., Mizel, D., Yang, T., Schnermann, J., Bachmann, S. (2006) Effect of apocynin treatment on renal expression of COX-2, NOS1, and renin in Wistar-Kyoto and spontaneously hypertensive rats. Am J Physiol Regul Integr Comp Physiol 290: R694-700.

[177] Takai, S., Jin, D., Ikeda, H., Sakonjo, H., Miyazaki, M. (2009) Significance of angiotensin II receptor blockers with high affinity to angiotensin II type 1 receptors for vascular protection in rats. Hypertens Res 32: 853-860.

[178] Asselin, C., Bouchard, B., Tardif, J.C., Des Rosiers, C. (2006) Circulating 4-hydroxynonenal-protein thioether adducts assessed by gas chromatography-mass spectrometry are increased with disease progression and aging in spontaneously hypertensive rats. Free Radic Biol Med 41: 97-105.

[179] de Cavanagh, E.M., Toblli, J.E., Ferder, L., Piotrkowski, B., Stella, I., Inserra, F. (2006) Renal mitochondrial dysfunction in spontaneously hypertensive rats is attenuated by losartan but not by amlodipine. Am J Physiol Regul Integr Comp Physiol 290: R1616-1625.

[180] Zhang, A., Jia, Z., Wang, N., Tidwell, T.J., Yang, T. (2011) Relative contributions of mitochondria and NADPH oxidase to deoxycorticosterone acetate-salt hypertension in mice. Kidney Int 80: 51-60.

[181] Yamamoto, E., Tamamaki, N., Nakamura, T., Kataoka, K., Tokutomi, Y., Dong, Y.F., Fukuda, M., Matsuba, S., Ogawa, H., Kim-Mitsuyama, S. (2008) Excess salt causes cerebral neuronal apoptosis and inflammation in stroke-prone hypertensive rats through angiotensin II-induced NADPH oxidase activation. Stroke 39: 3049-3056.

[182] Cifuentes, M.E., Rey, F.E., Carretero, O.A., Pagano, P.J. (2000) Upregulation of p67(phox) and gp91(phox) in aortas from angiotensin II-infused mice. Am J Physiol Heart Circ Physiol 279: H2234-2240.

[183] Akasaki, T., Ohya, Y., Kuroda, J., Eto, K., Abe, I., Sumimoto, H., Iida, M. (2006) Increased expression of gp91phox homologues of NAD(P)H oxidase in the aortic media

during chronic hypertension: involvement of the renin-angiotensin system. Hypertens Res 29: 813-820.

[184] Fukui, T., Ishizaka, N., Rajagopalan, S., Laursen, J.B., Capers, Q.t., Taylor, W.R., Harrison, D.G., de Leon, H., Wilcox, J.N., Griendling, K.K. (1997) p22phox mRNA expression and NADPH oxidase activity are increased in aortas from hypertensive rats. Circ Res 80: 45-51.

[185] Rajagopalan, S., Kurz, S., Munzel, T., Tarpey, M., Freeman, B.A., Griendling, K.K., Harrison, D.G. (1996) Angiotensin II-mediated hypertension in the rat increases vascular superoxide production via membrane NADH/NADPH oxidase activation. Contribution to alterations of vasomotor tone. J Clin Invest 97: 1916-1923.

[186] Virdis, A., Neves, M.F., Amiri, F., Touyz, R.M., Schiffrin, E.L. (2004) Role of NAD(P)H oxidase on vascular alterations in angiotensin II-infused mice. J Hypertens 22: 535-542.

[187] Suzuki, H., DeLano, F.A., Parks, D.A., Jamshidi, N., Granger, D.N., Ishii, H., Suematsu, M., Zweifach, B.W., Schmid-Schonbein, G.W. (1998) Xanthine oxidase activity associated with arterial blood pressure in spontaneously hypertensive rats. Proc Natl Acad Sci U S A 95: 4754-4759.

[188] DeLano, F.A., Parks, D.A., Ruedi, J.M., Babior, B.M., Schmid-Schonbein, G.W. (2006) Microvascular display of xanthine oxidase and NADPH oxidase in the spontaneously hypertensive rat. Microcirculation 13: 551-566.

[189] Breckwoldt, M.O., Chen, J.W., Stangenberg, L., Aikawa, E., Rodriguez, E., Qiu, S., Moskowitz, M.A., Weissleder, R. (2008) Tracking the inflammatory response in stroke in vivo by sensing the enzyme myeloperoxidase. Proc Natl Acad Sci U S A 105: 18584-18589.

[190] Ersahin, M., Sehirli, O., Toklu, H.Z., Suleymanoglu, S., Emekli-Alturfan, E., Yarat, A., Tatlidede, E., Yegen, B.C., Sener, G. (2009) Melatonin improves cardiovascular function and ameliorates renal, cardiac and cerebral damage in rats with renovascular hypertension. J Pineal Res 47: 97-106.

[191] Wang, X., Desai, K., Clausen, J.T., Wu, L. (2004) Increased methylglyoxal and advanced glycation end products in kidney from spontaneously hypertensive rats. Kidney Int 66: 2315-2321.

[192] Bayorh, M.A., Ganafa, A.A., Emmett, N., Socci, R.R., Eatman, D., Fridie, I.L. (2005) Alterations in aldosterone and angiotensin II levels in salt-induced hypertension. Clin Exp Hypertens 27: 355-367.

[193] Zhan, C.D., Sindhu, R.K., Vaziri, N.D. (2004) Up-regulation of kidney NAD(P)H oxidase and calcineurin in SHR: reversal by lifelong antioxidant supplementation. Kidney Int 65: 219-227.

[194] Tanito, M., Nakamura, H., Kwon, Y.W., Teratani, A., Masutani, H., Shioji, K., Kishimoto, C., Ohira, A., Horie, R., Yodoi, J. (2004) Enhanced oxidative stress and impaired thioredoxin expression in spontaneously hypertensive rats. Antioxid Redox Signal 6: 89-97.

[195] Tyther, R., Ahmeda, A., Johns, E., Sheehan, D. (2009) Protein carbonylation in the kidney medulla of the spontaneously hypertensive rat. Proteomics Clin Appl 3: 338-346.

[196] Rodriguez-Iturbe, B., Ferrebuz, A., Vanegas, V., Quiroz, Y., Mezzano, S., Vaziri, N.D. (2005) Early and sustained inhibition of nuclear factor-kappaB prevents hypertension in spontaneously hypertensive rats. J Pharmacol Exp Ther 315: 51-57.

[197] Diep, Q.N., El Mabrouk, M., Cohn, J.S., Endemann, D., Amiri, F., Virdis, A., Neves, M.F., Schiffrin, E.L. (2002) Structure, endothelial function, cell growth, and inflammation in blood vessels of angiotensin II-infused rats: role of peroxisome proliferator-activated receptor-gamma. Circulation 105: 2296-2302.

[198] Bayorh, M.A., Ganafa, A.A., Eatman, D., Walton, M., Feuerstein, G.Z. (2005) Simvastatin and losartan enhance nitric oxide and reduce oxidative stress in salt-induced hypertension. Am J Hypertens 18: 1496-1502.

[199] Hong, H.J., Hsiao, G., Cheng, T.H., Yen, M.H. (2001) Supplemention with tetrahydrobiopterin suppresses the development of hypertension in spontaneously hypertensive rats. Hypertension 38: 1044-1048.

[200] Elmarakby, A.A., Loomis, E.D., Pollock, J.S., Pollock, D.M. (2005) NADPH oxidase inhibition attenuates oxidative stress but not hypertension produced by chronic ET-1. Hypertension 45: 283-287.

[201] Schnackenberg, C.G., Wilcox, C.S. (1999) Two-week administration of tempol attenuates both hypertension and renal excretion of 8-Iso prostaglandin f2alpha. Hypertension 33: 424-428.

[202] Krug, S., Zhang, Y., Mori, T.A., Croft, K.D., Vickers, J.J., Langton, L.K., Whitworth, J.A. (2008) N-Acetylcysteine prevents but does not reverse dexamethasone-induced hypertension. Clin Exp Pharmacol Physiol 35: 979-981.

[203] Stefanska, J., Pawliczak, R. (2008) Apocynin: molecular aptitudes. Mediators Inflamm 2008: 106507.

[204] Selemidis, S., Sobey, C.G., Wingler, K., Schmidt, H.H., Drummond, G.R. (2008) NADPH oxidases in the vasculature: molecular features, roles in disease and pharmacological inhibition. Pharmacol Ther 120: 254-291.

[205] Heumuller, S., Wind, S., Barbosa-Sicard, E., Schmidt, H.H., Busse, R., Schroder, K., Brandes, R.P. (2008) Apocynin is not an inhibitor of vascular NADPH oxidases but an antioxidant. Hypertension 51: 211-217.

[206] Touyz, R.M. (2008) Apocynin, NADPH oxidase, and vascular cells: a complex matter. Hypertension 51: 172-174.

[207] Ximenes, V.F., Kanegae, M.P., Rissato, S.R., Galhiane, M.S. (2007) The oxidation of apocynin catalyzed by myeloperoxidase: proposal for NADPH oxidase inhibition. Arch Biochem Biophys 457: 134-141.

[208] Brandes, R.P. (2003) A radical adventure: the quest for specific functions and inhibitors of vascular NAPDH oxidases. Circ Res 92: 583-585.

[209] Rey, F.E., Cifuentes, M.E., Kiarash, A., Quinn, M.T., Pagano, P.J. (2001) Novel competitive inhibitor of NAD(P)H oxidase assembly attenuates vascular O(2)(-) and systolic blood pressure in mice. Circ Res 89: 408-414.

[210] George, J., Struthers, A.D. (2009) Role of urate, xanthine oxidase and the effects of allopurinol in vascular oxidative stress. Vasc Health Risk Manag 5: 265-272.

[211] Kelkar, A., Kuo, A., Frishman, W.H. (2011) Allopurinol as a cardiovascular drug. Cardiol Rev 19: 265-271.

[212] Moorhouse, P.C., Grootveld, M., Halliwell, B., Quinlan, J.G., Gutteridge, J.M. (1987) Allopurinol and oxypurinol are hydroxyl radical scavengers. FEBS Lett 213: 23-28.

[213] Schnackenberg, C.G., Welch, W.J., Wilcox, C.S. (1998) Normalization of blood pressure and renal vascular resistance in SHR with a membrane-permeable superoxide dismutase mimetic: role of nitric oxide. Hypertension 32: 59-64.

[214] Wilcox, C.S., Pearlman, A. (2008) Chemistry and antihypertensive effects of tempol and other nitroxides. Pharmacol Rev 60: 418-469.

[215] Dekhuijzen, P.N. (2004) Antioxidant properties of N-acetylcysteine: their relevance in relation to chronic obstructive pulmonary disease. Eur Respir J 23: 629-636.

[216] Moldeus, P., Cotgreave, I.A., Berggren, M. (1986) Lung protection by a thiol-containing antioxidant: N-acetylcysteine. Respiration 50 Suppl 1: 31-42.

[217] Heard, K.J. (2008) Acetylcysteine for acetaminophen poisoning. N Engl J Med 359: 285-292.

[218] Yilmaz, H., Sahin, S., Sayar, N., Tangurek, B., Yilmaz, M., Nurkalem, Z., Onturk, E., Cakmak, N., Bolca, O. (2007) Effects of folic acid and N-acetylcysteine on plasma homocysteine levels and endothelial function in patients with coronary artery disease. Acta Cardiol 62: 579-585.

[219] Beckman, J.S., Minor, R.L., Jr., White, C.W., Repine, J.E., Rosen, G.M., Freeman, B.A. (1988) Superoxide dismutase and catalase conjugated to polyethylene glycol increases endothelial enzyme activity and oxidant resistance. J Biol Chem 263: 6884-6892.

[220] Zhao, R., Masayasu, H., Holmgren, A. (2002) Ebselen: a substrate for human thioredoxin reductase strongly stimulating its hydroperoxide reductase activity and a superfast thioredoxin oxidant. Proc Natl Acad Sci U S A 99: 8579-8584.

[221] Schewe, T. (1995) Molecular actions of ebselen--an antiinflammatory antioxidant. Gen Pharmacol 26: 1153-1169.

[222] Arakawa, M., Ito, Y. (2007) N-acetylcysteine and neurodegenerative diseases: Basic and clinical pharmacology. Cerebellum: 1-7.

[223] Masumoto, H., Sies, H. (1996) The reaction of ebselen with peroxynitrite. Chem Res Toxicol 9: 262-267.

[224] Saito, I., Asano, T., Sano, K., Takakura, K., Abe, H., Yoshimoto, T., Kikuchi, H., Ohta, T., Ishibashi, S. (1998) Neuroprotective effect of an antioxidant, ebselen, in patients with delayed neurological deficits after aneurysmal subarachnoid hemorrhage. Neurosurgery 42: 269-277; discussion 277-268.

[225] Yamaguchi, T., Sano, K., Takakura, K., Saito, I., Shinohara, Y., Asano, T., Yasuhara, H. (1998) Ebselen in acute ischemic stroke: a placebo-controlled, double-blind clinical trial. Ebselen Study Group. Stroke 29: 12-17.

[226] Gilgun-Sherki, Y., Rosenbaum, Z., Melamed, E., Offen, D. (2002) Antioxidant therapy in acute central nervous system injury: current state. Pharmacol Rev 54: 271-284.

[227] Kizhakekuttu, T.J., Widlansky, M.E. (2010) Natural antioxidants and hypertension: promise and challenges. Cardiovasc Ther 28: e20-32.

[228] Zingg, J.M. (2007) Modulation of signal transduction by vitamin E. Mol Aspects Med 28: 481-506.

[229] Chen, X., Touyz, R.M., Park, J.B., Schiffrin, E.L. (2001) Antioxidant effects of vitamins C and E are associated with altered activation of vascular NADPH oxidase and superoxide dismutase in stroke-prone SHR. Hypertension 38: 606-611.

[230] Carr, A., Frei, B. (1999) Does vitamin C act as a pro-oxidant under physiological conditions? FASEB J 13: 1007-1024.

[231] Carr, A.C., Zhu, B.Z., Frei, B. (2000) Potential antiatherogenic mechanisms of ascorbate (vitamin C) and alpha-tocopherol (vitamin E). Circ Res 87: 349-354.

[232] Shay, K.P., Moreau, R.F., Smith, E.J., Smith, A.R., Hagen, T.M. (2009) Alpha-lipoic acid as a dietary supplement: molecular mechanisms and therapeutic potential. Biochim Biophys Acta 1790: 1149-1160.

[233] Kim, S.H., Han, S.I., Oh, S.Y., Chung, H.Y., Kim, H.D., Kang, H.S. (2001) Activation of heat shock factor 1 by pyrrolidine dithiocarbamate is mediated by its activities as pro-oxidant and thiol modulator. Biochem Biophys Res Commun 281: 367-372.

[234] Schreck, R., Meier, B., Mannel, D.N., Droge, W., Baeuerle, P.A. (1992) Dithiocarbamates as potent inhibitors of nuclear factor kappa B activation in intact cells. J Exp Med 175: 1181-1194.

[235] Min, Y.K., Park, J.H., Chong, S.A., Kim, Y.S., Ahn, Y.S., Seo, J.T., Bae, Y.S., Chung, K.C. (2003) Pyrrolidine dithiocarbamate-induced neuronal cell death is mediated by Akt, casein kinase 2, c-Jun N-terminal kinase, and IkappaB kinase in embryonic hippocampal progenitor cells. J Neurosci Res 71: 689-700.

[236] Knowles, R.G., Moncada, S. (1994) Nitric oxide synthases in mammals. Biochem J 298 (Pt 2): 249-258.

[237] Vasquez-Vivar, J., Kalyanaraman, B., Martasek, P., Hogg, N., Masters, B.S., Karoui, H., Tordo, P., Pritchard, K.A., Jr. (1998) Superoxide generation by endothelial nitric oxide synthase: the influence of cofactors. Proc Natl Acad Sci U S A 95: 9220-9225.

[238] Linscheid, P., Schaffner, A., Schoedon, G. (1998) Modulation of inducible nitric oxide synthase mRNA stability by tetrahydrobiopterin in vascular smooth muscle cells. Biochem Biophys Res Commun 243: 137-141.

[239] Kojima, S., Ona, S., Iizuka, I., Arai, T., Mori, H., Kubota, K. (1995) Antioxidative activity of 5,6,7,8-tetrahydrobiopterin and its inhibitory effect on paraquat-induced cell toxicity in cultured rat hepatocytes. Free Radic Res 23: 419-430.

[240] Park, Y.M., Park, M.Y., Suh, Y.L., Park, J.B. (2004) NAD(P)H oxidase inhibitor prevents blood pressure elevation and cardiovascular hypertrophy in aldosterone-infused rats. Biochem Biophys Res Commun 313: 812-817.

[241] Hu, L., Zhang, Y., Lim, P.S., Miao, Y., Tan, C., McKenzie, K.U., Schyvens, C.G., Whitworth, J.A. (2006) Apocynin but not L-arginine prevents and reverses dexamethasone-induced hypertension in the rat. Am J Hypertens 19: 413-418.

[242] Zhang, Y., Chan, M.M., Andrews, M.C., Mori, T.A., Croft, K.D., McKenzie, K.U., Schyvens, C.G., Whitworth, J.A. (2005) Apocynin but not allopurinol prevents and reverses adrenocorticotropic hormone-induced hypertension in the rat. Am J Hypertens 18: 910-916.

[243] Costa, C.A., Amaral, T.A., Carvalho, L.C., Ognibene, D.T., da Silva, A.F., Moss, M.B., Valenca, S.S., de Moura, R.S., Resende, A.C. (2009) Antioxidant treatment with tempol

and apocynin prevents endothelial dysfunction and development of renovascular hypertension. Am J Hypertens 22: 1242-1249.

[244] Pechanova, O., Jendekova, L., Vrankova, S. (2009) Effect of chronic apocynin treatment on nitric oxide and reactive oxygen species production in borderline and spontaneous hypertension. Pharmacol Rep 61: 116-122.

[245] Tian, N., Moore, R.S., Phillips, W.E., Lin, L., Braddy, S., Pryor, J.S., Stockstill, R.L., Hughson, M.D., Manning, R.D., Jr. (2008) NADPH oxidase contributes to renal damage and dysfunction in Dahl salt-sensitive hypertension. Am J Physiol Regul Integr Comp Physiol 295: R1858-1865.

[246] Armando, I., Wang, X., Villar, V.A., Jones, J.E., Asico, L.D., Escano, C., Jose, P.A. (2007) Reactive oxygen species-dependent hypertension in dopamine D2 receptor-deficient mice. Hypertension 49: 672-678.

[247] Kopkan, L., Huskova, Z., Vanourkova, Z., Thumova, M., Skaroupkova, P., Maly, J., Kramer, H.J., Dvorak, P., Cervenka, L. (2009) Reduction of oxidative stress does not attenuate the development of angiotensin II-dependent hypertension in Ren-2 transgenic rats. Vascul Pharmacol 51: 175-181.

[248] Liu, F., Wei, C.C., Wu, S.J., Chenier, I., Zhang, S.L., Filep, J.G., Ingelfinger, J.R., Chan, J.S. (2009) Apocynin attenuates tubular apoptosis and tubulointerstitial fibrosis in transgenic mice independent of hypertension. Kidney Int 75: 156-166.

[249] Ceravolo, G.S., Fernandes, L., Munhoz, C.D., Fernandes, D.C., Tostes, R.C., Laurindo, F.R., Scavone, C., Fortes, Z.B., Carvalho, M.H. (2007) Angiotensin II chronic infusion induces B1 receptor expression in aorta of rats. Hypertension 50: 756-761.

[250] Pech, V., Sikka, S.C., Sindhu, R.K., Vaziri, N.D., Majid, D.S. (2006) Oxidant stress and blood pressure responses to angiotensin II administration in rats fed varying salt diets. Am J Hypertens 19: 534-540.

[251] Zhou, M.S., Hernandez Schulman, I., Pagano, P.J., Jaimes, E.A., Raij, L. (2006) Reduced NAD(P)H oxidase in low renin hypertension: link among angiotensin II, atherogenesis, and blood pressure. Hypertension 47: 81-86.

[252] Viel, E.C., Benkirane, K., Javeshghani, D., Touyz, R.M., Schiffrin, E.L. (2008) Xanthine oxidase and mitochondria contribute to vascular superoxide anion generation in DOCA-salt hypertensive rats. Am J Physiol Heart Circ Physiol 295: H281-288.

[253] Wallwork, C.J., Parks, D.A., Schmid-Schonbein, G.W. (2003) Xanthine oxidase activity in the dexamethasone-induced hypertensive rat. Microvasc Res 66: 30-37.

[254] Szasz, T., Linder, A.E., Davis, R.P., Burnett, R., Fink, G.D., Watts, S.W. (2010) Allopurinol does not decrease blood pressure or prevent the development of hypertension in the deoxycorticosterone acetate-salt rat model. J Cardiovasc Pharmacol 56: 627-634.

[255] Ong, S.L., Vickers, J.J., Zhang, Y., McKenzie, K.U., Walsh, C.E., Whitworth, J.A. (2007) Role of xanthine oxidase in dexamethasone-induced hypertension in rats. Clin Exp Pharmacol Physiol 34: 517-519.

[256] Usui, M., Egashira, K., Kitamoto, S., Koyanagi, M., Katoh, M., Kataoka, C., Shimokawa, H., Takeshita, A. (1999) Pathogenic role of oxidative stress in vascular angiotensin-converting enzyme activation in long-term blockade of nitric oxide synthesis in rats. Hypertension 34: 546-551.

[257] Laakso, J.T., Teravainen, T.L., Martelin, E., Vaskonen, T., Lapatto, R. (2004) Renal xanthine oxidoreductase activity during development of hypertension in spontaneously hypertensive rats. J Hypertens 22: 1333-1340.

[258] Nakazono, K., Watanabe, N., Matsuno, K., Sasaki, J., Sato, T., Inoue, M. (1991) Does superoxide underlie the pathogenesis of hypertension? Proc Natl Acad Sci U S A 88: 10045-10048.

[259] Park, J.B., Touyz, R.M., Chen, X., Schiffrin, E.L. (2002) Chronic treatment with a superoxide dismutase mimetic prevents vascular remodeling and progression of hypertension in salt-loaded stroke-prone spontaneously hypertensive rats. Am J Hypertens 15: 78-84.

[260] Beswick, R.A., Zhang, H., Marable, D., Catravas, J.D., Hill, W.D., Webb, R.C. (2001) Long-term antioxidant administration attenuates mineralocorticoid hypertension and renal inflammatory response. Hypertension 37: 781-786.

[261] Bayorh, M.A., Mann, G., Walton, M., Eatman, D. (2006) Effects of enalapril, tempol, and eplerenone on salt-induced hypertension in dahl salt-sensitive rats. Clin Exp Hypertens 28: 121-132.

[262] Dobrian, A.D., Schriver, S.D., Prewitt, R.L. (2001) Role of angiotensin II and free radicals in blood pressure regulation in a rat model of renal hypertension. Hypertension 38: 361-366.

[263] Zhang, Y., Jang, R., Mori, T.A., Croft, K.D., Schyvens, C.G., McKenzie, K.U., Whitworth, J.A. (2003) The anti-oxidant Tempol reverses and partially prevents adrenocorticotrophic hormone-induced hypertension in the rat. J Hypertens 21: 1513-1518.

[264] Elmarakby, A.A., Williams, J.M., Imig, J.D., Pollock, J.S., Pollock, D.M. (2007) Synergistic actions of enalapril and tempol during chronic angiotensin II-induced hypertension. Vascul Pharmacol 46: 144-151.

[265] Pechanova, O., Zicha, J., Kojsova, S., Dobesova, Z., Jendekova, L., Kunes, J. (2006) Effect of chronic N-acetylcysteine treatment on the development of spontaneous hypertension. Clin Sci (Lond) 110: 235-242.

[266] Mondo, C.K., Zhang, Y., de Macedo Possamai, V., Miao, Y., Schyvens, C.G., McKenzie, K.U., Hu, L., Guo, Z., Whitworth, J.A. (2006) N-acetylcysteine antagonizes the development but does not reverse ACTH-induced hypertension in the rat. Clin Exp Hypertens 28: 73-84.

[267] Tian, N., Rose, R.A., Jordan, S., Dwyer, T.M., Hughson, M.D., Manning, R.D., Jr. (2006) N-Acetylcysteine improves renal dysfunction, ameliorates kidney damage and decreases blood pressure in salt-sensitive hypertension. J Hypertens 24: 2263-2270.

[268] Rauchova, H., Pechanova, O., Kunes, J., Vokurkova, M., Dobesova, Z., Zicha, J. (2005) Chronic N-acetylcysteine administration prevents development of hypertension in N(omega)-nitro-L-arginine methyl ester-treated rats: the role of reactive oxygen species. Hypertens Res 28: 475-482.

[269] Jung, O., Marklund, S.L., Xia, N., Busse, R., Brandes, R.P. (2007) Inactivation of extracellular superoxide dismutase contributes to the development of high-volume hypertension. Arterioscler Thromb Vasc Biol 27: 470-477.

[270] Weber, D.S., Rocic, P., Mellis, A.M., Laude, K., Lyle, A.N., Harrison, D.G., Griendling, K.K. (2005) Angiotensin II-induced hypertrophy is potentiated in mice overexpressing p22phox in vascular smooth muscle. Am J Physiol Heart Circ Physiol 288: H37-42.

[271] Nishikawa, Y., Tatsumi, K., Matsuura, T., Yamamoto, A., Nadamoto, T., Urabe, K. (2003) Effects of vitamin C on high blood pressure induced by salt in spontaneously hypertensive rats. J Nutr Sci Vitaminol (Tokyo) 49: 301-309.

[272] Ettarh, R.R., Odigie, I.P., Adigun, S.A. (2002) Vitamin C lowers blood pressure and alters vascular responsiveness in salt-induced hypertension. Can J Physiol Pharmacol 80: 1199-1202.

[273] Tian, N., Moore, R.S., Braddy, S., Rose, R.A., Gu, J.W., Hughson, M.D., Manning, R.D., Jr. (2007) Interactions between oxidative stress and inflammation in salt-sensitive hypertension. Am J Physiol Heart Circ Physiol 293: H3388-3395.

[274] Schyvens, C.G., Andrews, M.C., Tam, R., Mori, T.A., Croft, K.D., McKenzie, K.U., Whitworth, J.A., Zhang, Y. (2007) Antioxidant vitamins and adrenocorticotrophic hormone-induced hypertension in rats. Clin Exp Hypertens 29: 465-478.

[275] Noguchi, T., Ikeda, K., Sasaki, Y., Yamamoto, J., Yamori, Y. (2004) Effects of vitamin E and sesamin on hypertension and cerebral thrombogenesis in stroke-prone spontaneously hypertensive rats. Clin Exp Pharmacol Physiol 31 Suppl 2: S24-26.

[276] Vasdev, S., Ford, C.A., Parai, S., Longerich, L., Gadag, V. (2000) Dietary alpha-lipoic acid supplementation lowers blood pressure in spontaneously hypertensive rats. J Hypertens 18: 567-573.

[277] Vasdev, S., Ford, C.A., Parai, S., Longerich, L., Gadag, V. (2000) Dietary lipoic acid supplementation prevents fructose-induced hypertension in rats. Nutr Metab Cardiovasc Dis 10: 339-346.

[278] Vasdev, S., Gill, V., Longerich, L., Parai, S., Gadag, V. (2003) Salt-induced hypertension in WKY rats: prevention by alpha-lipoic acid supplementation. Mol Cell Biochem 254: 319-326.

[279] Takaoka, M., Kobayashi, Y., Yuba, M., Ohkita, M., Matsumura, Y. (2001) Effects of alpha-lipoic acid on deoxycorticosterone acetate-salt-induced hypertension in rats. Eur J Pharmacol 424: 121-129.

[280] Koeners, M.P., Braam, B., Joles, J.A. (2011) Perinatal inhibition of NF-kappaB has long-term antihypertensive effects in spontaneously hypertensive rats. J Hypertens 29: 1160-1166.

[281] Fortepiani, L.A., Reckelhoff, J.F. (2005) Treatment with tetrahydrobiopterin reduces blood pressure in male SHR by reducing testosterone synthesis. Am J Physiol Regul Integr Comp Physiol 288: R733-736.

[282] Zhang, Y., Pang, T., Earl, J., Schyvens, C.G., McKenzie, K.U., Whitworth, J.A. (2004) Role of tetrahydrobiopterin in adrenocorticotropic hormone-induced hypertension in the rat. Clin Exp Hypertens 26: 231-241.

[283] Kris-Etherton, P.M., Lichtenstein, A.H., Howard, B.V., Steinberg, D., Witztum, J.L. (2004) Antioxidant vitamin supplements and cardiovascular disease. Circulation 110: 637-641.

[284] Bjelakovic, G., Nikolova, D., Gluud, L.L., Simonetti, R.G., Gluud, C. (2007) Mortality in randomized trials of antioxidant supplements for primary and secondary prevention: systematic review and meta-analysis. JAMA 297: 842-857.

[285] Dotan, Y., Pinchuk, I., Lichtenberg, D., Leshno, M. (2009) Decision analysis supports the paradigm that indiscriminate supplementation of vitamin E does more harm than good. Arterioscler Thromb Vasc Biol 29: 1304-1309.

[286] Bates, C.J., Walmsley, C.M., Prentice, A., Finch, S. (1998) Does vitamin C reduce blood pressure? Results of a large study of people aged 65 or older. J Hypertens 16: 925-932.

[287] McMackin, C.J., Widlansky, M.E., Hamburg, N.M., Huang, A.L., Weller, S., Holbrook, M., Gokce, N., Hagen, T.M., Keaney, J.F., Jr., Vita, J.A. (2007) Effect of combined treatment with alpha-Lipoic acid and acetyl-L-carnitine on vascular function and blood pressure in patients with coronary artery disease. J Clin Hypertens (Greenwich) 9: 249-255.

[288] Rahman, S.T., Merchant, N., Haque, T., Wahi, J., Bhaheetharan, S., Ferdinand, K.C., Khan, B.V. (2012) The Impact of Lipoic Acid on Endothelial Function and Proteinuria in Quinapril-Treated Diabetic Patients With Stage I Hypertension: Results From the QUALITY Study, J Cardiovasc Pharmacol Ther 17: 139-145.

[289] John, J.H., Ziebland, S., Yudkin, P., Roe, L.S., Neil, H.A. (2002) Effects of fruit and vegetable consumption on plasma antioxidant concentrations and blood pressure: a randomised controlled trial. Lancet 359: 1969-1974.

[290] Lopes, H.F., Martin, K.L., Nashar, K., Morrow, J.D., Goodfriend, T.L., Egan, B.M. (2003) DASH diet lowers blood pressure and lipid-induced oxidative stress in obesity. Hypertension 41: 422-430.

[291] Griendling, K.K., FitzGerald, G.A. (2003) Oxidative stress and cardiovascular injury: Part II: animal and human studies. Circulation 108: 2034-2040.

[292] Steinhubl, S.R. (2008) Why have antioxidants failed in clinical trials? Am J Cardiol 101: 14D-19D.

[293] Suzuki, K. (2009) Anti-oxidants for therapeutic use: why are only a few drugs in clinical use? Adv Drug Deliv Rev 61: 287-289.

[294] Violi, F., Loffredo, L., Musella, L., Marcoccia, A. (2004) Should antioxidant status be considered in interventional trials with antioxidants? Heart 90: 598-602.

[295] Yusuf, S., Dagenais, G., Pogue, J., Bosch, J., Sleight, P. (2000) Vitamin E supplementation and cardiovascular events in high-risk patients. The Heart Outcomes Prevention Evaluation Study Investigators. N Engl J Med 342: 154-160.

[296] Montezano, A.C., Touyz, R.M. (2012) Molecular Mechanisms of Hypertension-Reactive Oxygen Species and Antioxidants: A Basic Science Update for the Clinician. Can J Cardiol.

[297] Violi, F., Cangemi, R. (2008) Statin treatment as a confounding factor in human trials with vitamin E. J Nutr 138: 1179-1181.

[298] George, J., Carr, E., Davies, J., Belch, J.J., Struthers, A. (2006) High-dose allopurinol improves endothelial function by profoundly reducing vascular oxidative stress and not by lowering uric acid. Circulation 114: 2508-2516.

[299] Kostka-Jeziorny, K., Uruski, P., Tykarski, A. (2011) Effect of allopurinol on blood pressure and aortic compliance in hypertensive patients. Blood Press 20: 104-110.

[300] Feig, D.I., Soletsky, B., Johnson, R.J. (2008) Effect of allopurinol on blood pressure of adolescents with newly diagnosed essential hypertension: a randomized trial. JAMA 300: 924-932.

[301] Kanbay, M., Ozkara, A., Selcoki, Y., Isik, B., Turgut, F., Bavbek, N., Uz, E., Akcay, A., Yigitoglu, R., Covic, A. (2007) Effect of treatment of hyperuricemia with allopurinol on blood pressure, creatinine clearence, and proteinuria in patients with normal renal functions. Int Urol Nephrol 39: 1227-1233.

[302] Garrido, A.M., Griendling, K.K. (2009) NADPH oxidases and angiotensin II receptor signaling. Mol Cell Endocrinol 302: 148-158.

[303] Griendling, K.K., Ushio-Fukai, M. (2000) Reactive oxygen species as mediators of angiotensin II signaling. Regul Pept 91: 21-27.

[304] Tamba, M., Torreggiani, A. (2000) Free radical scavenging and copper chelation: a potentially beneficial action of captopril. Free Radic Res 32: 199-211.

[305] Chen, S., Ge, Y., Si, J., Rifai, A., Dworkin, L.D., Gong, R. (2008) Candesartan suppresses chronic renal inflammation by a novel antioxidant action independent of AT1R blockade. Kidney Int 74: 1128-1138.

[306] Kadowaki, D., Anraku, M., Tasaki, Y., Taguchi, K., Shimoishi, K., Seo, H., Hirata, S., Maruyama, T., Otagiri, M. (2009) Evaluation for antioxidant and renoprotective activity of olmesartan using nephrectomy rats. Biol Pharm Bull 32: 2041-2045.

[307] Miyata, T., van Ypersele de Strihou, C., Ueda, Y., Ichimori, K., Inagi, R., Onogi, H., Ishikawa, N., Nangaku, M., Kurokawa, K. (2002) Angiotensin II receptor antagonists and angiotensin-converting enzyme inhibitors lower in vitro the formation of advanced glycation end products: biochemical mechanisms. J Am Soc Nephrol 13: 2478-2487.

[308] Dandona, P., Ghanim, H., Brooks, D.P. (2007) Antioxidant activity of carvedilol in cardiovascular disease. J Hypertens 25: 731-741.

[309] Evangelista, S., Garbin, U., Pasini, A.F., Stranieri, C., Boccioletti, V., Cominacini, L. (2007) Effect of DL-nebivolol, its enantiomers and metabolites on the intracellular production of superoxide and nitric oxide in human endothelial cells. Pharmacol Res 55: 303-309.

[310] Gupta, S., Wright, H.M. (2008) Nebivolol: a highly selective beta1-adrenergic receptor blocker that causes vasodilation by increasing nitric oxide. Cardiovasc Ther 26: 189-202.

[311] Cominacini, L., Fratta Pasini, A., Garbin, U., Pastorino, A.M., Davoli, A., Nava, C., Campagnola, M., Rossato, P., Lo Cascio, V. (2003) Antioxidant activity of different dihydropyridines. Biochem Biophys Res Commun 302: 679-684.

[312] Godfraind, T. (2005) Antioxidant effects and the therapeutic mode of action of calcium channel blockers in hypertension and atherosclerosis. Philos Trans R Soc Lond B Biol Sci 360: 2259-2272.

Chapter 6

Lipid Peroxidation by-Products and the Metabolic Syndrome

Nicolas J. Pillon and Christophe O. Soulage

Additional information is available at the end of the chapter

1. Introduction

About twenty-one percent of the air we breathe is composed of oxygen, and our life will not be possible without it. Oxygen is however a toxic, highly reactive molecule which was originally released in atmosphere as a waste product of the first photosynthetic organisms. Its accumulation on Earth indeed led to a massive extinction of living species. Few organisms survived and some developed the ability to use this toxic oxygen to improve the production of energy from carbohydrates. This "oxidative metabolism" was however a double-edge sword as the use of intracellular oxygen generates deleterious oxidative damages. To protect themselves toward this toxicity, those organisms consequently developed several "antioxidants" protection mechanisms which helped them maintain a balance between oxidative damage and efficient use of oxygen to produce energy.

When antioxidant defences are reduced and/or oxidative mechanisms increased, uncontrolled oxidation of cell targets leads to the accumulation of reactive oxygen species (ROS) and a state of "oxidative stress", often deleterious for the cells. This stress is involved in the pathophysiology of several human diseases, and especially in the development of metabolic diseases, even if its causative role remains questionable. A definite increase in oxidative stress biomarkers can be found in obese and diabetic humans as well as in animal model of these diseases. Accumulation of ROS can be deleterious by itself or can induce the oxidation of proteins, nucleic acids and lipids, generating secondary by-products. The specific reactivity of ROS towards polyunsaturated fatty acids (PUFAs) present in cell membranes induces lipid peroxidation, a noxious mechanism producing toxic aldehydes. Among them, malondialdehyde (MDA) and 4-hydroxy-2-nonenal (HNE) have been extensively studied. Originally simple markers of lipid peroxidation, these aldehydes have demonstrated causative roles in the impairment of cellular functions: activation of signalling pathways, apoptosis, and modification of enzyme function. In addition to being hallmarks of oxidative damage, lipid aldehydes could be mediators of oxidative insults, propagating tissue injury and activating cellular stress signalling pathways. Several studies demonstrated the association of obesity and diabetes with lipid peroxidation by-products, and the role of aldehydes in impairment of insulin function and signalling was recently pointed out.

This chapter aims to review the diverse implications of lipid peroxidation by-products in the pathophysiology of metabolic diseases, from evidence of their production during obesity and diabetes to the cellular mechanisms of their toxicity and protection against their deleterious effects.

2. Lipid peroxidation by-products

Under conditions of oxidative stress, excessive production of reactive oxygen species promotes the peroxidation of polyunsaturated fatty acids (PUFA). The resulting accumulation of hydroperoxides, unstable molecules, leads to their non-enzymatic degradation in many compounds, including aldehydes. Quantification of lipid peroxidation in biological samples has been extensively performed with the thiobarbituric acid (TBA) test. TBA detects malondialdehyde (MDA), an end-product of nonenzymatic PUFA oxidative degradation, which has therefore been used for decades as a marker of lipid peroxidation (Gutteridge, 1982). Another aldehyde: acrolein, first attracted attention because of its formation during tobacco combustion and its ubiquitous presence in the environment (Dong & Moldoveanu, 2004). Because of its carcinogenic potential, its role in smoking-related diseases has received extensive attention; however, acrolein is also produced endogenously though lipid peroxidation and its link with oxidative-associated pathologies is now well established. 4-hydroxy-2-alkenals are specific by-products of the oxidation of omega-3 and omega-6 fatty acids. 4-hydroxy-2-nonenal (HNE) is derived from the oxidation of polyunsaturated fatty acids of the n-6 series, mainly linoleic and arachidonic acids, while 4-hydroxy-2-hexenal (HHE) results from the peroxidation of polyunsaturated fatty acids of the n-3 series (mainly docosahexaenoic, eicosapentaenoic and linolenic acid). The peroxidation of arachidonic acid via 12-lipoxygenase leads to the formation of 4-hydroxy-2-dodecadienal (HDDE) (Guichardant et al., 2006).

Name	Molecular Weight (Da)	Molecular Formula	Skeletal Formula
Acrolein (prop-2-enal)	56.1	C_3H_4O	
Malondialdehyde (propanedial, MDA)	72.1	$C_3H_4O_2$	
4-hydroxy-2-hexenal (HHE)	114.1	$C_6H_{10}O_2$	
4-hydroxy-2-nonenal (HNE)	156.2	$C_9H_{16}O_2$	
4-hydroxy-2-dodecadienal (HDDE)	196.0	$C_{12}H_{20}O_2$	

Table 1. α,β-Unsaturated aldehydes produced during polyunsaturated fatty acids oxidation

2.1. Chemistry and reactivity

Acrolein, MDA and 4-hydroxy-alkenals are α,β-unsaturated aldehydes, a class of compounds sharing the general structure C=C–C=O. They are characterised by an aldehyde group (C=O) on carbon 1 and a conjugated double bond (C=C) between carbons 2 and 3 (Table 1). In this structure, the oxygen atom of the carbonyl group increases the polarity of the double bond, which makes α,β-unsaturated aldehydes potent electrophiles. Acrolein has the simplest structure composed of 3 carbons, MDA is a dicarbonyl compound and 4-hydroxy-2-alkenals are characterized by the presence of a hydroxyl group on carbon 4. HHE, HNE and HDDE only differ by the length of their carbon chain and the presence of an additional double bond for the HDDE. In the case of MDA and 4-hydroxy-alkenals, the presence of a second oxygen atom makes the double bond even more reactive.

These aldehydes are part of the *"reactive electrophile species"* able to form covalent adducts with the nucleophilic groups present in DNA, proteins and phospholipids. In physiological conditions, they spontaneously react with the thiol group of glutathione to form Michael adduct by attack of the nucleophilic group of glutathione to the double bond of aldehydes. They can also react with thiol groups present on cysteine residues of certain proteins, leading to impairment of their biological activity. Under certain conditions, especially alkaline pH, aldehydes react with the amine groups present in proteins, nucleic acids and aminophospholipids, leading to Michael adducts. On the other hand, the reaction between a primary amine group and the carbonyl group of the aldehyde leads to the formation of Schiff bases (Schaur, 2003).

Aldehydes produced during lipid peroxidation are precursors of Advanced Lipoxidation End products (ALEs). Together with Advanced Glycation End products (AGEs) generated during glycoxidation, they accumulate in cells and tissues. The "carbonyl stress" is a result of this adduct accumulation, which induces protein dysfunctions and consequent pathological events such as inflammation and apoptosis (Negre-Salvayre *et al.*, 2008).

2.2. Cellular effects

2.2.1. Cytotoxicity

Since α,β-unsaturated compounds are strong electrophiles, they exhibit a high cytotoxic and mutagenic potential and have consequently been extensively studied for their effects on cell viability. HNE, HHE and acrolein indeed induce cell death, but the lethal concentration 50 (LC50), concentration that induces the death of 50% of the cells, is subject to variation, depending on the aldehyde, exposure duration and cell type (Table 2). The LC50 for a long exposure (>16 hours) to HNE or HHE is however consistently found 20-60 μM in several cell types, including human lymphoma Jurkat cells, lens epithelial cells, hamster V79-4 cells and muscle cells (Table 2). For acrolein, the same treatment gives a range of LC50 of 5-100 μM in human fibroblasts, human neuroblastoma cells, PC12 chromaffin cells and lymphocytes. MDA-induced cell death is less documented, even if the LC50 range for MDA is found around 1 mM in cortical, endothelial cells and fibroblasts. MDA was also reported to induce cell cycle arrest, which is to relate to cell damage and death (Ji *et al.*, 1998). HDDE appears to be the most toxic lipid aldehyde with a LC50 in endothelial cells in the submicromolar range.

Name	LC50, μM	References
Acrolein (prop-2-enal)	5-100	Poirier *et al.*, 2002; Luo *et al.*, 2005; Jia *et al.*, 2009*b*, 2009*a*
Malondialdehyde (propanedial, MDA)	600-3000	Michiels & Remacle, 1991; Hipkiss *et al.*, 1997; Cheng *et al.*, 2011
4-hydroxy-2-hexenal (HHE)	20-60	Liu *et al.*, 2000; Choudhary *et al.*, 2002; Pillon *et al.*, 2010; Li *et al.*, 2011
4-hydroxy-2-nonenal (HNE)	20-60	Liu *et al.*, 2000; Choudhary *et al.*, 2002; Pillon *et al.*, 2010; Li *et al.*, 2011
4-hydroxy-2-dodecadienal (HDDE)	0.22	Riahi *et al.*, 2010

Table 2. Range of lethal concentration 50 for long-term treatment (>16 hours) with α,β-unsaturated aldehydes. LC50 values were calculated from the indicated references.

On the opposite, very little cell death is detectable for short term treatments (<4 hours), likely because aldehyde-induced cell death involves apoptosis mechanisms not yet occurring during this short period of time. Several studies indeed reported that cell death is induced by aldehydes through apoptosis for low concentration and both apoptosis and necrosis for high doses (Luo *et al.*, 2005; Liu *et al.*, 2010). Acrolein-induced necrosis was described in few studies (Luo *et al.*, 2005), but mitochondrial-driven cell death seems to be the canonical road and was widely studied. Acrolein-induced apoptosis was indeed confirmed in several cell types through DNA fragmentation, phosphatidylserine externalization, poly(ADP-ribose) polymerase cleavage and activation of caspases (Pan *et al.*, 2009; Roy *et al.*, 2010). Last but not least, hydroxyalkenals are potent activators of apoptosis. They both induce DNA fragmentation and activation of caspases in very different cell types (Choudhary *et al.*, 2002; Vaillancourt *et al.*, 2008). In addition, HHE has been shown to decrease the expression/phosphorylation of Bcl-2, while increasing that of Bax, leading to apoptosis of human renal epithelial cells (Bae *et al.*, 2011; Bodur *et al.*, 2012).

Interestingly, the toxicity of aldehydes is highly correlated to their ability to form covalent adducts on proteins. In muscle cells, the lethal concentration 50 (LC50) for 12 different aldehydes was calculated, including HHE and HNE (Pillon *et al.*, 2010). This LC50 was strongly correlated with their respective potency to form covalent adducts on albumin *in vitro* (Figure 1). This demonstrates that the cytotoxicity and likely other biological effects of aldehydes mainly occur through chemical adduction of other biomolecules.

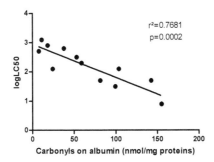

Figure 1. Toxicity is correlated to the adduction ability. Viability of muscle cells was measured in response to 12 different aldehydes[1]. The calculated LC50 was then correlated to their ability to form covalent adducts on bovine serumalbumin (Pillon et al., 2010).

2.2.2. Oxidative stress and ROS production

The classical sequence of events is that oxidative stress triggers lipid peroxidation which in turns produces aldehydes by-products. However, an interesting paradigm was pointed out by several groups: lipid aldehydes are able to induce the production of ROS, and this is thought to be of importance in their deleterious effects. For instance, acrolein treatment produces nitric oxide (Misonou et al., 2006) and induces generation of intracellular oxidants (Luo et al., 2005; Wang et al., 2011). Similarly, accumulation of intracellular ROS was described in cells treated with MDA (Cheng et al., 2011); and unsurprisingly, the 4-hydroxyalkenals HHE and HNE share the same ability. HHE induces ROS in neurons and tubular epithelial cells (Long et al., 2008; Bae et al., 2011) and HNE induces mitochondrial oxidative stress in neurons, vascular muscle, liver and skeletal muscle cells (Uchida et al., 1999; Lee et al., 2006; Pillon et al., 2012). The source of ROS was suggested to be mitochondria, as several aldehydes have been shown to induce a significant decrease in mitochondrial membrane potential (Uchida et al., 1999; Luo et al., 2005).

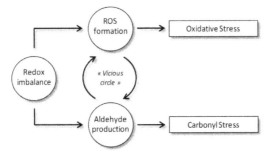

Figure 2. The vicious circle linking ROS and lipid aldehyde production

---

[1] 4-hydroxy-2-hexenal, 4-hydroxy-2-nonenal, 4-hydroxy-2-dodecenal, 4-hydroxy-2-hexenal dimethylacetal, 4-hydroxy-2-nonenal dimethylacetal, 4-hydroxy-2-dodecenal dimethylacetal, trans-2-hexenal, trans-2-nonenal, trans-2-dodecenal, hexanal, nonanal, dodecanal

This concept is reinforced by the fact that an increase in glutathione pool can prevent the deleterious effects of HNE on both adduct formation and ROS production (Pillon *et al.*, 2012) and that dysfunction of glutathione S-transferase, a major enzyme for aldehyde detoxification, leads to excess 4-hydroxy-2-nonenal and oxidative stress (Kostyuk *et al.*, 2010; Curtis *et al.*, 2010). Based on this body of evidence, lipid by-products can be seen as parts of a vicious circle in which increased ROS production generates aldehydes which further amplify the generation of oxidative species and so on (Figure 2).

2.2.3. Signalling pathways and transcription factors

Consistent with the extensive work done on cytotoxicity, the stress signalling pathways have been primarily pointed out as the main intracellular route activated by aldehydes. The mitogen-activated protein kinases (MAPKs) are indeed activated by several aldehydes including MDA, acrolein and HNE. MDA activates c-Jun N-terminal kinases (JNK) and extracellular signal-regulated kinases (ERK) (Cheng *et al.*, 2011) and acrolein-induced apoptosis occurs through activation of p38 and ERK (Tanel & Averill-Bates, 2007). The activation of these three MAPK has also been described following treatment with HNE (Uchida *et al.*, 1999; Zarrouki *et al.*, 2007; Pillon *et al.*, 2012) and HHE (Je *et al.*, 2004; Bae *et al.*, 2011). Overall, most studies investigated the cytotoxic effects of aldehydes and thus focused on cellular stress pathways; therefore, very little data is available regarding their potential effects on other pathways. Only very recent work shows that HNE interfere with insulin signalling pathway through oxidative stress and adduction of IRS1 and Akt (Demozay *et al.*, 2008; Shearn *et al.*, 2011; Pillon *et al.*, 2012).

Aldehydes regulate gene expression by activating the signalling pathways described above, or by direct modification of transcription factors. Unsurprisingly, aldehyde production regulates the expression of several antioxidant enzymes such as NAD(P)H quinone oxidoreductase-1, Heme oxygenase-1 and glutathione S-transferase (GST). This occurs through the activation of the Nuclear factor (erythroid-derived 2)-like 2 (Nrf2), which drives the expression of these antioxidant enzymes. Aldehyde-induced Heme Oxygenase-1 expression is indeed mediated by the Nrf2 pathway in HUVECs (Lee *et al.*, 2011) and Nrf2 silencing significantly attenuates the induction of this same gene by acrolein (Zhang & Forman, 2008).

Beyond the expression of antioxidant enzymes, cellular growth, apoptosis and inflammatory responses can be induced by aldehydes, involving the activation of the nuclear factor kappa B (NF-κB) family. The effects of acrolein on NF-κB activation are controversial, but its activation by HNE and HHE is well documented. HHE and HNE induce NF-kappaB activation through IKK/NIK pathway, leading to IκB phosphorylation and subsequent proteolysis (Page *et al.*, 1999; Je *et al.*, 2004; Lee *et al.*, 2004). HNE has moreover been shown to induce DNA-binding of NF-κB in vascular smooth muscle cells (Ruef *et al.*, 2001).

The peroxisome proliferator-activated receptor (PPAR) family regulate the expression of genes that encode proteins involved in energy balance. They act as ligand-activated transcription factors and are responsive to the lipid status of the cell, therefore important during high fat diets and obesity. HNE is an intracellular agonist of PPARβ/δ while HHE do not activate this receptor (Coleman *et al.*, 2007). Through this pathway, HNE significantly elevates adiponectin gene expression, concomitant with increased PPAR-γ gene expression and transactivity. Meanwhile, HDDE acts through PPARδ signalling pathways to regulate glucose transport in vascular endothelial cells subjected to hyperglycemia (Riahi *et al.*, 2010) and HNE stimulates insulin secretion from Beta cells through interaction with PPARδ (Cohen *et al.*, 2011).

2.3. Concentration in plasma and tissues

Little data is available concerning aldehydes levels in biological fluids, except for MDA and HNE which have been widely used as lipid peroxidation markers. The concentration of MDA in the plasma of healthy subjects is around 2-5 µM and increases up to 2-fold in type 2 diabetic patients (Figure 4). In the specific context of metabolic diseases, MDA is positively correlated with BMI and waist circumference in obese patients (Furukawa *et al.*, 2004). Depending on the study, HNE concentration has been found to range from 50 nM to 10 µM under normal conditions. This significant variability in concentration according to the authors could be explained partly by the method used (LC/MS, GC/MS ...) and also by the difficulty to measure such reactive derivatives. If HNE has been widely studied, there is however scarce data in the literature regarding the pathophysiological concentrations of HHE and HDDE. Plasma HHE concentration was however found to be around 9 nM in human, and dramatically increases to reach 90 nM after several weeks of a diet rich in omega-3 fatty acids (Calzada *et al.*, 2010). Our group recently showed that HHE concentration was 20 nM in humans and 7 nM in rats and that it increases in both type-2 diabetes patients and type-1 diabetic rats, reinforcing existing evidence for a role of lipid aldehydes in metabolic diseases.

Tissue	Aldehyde	Concentration (µM)	References
Plasma	MDA	1 – 6	See figure 4
	HHE	0.006 – 0.090	(Calzada et al., 2010; Pillon et al, unpublished)
	HNE	0.007 – 11	(McGrath *et al.*, 2001; Selley, 2004; Syslova *et al.*, 2009)
Exhaled breath condensate	HNE	0.25 - 5	(Syslova *et al.*, 2009; Manini *et al.*, 2010)
	MDA	0.63 - 14	(Syslova *et al.*, 2009; Manini *et al.*, 2010)
Ventricular Fluid	HNE	0.2 - 120	(Lovell *et al.*, 1997)
Pancreatic Islets	HNE	23 – 35	(Miwa *et al.*, 2000)

Table 3. Concentration range of aldehydes in healthy plasma and tissues

One should keep in mind that all current quantification methods (HPLC, GC) only assay the free fraction (*i.e.* unreacted) of lipid aldehydes present in samples. Being very reactive, this is certainly not representative of the amount indeed produced from lipid peroxidation, which rapidly react with neighbour targets to form covalent adducts and thus, other non-quantitative methods estimating the amount of HNE have been used. For example, detection of protein adducts by immunohistochemistry has shown a significant increase in HHE and HNE proteins adducts in Parkinson's disease and in chronic liver disease (Yoritaka *et al.*, 1996; Paradis *et al.*, 1997). Of particular interest for this chapter, HNE adducts on plasma albumin are increased in type 2 diabetes (Toyokuni *et al.*, 2000).

3. Insulin secretion and type-1 diabetes

3.1. Lipid aldehydes and insulin secretion

Oxidative stress associated with hyperglycemia is suspected to participate in beta cell dysfunction in terms of insulin synthesis and/or secretion. Diabetic animals display increased levels of lipoperoxidation in pancreas, and HNE concentration was reported to reach up to 35 μM in pancreatic islets of diabetic rats (Miwa *et al.*, 2000). In addition, HNE-modified proteins are increased in the pancreatic beta-cells of Goto-Kakizaki rats, a genetic model of non obese type 2 diabetes (Ihara *et al.*, 1999), as well as in type 2 diabetic patients (Sakuraba *et al.*, 2002). HNE and other lipid peroxidation by-products such as 2-hexenal and 2-butenal inhibit glucose-induced insulin secretion in isolated rat islets. Both glucose utilization and glucose oxidation are blunted in islets after treatment with aldehydes suggesting that they impair glucose-induced insulin secretion through an interference with glycolytic pathway and citric acid cycle (Miwa *et al.*, 2000). Another piece of evidence comes from the exposure of beta cells to NO donors and to interleukin-1 beta, which leads to generation of oxidative stress and lipoperoxidation by-products. MDA and HNE produced under this condition are involved in the activation of an apoptotic program, contributing to the reduction in the beta cell mass (Cahuana *et al.*, 2003). Peroxynitrite indeed triggers lipoperoxidation in the beta-cell line RIN-5-F, and the resulting protein carbonylation is a key factor linking NO-dependent lipoperoxidation and apoptosis (Cahuana *et al.*, 2003). Alloxan, a toxic glucose analogue, has been widely used to generate rodent models of type-1 diabetes, as it selectively destroys insulin-producing cells in the pancreas. Alloxan-induced diabetic rats exhibit increased lipid peroxidation associated with defects in insulin secretion, which can be prevented by the antioxidant S-allyl cysteine therefore restoring insulin secretion and ameliorating the glycaemic control (Augusti & Sheela, 1996).

3.2. Lipid aldehydes and the beta cells: Doctor Jeckyll or Mr Hyde?

HNE was shown to elicit various physiological or physiopathological responses: high concentrations elicit beta cell death and defect in insulin secretion, while lower concentrations act as signalling mediators. In INS-1E beta-cells, elevated glucose levels increase the release of arachidonic acid and linoleic acid from membrane phospholipids and promote their peroxidation to HNE. At non cytotoxic concentrations, HNE behaves as an endogenous ligand for nuclear receptor PPAR-δ, stimulating insulin secretion in beta-cells (Cohen *et al.*, 2011). In rat islet beta-cell-derived RINm5F cells, a recent report demonstrates the involvement of the transient receptor potential (TRP) cation channels in the HNE-induced insulin secretion. Short-term (1h) exposure to HNE induces a transient increase in intracellular calcium concentration and triggers insulin secretion. HNE induces calcium influx through activation of TRP channels (amongst which TRPA1) which appears to be coupled with the L-type voltage-dependent calcium channel, and ultimately insulin secretion (Numazawa *et al.*, 2012). Lipid aldehydes should therefore be considered either as detrimental (>10 μM) or as beneficial (sub micromolar range) depending on their actual tissue concentration.

3.3. Direct adduction of the insulin polypeptide

Under conditions of oxidative stress, insulin, a polypeptidic hormone composed of 51 amino acid residues, is exposed to direct oxidative insult or to modification by lipoperoxidation by-products. Several amino acids are putative sites of adduction, and thus, covalent binding of lipid aldehydes affect the biological actions of this hormone. This applies to acrolein and methylglyoxal, whose fixation on insulin has been shown to reduce both hypoglycemic effects in rats and glucose uptake in 3T3-L1 adipocytes (Jia *et al.*, 2006; Medina-Navarro *et al.*, 2007). HHE and HNE, toxic aldehydes generated during lipid peroxidation, also modify the B-chain of human insulin *in vitro*, predominantly at the His B5 and His B10 residues via Michael adduction (Figure 3). Adduct formation affects the biological activity of insulin *in vivo*, decreasing its hypoglycemic effect in mice and stimulation of glucose uptake in adipose and muscle cells (Pillon *et al.*, 2011).

Figure 3. Structure of insulin monomer displaying two HHE adducts on histidine residues. From Pillon et al, 2011.

4. Obesity, insulin resistance and type-2 diabetes

4.1. Lipid peroxidation by-products association with obesity

Obesity is a major factor in the development of metabolic syndrome. After consumption of an energy-dense (i.e. high-fat) diet, plasma HNE levels increase rapidly and significantly within minutes (Devaraj *et al.*, 2008). When consumed regularly, this diet promotes obesity, which suggests a role for HNE very early in the development of obesity. On the other hand, levels of circulating HNE tend to decrease when obese people are maintained on calorie restriction (Johnson et al., 2007), demonstrating that lipid peroxidation is tightly linked to high fat diet and obesity. Furukawa et al. reported that increased oxidative stress in accumulated fat is an important pathogenic mechanism of obesity-associated metabolic syndrome. Production of ROS is indeed selectively increased in white adipose tissue of obese mice and associated with a blunted expression of antioxidant enzymes. In good agreement, fat accumulation correlates with systemic lipid peroxidation in humans (Furukawa et al., 2004), and the plasma concentration of MDA is 1.8 fold higher in subjects with a BMI above 40 kg/m² compared to lean individuals (Olusi, 2002). Diet-induced obesity increases tissue and plasma accumulation of ALEs (protein–acrolein and protein–HNE adducts for example), suggesting that obesity is associated with an increase in the formation of lipid peroxidation-derived aldehydes (Baba et al., 2011). A significant accumulation of HNE was noticed in the white adipose tissue of obese mice, where the adipocyte fatty acid binding protein (AFABP also known as aP2) is the soluble protein most highly modified by HNE in this tissue (Grimsrud et al., 2007). In obese mice roughly 7% of the AFABP in adipose tissue is covalently modified by HNE resulting in a decreased binding affinity for fatty acids. Lipid peroxidation is however not restricted to adipose tissue since HNE is also elevated in skeletal muscles of Otsuka Long-Evans Tokushima fatty (OLETF) rat, a model for hyperphagic obesity (Morris et al., 2008). Intracellular triglyacylglycerols accumulate in the muscle of obese humans where it is considered as a pathogenic factor in the development of insulin resistance. In obese compared to endurance-trained subjects, the lipid peroxidation to intracellular triacylglycerols ratio was 4-fold higher suggesting that obesity is associated with increased muscle lipid peroxidation (Russell et al., 2003; Vincent et al., 2006).

Chronic, low grade inflammation of white adipose tissue is a hallmark of obesity and a major contributor to oxidative stress and lipid peroxidation (Wellen & Hotamisligil, 2003). In the expanding adipose tissue, hypertrophied adipocytes contribute to the inflammation by up-regulating the expression and release of pro-inflammatory cytokines. In 3T3-L1 adipocytes, HNE can dose-dependently increase the expression of the inducible cyclooxygenase (COX-2) (Zarrouki et al., 2007) and that of the plasminogen activator inhibitor-1 (PAI-1). In the meantime, HNE decreases the expression of the anti-inflammatory, insulin-sensitizing hormone adiponectin (Soares et al., 2005; Wang et al., 2012), therefore linking lipid peroxidation by-products and chronic inflammation.

4.2. Lipid peroxidation by-products association with type-2 diabetes

Epidemiological studies demonstrates that fasting glycemia is positively correlated with oxidative stress markers such as 8-epi-PGF2α and TBARs and negatively correlated with plasma glutathione (Trevisan et al., 2001; Menon et al., 2004). In type-2 diabetic individuals, 8-epi-PGF2α is positively correlated to the HOMA index for insulin resistance (Gopaul et al., 2001) and urinary acrolein correlates with glycated haemoglobin HbA1c (Daimon et al., 2003). This was confirmed in animal models of insulin resistance which exhibit increased markers of oxidative stress, such as plasma F2-isoprostanes (Laight et al., 1999a). In parallel, antioxidant defences are reduced during an oral glucose tolerance test in normal and non-insulin-dependent diabetic subjects (Ceriello *et al.*, 1998), and diabetes is associated with decreased vitamin C and glutathione (Maxwell *et al.*, 1997; Dierckx *et al.*, 2003). Reciprocally, an intensive treatment of diabetes improves circulating levels of H_2O_2 and MDA (Wierusz-Wysocka *et al.*, 1995); and improved insulin sensitivity resulting from exercise and/or dietary restriction is associated with reduced levels of lipid peroxidation products (Reviewed by Vincent *et al.*, 2007). On the other hand, insulin sensitivity can be improved through antioxidant or carbonyl scavenging treatment (Kamenova, 2006; Vincent *et al.*, 2009), demonstrating the tight link existing between oxidative stress, oxidation by-products and insulin resistance.

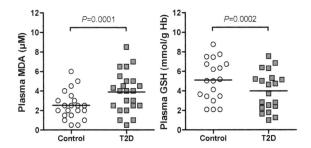

Figure 4. Plasma MDA concentration (μM) and blood GSH content (mmol/g haemoglobin) in healthy and type-2 diabetic (T2D) individuals. Results are a meta-analysis from 23 different publications; one dot represents the mean value obtained in one study[2]. In average, MDA is significantly increased by 60% while GSH is decreased by 25% in T2D compared to healthy subjects (paired student t-test, n=21).

Among α,β-unsaturated aldehydes, only MDA and to some extend HNE have been studied as oxidative stress biomarkers in diabetes, and they are indeed both increased up to 2-fold in both human (Figure 4) and animal models of type-2 diabetes (Wierusz-Wysocka *et al.*, 1995; Dierckx *et al.*, 2003). It has moreover been demonstrated that Type-2 diabetes duration is independently associated with increased levels of lipid peroxidation (Nakhjavani *et al.*, 2010), and our group recently showed an increase in HHE concentration in type-2 diabetes patients, reinforcing existing evidence for the specific role of lipid aldehydes in insulin resistance.

---

[2] Results from Wierusz-Wysocka et al., 1995; Vijayalingam et al., 1996; Feillet-Coudray et al., 1999; Rábago-Velasco et al., 2000; Rizvi & Zaid, 2001; Seghrouchni et al., 2002; Dinçer et al., 2002; Dierckx et al., 2003; Duman et al., 2003; Pasaoglu et al., 2004; Memişoğullari & Bakan, 2004; Skrha et al., 2005; Kurtul et al., 2005; Ozdemir et al., 2005; Mahboob et al., 2005; Saxena et al., 2005; Sampathkumar et al., 2005; Kuppusamy et al., 2005; Lapolla et al., 2007; Sathiyapriya et al., 2007; Singhania et al., 2008; Jain et al., 2009; Nakhjavani et al., 2010; Narasimhan et al., 2010; Shinde et al., 2011; Huang et al., 2011; Bahadoran et al., 2011; Zhang et al., 2011; Pácal et al., 2011; Calabrese et al., 2011 and Rasic-Milutinovic et al., 2012

Aldehydes concentration is increased in several tissues during diabetes, as revealed by increased levels of HNE in pancreas, liver, brain and heart. Pancreatic islets from type-2 diabetic patients are positively stained with HNE, suggesting that oxidative stress through lipid peroxidation could contribute to the reduced beta-cell mass and islet-cell injury (Sakuraba *et al.*, 2002). An increased level of HNE-modified proteins is reported in the pancreatic beta cells of Goto Kakizaki rats as a result of hyperglycaemia (Ihara *et al.*, 1999); and accumulation of HNE is observed in liver of diabetic rats due to the impairment of HNE-metabolizing enzymes (Traverso *et al.*, 1998, 2002). Diabetic mice under a high fat diet exhibit increased HNE adducts levels in temporal lobes relative to control (Lyn-Cook *et al.*, 2009), and HNE conjugation of GLUT3, the glucose transporter present in neurons, is increased in the hippocampus of diabetic rats subjected to stress (Reagan *et al.*, 2000). When db/db obese mice are fed a Western diet containing 21% fat and 0.15% cholesterol, they develop obesity, hyperglycemia, and insulin resistance. In this situation, HNE is significantly elevated in the left ventricular myocardium of diabetic mice compared to their lean littermates (Yamashita *et al.*, 2010). Finally, hyperglycemic Zucker Diabetic Fatty (ZDF) rats exhibited a 8-fold increase in plasma HDDE concentration compared to their lean non diabetic counterparts (Riahi *et al.*, 2010).

4.3. Causative role for lipid peroxidation by-products in the metabolic syndrome

This body of evidence linking oxidative stress with the metabolic syndrome is however only based on correlations and do not decipher the mechanisms and/or the causative role of oxidative stress in diabetes. There are consequently two main hypotheses:

1. First is chronic hyperglycemia which then leads to oxidative stress. In this case, lipid peroxidation products would be by-products of this oxidative stress and a consequence of diabetes, even if contributing to the progression of the disease and its complications.
2. First is oxidative stress which precedes the development of diabetes and plays a causative role in its development. In this hypothesis, deregulation of antioxidant defences and increased oxidative stress would lead to accumulation of secondary by-products consequently inducing insulin resistance.

We will focus in this chapter on the second hypothesis for which evidences have been recently accumulating from human, animal and cell culture studies.

Oxidative stress, through reactive oxygen species can positively and negatively regulate insulin signalling, depending on time, dose, model and free radical used (for review, see Bashan *et al.*, 2009). It is however admitted that prolonged oxidative stress impairs insulin signalling, insulin-induced GLUT4 translocation and glucose uptake. This occurs through several mechanisms including, but not limited to IRS inhibitory phosphorylation, MAPK activation and endoplasmic reticulum stress. This body of evidence for the role of oxidative stress in insulin resistance has been demonstrated in adipocytes, muscle, liver and cardiac cells (Rudich *et al.*, 1998; Bloch-Damti *et al.*, 2006; JeBailey *et al.*, 2007; Singh *et al.*, 2008a; Shibata *et al.*, 2010; Tan *et al.*, 2011). Even *in vivo*, a pro-oxidant challenge provokes the onset of type-2 diabetes in insulin resistant rats (Laight *et al.*, 1999b) and chronic methylglyoxal infusion by minipump causes pancreatic beta-cell dysfunction and induces type-2 diabetes in Sprague-Dawley rats (Dhar *et al.*, 2011).

The role of lipid aldehydes *in vivo* during metabolic diseases is often indirectly assessed, and the most compelling evidence for a causative role of aldehydes comes from polymorphisms in the glutathione-S-transferase (GST) gene. This family of enzymes is responsible for the detoxification of aldehydes through conjugation to glutathione; and several deletion polymorphisms leading to blunted enzyme activity have been described. Patients carrying certain null GST polymorphisms had up to 3-fold increased incidence of type-2 diabetes mellitus compared to those with normal genotypes (Amer *et al.*, 2011). Accordingly, the expression of the GST4A is blunted in the adipose tissue of obese insulin resistant subjects (Curtis *et al.*, 2010).

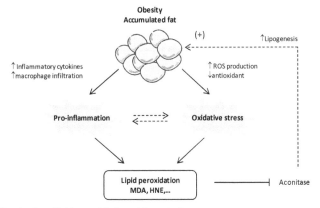

Figure 5. The obesity – lipid peroxidation vicious circle

This same observation was made in animals model, were it has been suggested that excessive production of HNE might be sufficient to cause obesity and the metabolic syndrome. Mice lacking the gene encoding the HNE-conjugating enzyme mGSTA4-4 develop obesity and insulin resistance, unsurprisingly associated with HNE accumulation in multiple tissues (Singh *et al.*, 2008*b*). Cell culture work further shows that dysfunction of glutathione S-transferase or its ablation by siRNA leads to excess HNE accumulation, increased protein carbonylation, oxidative stress, and mitochondrial dysfunction (Kostyuk *et al.*, 2010; Curtis *et al.*, 2010). On the other hand, overexpression of glutathione-S-transferase A4-4 protects against oxidative stress and HNE-induced apoptosis (Vaillancourt *et al.*, 2008). Additional findings suggest that HNE causes fat accumulation by promoting fatty acid synthesis and suppressing fatty acid beta-oxidation. Interestingly, the phenotype of mGSTA4-4 null mice is strain dependent: mGSTA4-4 null mice with 129/sv genetic background exhibit both increased accumulation of HNE and obesity while those with C57Bl6 genetic background are lean and HNE is unchanged (Singh *et al.*, 2008*b*). In good agreement, silencing of the mGSTA4-4 gene in the nematode Caenorhabditis elegans also results in an accumulation of lipid peroxidation by-products and a fatty phenotype (Singh *et* *al.*, 2009). When HNE is experimentally increased in the nematode, either by genetic deletion or through direct exposure, it promotes fat accumulation. The mechanism involves HNE inhibition of mitochondrial aconitase leading to an accumulation of malonyl CoA, precursor of fatty acid synthesis (Zimniak, 2010). Taken together, these data suggest that accumulation of lipid aldehydes and fat deposition could be mutually inductive leading to a vicious circle promoting fat accretion (Figure 5).

The direct causative role of lipid peroxidation by-products in insulin resistance has also been recently evidenced in cell cultures studies. A first study identified that methylglyoxal, an aldehyde by-product of glucose oxidation, can impair the insulin signalling pathways independently of the formation of intracellular reactive oxygen species (Riboulet-Chavey *et al.*, 2006). Then, focusing on lipid peroxidation products, two studies demonstrated that HNE can induce insulin resistance in adipocytes and muscle cell through inhibition of IRS and Akt signalling, as well as insulin-induced glucose uptake (Demozay *et al.*, 2008; Pillon *et al.*, 2012). These two studies identified carbonyl stress (notably IRS1 adduction) and ROS production as the possible mechanisms responsible for HNE effects; a third one being possibly the adduction of Akt2, which inhibits insulin-dependent Akt signalling in HepG2 cells (Shearn *et al.*, 2011). Altogether, these studies strongly suggest that excessive production of aldehydes might be sufficient to cause obesity, diabetes and the metabolic syndrome.

5. Preventing the deleterious effects of aldehydes

It has been known for decades that supplementation in α-lipoic acid in type-2 diabetic subject improves glucose tolerance and insulin sensitivity (Jacob et al., 1996). Similarly, increased intake of Vitamin E in obese and insulin resistant patients reduces fasting glycemia and the HOMA index for insulin resistance, and this is correlated with a decrease in the concentration of peroxides in plasma (Manning et al., 2004). This adds further evidence for a causative role of oxidative stress in the metabolic syndrome, and opens new therapeutic perspectives using antioxidant and/or scavenging of toxic aldehydes.

5.1. Glutathione and enzymatic detoxification

Glutathione (GSH) is a tripeptide (L-γ-glutamyl-L-cysteinyl-glycine) which is present in high concentration in the cytoplasm of living cells: 5-50 nmol/mg proteins (Jungas *et al.*, 2002; Dominy *et al.*, 2007). GSH is an important coenzyme of several enzymatic reactions, and exerts its antioxidant activity mainly through regeneration of vitamin E, and also through direct interaction with free radical and aldehydes. GSH is obviously necessary for the activity of glutathione-S-transferases (GSTs), a family of enzymes responsible for detoxification of electrophile by-products, such as the ones derived from lipid peroxidation, and GSH exhibits indeed a high reactivity for HNE. Though HNE is electrophile enough to spontaneously react with GSH, this reaction is dramatically accelerated via the conjugation process catalyzed by GSTs (Alin *et al.*, 1985). As described above, mice lacking GSTs develop obesity and insulin resistance (Singh *et al.*, 2008*b*), and humans carrying null polymorphisms for GST have a 3-fold increased risk of having type-2 diabetes (Amer *et al.*, 2011), pointing out the important role of GSH in the metabolic syndrome. As addition reaction to GSH contributes to the detoxification of aldehydes, pharmacological strategies to increase glutathione pools or GST activity should be protective against aldehydes. Several studies were indeed successful in protecting cultured cells from the deleterious effects of oxidative stress through an increase in intracellular pools of reduced glutathione. This strategy was particularly efficient to protect the cells against the deleterious effects of HNE (Yadav *et al.*, 2008; Jia *et al.*, 2009*b*; Pillon *et al.*, 2012).

In addition to glutathione-S-transferases, several enzymes are also responsible for the detoxification of aldehydes: aldehyde dehydrogenase, alcohol dehydrogenase (Hartley *et al.*, 1995), aldose reductase (Srivastava *et al.*, 2000) and fatty aldehyde dehydrogenase (Demozay *et al.*, 2008). These enzymes however participate to a lower extent in the metabolism of aldehydes, compared to glutathione and GST. Recent data indicate that if HHE and HNE are both metabolized via glutathione, the effectiveness of detoxification differs for these two molecules. By extension, aldehydes may be metabolized with different affinities and efficiencies by detoxification enzymes, which could explain some differences in their respective toxicities (Long *et al.*, 2010).

5.2. Scavenging

In chemistry, a scavenger is a chemical substance able to remove or inactivate impurities or unwanted reaction products. In living cells, a scavenger is, by extension, a molecule able to inactivate toxic compounds such as ROS and aldehydes, therefore preventing their deleterious effects. In the case of aldehydes, a scavenger would be a strong nucleophile molecule on which HHE, HNE or any other aldehyde would form a covalent adduct. Consequently, most of the aldehyde scavengers are amino- or sulphur-containing drugs such as N-acetyl-cysteine (NAC), hydralazine, S-adenosyl-methionine (SAM) aminoguanidine (AGD) and α-lipoic acid, the latter being tested in several human studies for treatment of type-2 diabetes.

5.2.1. α-lipoic acid

α-lipoic acid is a natural compound found in many foodstuffs (such as potatoes, broccoli and meat), but in rather low amount. The effect of lipoic acid was demonstrated in animal models, where it enhances insulin-stimulated glucose metabolism in skeletal muscle from insulin-resistant rat (Jacob et al., 1996). It was rapidly tested in several clinical studies which demonstrated its beneficial effects in type-2 diabetes through a decrease in fasted blood glucose, enhancement of glucose disposal, improved insulin sensitivity and decreased insulin resistance (Jacob *et al.*, 1995). These results were confirmed *in vitro*, where α-lipoic acid prevents the development of glucose-induced insulin resistance in adipocytes (Greene et al., 2001) and were reproduced by many independent studies. Treatment with lipoic acid decreases oxidative stress in both adipocytes and muscle cells (Rudich et al., 1999; Maddux et al., 2001) and also decreases lipid peroxidation markers in insulin resistant rats (Thirunavukkarasu & Anuradha, 2004), suggesting that the improvement of insulin sensitivity is due to its antioxidant properties. Hence, the current literature supports the use of alpha lipoic acid for the treatment of diabetes complications and it consequently became the first antioxidant supplement used for the treatment of diabetes complications, being already approved in Germany for the treatment of diabetic neuropathy.

5.2.2. N-acetylcysteine (NAC)

N-acetylcysteine is a cysteine derivative and a potent antioxidant. Its properties are mainly due to its thiol group able to reduce free radicals as well as its role as a precursor in the formation of glutathione (Zafarullah et al., 2003). NAC exhibits highly protective scavenging properties against aldehydes and protects against MDA increase and GSH decrease in animal models of insulin resistance. NAC is able to improve insulin sensitivity in healthy rats (Figure 6) and reverses insulin resistance and aldehyde-induced hypertension in rats (Haber et al., 2003). In cell culture studies, NAC can prevent the insulin resistance induced by HNE in muscle cells (Pillon et al., 2012), as well as the one induced by advanced glycation end products in adipocytes (Unoki et al., 2007), thus confirming the important role NAC can play in improving both oxidative stress parameters and insulin resistance.

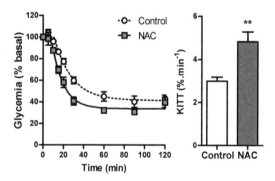

Figure 6. Insulin sensitizing effect of N-acetylcysteine. Wistar rats were given NAC in drinking water for one week (total intake was 225 mg.kg^{-1}.day^{-1}). Insulin sensitivity was calculated using a standard insulin tolerance test[3]. Results are average ± SEM from 5 different animals per group, expressed as percent of basal glycemia. From Pillon et al. unpublished results.

---

[3] Animals fasted overnight were then injected intraperitoneally with 0.5 UI/kg body weight of insulin. Plasma glucose was measured from tail vein blood using a glucometer at 0, 20, 40, 60, and 120 min following the injection. Glucose disappearance rate for ITT (K_{ITT}; %/min) was calculated as: $K_{ITT}=(0.693 \times 100)/t_{1/2}$, where $t_{1/2}$ was calculated from the slope of the plasma glucose concentration, considering an exponential decrement of glucose concentration during the 20 min after insulin administration. Higher insulin sensitivity index (K_{ITT}) scores mean higher response of tissues to insulin.

Results from clinical trials in type-2 diabetic subjects show that NAC is able to decrease oxidative stress parameters, increase GSH and decrease plasma VCAM-1 (De Mattia et al., 1998); moreover, long-term N-acetylcysteine administration reduces endothelial activation and is proposed as a potential antiatherogenic therapy (Martina et al., 2008). Despite the fact that NAC can improve insulin sensitivity in women with polycystic ovary syndrome (Fulghesu *et al.*, 2002) and this body of evidence suggesting that NAC may slow down the progression of diabetic complications, to date no clinical trial demonstrated any significant benefit of its supplementation in diabetes.

5.2.3. Aminoguanidine (AGD)

Aminoguanidine is a highly nucleophilic agent which reacts *in vitro* and *in vivo* with aldehydes, therefore protecting against the deleterious effects of ALE precursors (Peyroux & Sternberg, 2006). It is also an antioxidant able to quench hydroxyl radicals and *in vivo*. AGD in drinking water decreases lipid peroxidation in type-1 diabetic rats and rabbits (Ihm *et al.*, 1999). In experimental animal models of diabetes, AGD demonstrates significant effects in protecting against pathological complications, such as diabetic nephropathies, atherosclerosis and neurovascular complications (El Shazly *et al.*, 2009). Consequently, several clinical trials in humans have been designed to evaluate AGD efficiency but they demonstrate only mild effects and were not conclusive, partly because of side-effects, and of weak carbonyl scavenger effects in human vascular tissues (Bolton *et al.*, 2004).

5.2.4. Hydralazine

Primarily used as an antihypertensive drug, hydralazine exhibits a pronounced nucleophilicity and is consequently very efficient in scavenging several aldehydes (acrolein, HNE) and ketones, as well as aldehyde-adducted proteins (Burcham et al., 2002). It is also a powerful antioxidant, able to inhibit the generation of ROS (Münzel *et al.*, 1996). Its scavenging activity *in vivo* was demonstrated by its ability to reverse the formation of HNE and acrolein adducts on tissue proteins in atherosclerotic aortas of hypercholesterolemic animals (Vindis et al., 2006), but its effects on insulin resistance and diabetes are to date uncharacterized.

5.2.5. S-adenosyl-methionine (SAM)

In living organisms, SAM is endogenously synthesized from methionine in every cell, but the liver is the major site of its synthesis and degradation. SAM is an important precursor for cysteine and glutathione production (Lu, 2000) and its involvement in several metabolic pathways makes it essential for a wide spectrum of cellular processes. SAM inhibits both HNE production and adducts formation and efficiently prevents high-fat diet-induced non-alcoholic steatohepatitis in rats (Lieber et al., 2007), even if its direct effects on insulin resistance are still unknown.

6. Conclusion

Lipid peroxidation by-products are associated with metabolic diseases, but their primary role during obesity and diabetes is subject to debate. *In vitro* and *in vivo* animal studies highlighted that an aldehyde challenge, through a state of carbonyl stress, affects several steps involved in the development of obesity and type-2 diabetes (Figure 7). On insulin-sensitive tissues (muscle, adipose tissue), aldehydes lead to insulin resistance, while in pancreas, aldehydes impair insulin secretion. Together with the carbonylation of the insulin peptide itself, aldehydes could contribute to the defects in insulin action, leading to the metabolic syndrome.

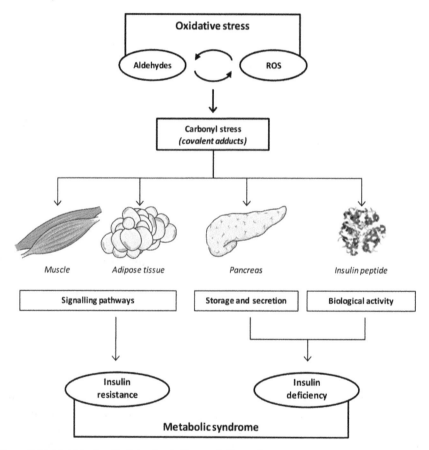

Figure 7. Lipid aldehydes effects leading to the metabolic syndrome.

Despite this evidence, the role of aldehydes is mitigated by the mild effect obtained with antioxidants and/or aldehyde scavengers in the treatment of diabetic complications in human. "Oxidative stress" and its derivatives are nevertheless important in the metabolic syndrome, and prevention or treatment of some of its associated complications could be accessed through reduction of both ROS and toxic aldehydes by-products, as it is the case with lipoic acid.

Author details

Nicolas J. Pillon and Christophe O. Soulage
Université de Lyon, INSA de Lyon, CarMeN, INSERM U1060, Univ Lyon-1, F-69621, Villeurbanne, France

7. References

Alin P, Danielson UH & Mannervik B (1985). 4-Hydroxyalk-2-enals are substrates for glutathione transferase. *FEBS Lett* 179, 267–270.

Amer MA, Ghattas MH, Abo-Elmatty DM & Abou-El-Ela SH (2011). Influence of glutathione S-transferase polymorphisms on type-2 diabetes mellitus risk. *Genet Mol Res*; DOI: 10.4238/2011.October.31.14.

Augusti KT & Sheela CG (1996). Antiperoxide effect of S-allyl cysteine sulfoxide, an insulin secretagogue, in diabetic rats. *Experientia* 52, 115–120.

Baba SP, Hellmann J, Srivastava S & Bhatnagar A (2011). Aldose reductase (AKR1B3) regulates the accumulation of advanced glycosylation end products (AGEs) and the expression of AGE receptor (RAGE). *Chem Biol Interact* 191, 357–363.

Bae EH, Cho S, Joo SY, Ma SK, Kim SH, Lee J & Kim SW (2011). 4-Hydroxy-2-hexenal-induced apoptosis in human renal proximal tubular epithelial cells. *Nephrol Dial Transplant* 26, 3866–3873.

Bashan N, Kovsan J, Kachko I, Ovadia H & Rudich A (2009). Positive and negative regulation of insulin signaling by reactive oxygen and nitrogen species. *Physiol Rev* 89, 27–71.

Bloch-Damti A, Potashnik R, Gual P, Le Marchand-Brustel Y, Tanti JF, Rudich A & Bashan N (2006). Differential effects of IRS1 phosphorylated on Ser307 or Ser632 in the induction of insulin resistance by oxidative stress. *Diabetologia* 49, 2463–2473.

Bodur C, Kutuk O, Tezil T & Basaga H (2012). Inactivation of Bcl-2 through IκB kinase (IKK)-dependent phosphorylation mediates apoptosis upon exposure to 4-hydroxynonenal (HNE). *J Cell Physiol*; DOI: 10.1002/jcp.24057.

Bolton WK, Cattran DC, Williams ME, Adler SG, Appel GB, Cartwright K, Foiles PG, Freedman BI, Raskin P, Ratner RE, Spinowitz BS, Whittier FC & Wuerth J-P (2004). Randomized trial of an inhibitor of formation of advanced glycation end products in diabetic nephropathy. *Am J Nephrol* 24, 32–40.

Burcham PC, Kaminskas LM, Fontaine FR, Petersen DR & Pyke SM (2002). Aldehyde-sequestering drugs: tools for studying protein damage by lipid peroxidation products. *Toxicology* 181-182, 229–236.

Cahuana GM, Tejedo JR, Jimenez J, Ramirez R, Sobrino F & Bedoya FJ (2003). Involvement of advanced lipooxidation end products (ALEs) and protein oxidation in the apoptotic actions of nitric oxide in insulin secreting RINm5F cells. *Biochem Pharmacol* 66, 1963–1971.

Calzada C, Colas R, Guillot N, Guichardant M, Laville M, Véricel E & Lagarde M (2010). Subgram daily supplementation with docosahexaenoic acid protects low-density lipoproteins from oxidation in healthy men. *Atherosclerosis* 208, 467–472.

Ceriello A, Bortolotti N, Crescentini A, Motz E, Lizzio S, Russo A, Ezsol Z, Tonutti L & Taboga C (1998). Antioxidant defences are reduced during the oral glucose tolerance test in normal and non-insulin-dependent diabetic subjects. *Eur J Clin Invest* 28, 329–333.

Cheng J, Wang F, Yu D-F, Wu P-F & Chen J-G (2011). The cytotoxic mechanism of malondialdehyde and protective effect of carnosine via protein cross-linking/mitochondrial dysfunction/reactive oxygen species/MAPK pathway in neurons. *Eur J Pharmacol* 650, 184–194.

Choudhary S, Zhang W, Zhou F, Campbell GA, Chan LL, Thompson EB & Ansari NH (2002). Cellular lipid peroxidation end-products induce apoptosis in human lens epithelial cells. *Free Radic Biol Med* 32, 360–369.

Cohen G, Riahi Y, Shamni O, Guichardant M, Chatgilialoglu C, Ferreri C, Kaiser N & Sasson S (2011). Role of lipid peroxidation and PPAR-δ in amplifying glucose-stimulated insulin secretion. *Diabetes* 60, 2830–2842.

Coleman JD, Prabhu KS, Thompson JT, Reddy PS, Peters JM, Peterson BR, Reddy CC & Vanden Heuvel JP (2007). The oxidative stress mediator 4-hydroxynonenal is an intracellular agonist of the nuclear receptor peroxisome proliferator-activated receptor-beta/delta (PPARbeta/delta). *Free Radic Biol Med* 42, 1155–1164.

Curtis JM, Grimsrud PA, Wright WS, Xu X, Foncea RE, Graham DW, Brestoff JR, Wiczer BM, Ilkayeva O, Cianflone K, Muoio DE, Arriaga EA & Bernlohr DA (2010). Downregulation of adipose glutathione S-transferase A4 leads to increased protein carbonylation, oxidative stress, and mitochondrial dysfunction. *Diabetes* 59, 1132–1142.

Daimon M, Sugiyama K, Kameda W, Saitoh T, Oizumi T, Hirata A, Yamaguchi H, Ohnuma H, Igarashi M & Kato T (2003). Increased urinary levels of pentosidine, pyrraline and acrolein adduct in type 2 diabetes. *Endocr J* 50, 61–67.

Demozay D, Mas J-C, Rocchi S & Van Obberghen E (2008). FALDH reverses the deleterious action of oxidative stress induced by lipid peroxidation product 4-hydroxynonenal on insulin signaling in 3T3-L1 adipocytes. *Diabetes* 57, 1216–1226.

Devaraj S, Wang-Polagruto J, Polagruto J, Keen CL & Jialal I (2008). High-fat, energy-dense, fast-food-style breakfast results in an increase in oxidative stress in metabolic syndrome. *Metab Clin Exp* 57, 867–870.

Dhar A, Dhar I, Jiang B, Desai KM & Wu L (2011). Chronic methylglyoxal infusion by minipump causes pancreatic beta-cell dysfunction and induces type 2 diabetes in Sprague-Dawley rats. *Diabetes* 60, 899–908.

Dierckx N, Horvath G, van Gils C, Vertommen J, van de Vliet J, De Leeuw I & Manuel-y-Keenoy B (2003). Oxidative stress status in patients with diabetes mellitus: relationship to diet. *Eur J Clin Nutr* 57, 999–1008.

Dominy JE, Hwang J & Stipanuk MH (2007). Overexpression of cysteine dioxygenase reduces intracellular cysteine and glutathione pools in HepG2/C3A cells. *Am J Physiol Endocrinol Metab* 293, E62–69.

Dong J-Z & Moldoveanu SC (2004). Gas chromatography-mass spectrometry of carbonyl compounds in cigarette mainstream smoke after derivatization with 2,4-dinitrophenylhydrazine. *J Chromatogr A* 1027, 25–35.

El Shazly AHM, Mahmoud AM & Darwish NS (2009). Potential prophylactic role of aminoguanidine in diabetic retinopathy and nephropathy in experimental animals. *Acta Pharm* 59, 67–73.

Fulghesu AM, Ciampelli M, Muzj G, Belosi C, Selvaggi L, Ayala GF & Lanzone A (2002). N-acetyl-cysteine treatment improves insulin sensitivity in women with polycystic ovary syndrome. *Fertil Steril* 77, 1128–1135.

Furukawa S, Fujita T, Shimabukuro M, Iwaki M, Yamada Y, Nakajima Y, Nakayama O, Makishima M, Matsuda M & Shimomura I (2004). Increased oxidative stress in obesity and its impact on metabolic syndrome. *J Clin Invest* 114, 1752–1761.

Gopaul NK, Manraj MD, Hebe A, Lee Kwai Yan S, Johnston A, Carrier MJ & Anggard EE (2001). Oxidative stress could precede endothelial dysfunction and insulin resistance in Indian Mauritians with impaired glucose metabolism. *Diabetologia* 44, 706–712.

Greene EL, Nelson BA, Robinson KA & Buse MG (2001). alpha-Lipoic acid prevents the development of glucose-induced insulin resistance in 3T3-L1 adipocytes and accelerates the decline in immunoreactive insulin during cell incubation. *Metab Clin Exp* 50, 1063–1069.

Grimsrud PA, Picklo MJ, Griffin TJ & Bernlohr DA (2007). Carbonylation of adipose proteins in obesity and insulin resistance: identification of adipocyte fatty acid-binding protein as a cellular target of 4-hydroxynonenal. *Mol Cell Proteomics* 6, 624–637.

Guichardant M, Bacot S, Moliere P & Lagarde M (2006). Hydroxy-alkenals from the peroxidation of n-3 and n-6 fatty acids and urinary metabolites. *Prostaglandins Leukot Essent Fatty Acids* 75, 179–182.

Gutteridge JM (1982). Free-radical damage to lipids, amino acids, carbohydrates and nucleic acids determined by thiobarbituric acid reactivity. *Int J Biochem* 14, 649–653.

Haber CA, Lam TKT, Yu Z, Gupta N, Goh T, Bogdanovic E, Giacca A & Fantus IG (2003). N-acetylcysteine and taurine prevent hyperglycemia-induced insulin resistance in vivo: possible role of oxidative stress. *Am J Physiol Endocrinol Metab* 285, E744–753.

Hartley DP, Ruth JA & Petersen DR (1995). The hepatocellular metabolism of 4-hydroxynonenal by alcohol dehydrogenase, aldehyde dehydrogenase, and glutathione S-transferase. *Arch Biochem Biophys* 316, 197–205.

Hipkiss AR, Preston JE, Himsworth DT, Worthington VC & Abbot NJ (1997). Protective effects of carnosine against malondialdehyde-induced toxicity towards cultured rat brain endothelial cells. *Neurosci Lett* 238, 135–138.

Ihara Y, Toyokuni S, Uchida K, Odaka H, Tanaka T, Ikeda H, Hiai H, Seino Y & Yamada Y (1999). Hyperglycemia causes oxidative stress in pancreatic beta-cells of GK rats, a model of type 2 diabetes. *Diabetes* 48, 927–932.

Ihm SH, Yoo HJ, Park SW & Ihm J (1999). Effect of aminoguanidine on lipid peroxidation in streptozotocin-induced diabetic rats. *Metab Clin Exp* 48, 1141–1145.

Jacob S, Henriksen EJ, Schiemann AL, Simon I, Clancy DE, Tritschler HJ, Jung WI, Augustin HJ & Dietze GJ (1995). Enhancement of glucose disposal in patients with type 2 diabetes by alpha-lipoic acid. *Arzneimittelforschung* 45, 872–874.

Jacob S, Streeper RS, Fogt DL, Hokama JY, Tritschler HJ, Dietze GJ & Henriksen EJ (1996). The antioxidant alpha-lipoic acid enhances insulin-stimulated glucose metabolism in insulin-resistant rat skeletal muscle. *Diabetes* 45, 1024–1029.

Je JH, Lee JY, Jung KJ, Sung B, Go EK, Yu BP & Chung HY (2004). NF-kappaB activation mechanism of 4-hydroxyhexenal via NIK/IKK and p38 MAPK pathway. *FEBS Lett* 566, 183–189.

JeBailey L, Wanono O, Niu W, Roessler J, Rudich A & Klip A (2007). Ceramide- and oxidant-induced insulin resistance involve loss of insulin-dependent Rac-activation and actin remodeling in muscle cells. *Diabetes* 56, 394–403.

Ji C, Rouzer CA, Marnett LJ & Pietenpol JA (1998). Induction of cell cycle arrest by the endogenous product of lipid peroxidation, malondialdehyde. *Carcinogenesis* 19, 1275–1283.

Jia L, Zhang Z, Zhai L & Bai Y (2009a). Protective effect of lipoic acid against acrolein-induced cytotoxicity in IMR-90 human fibroblasts. *J Nutr Sci Vitaminol* 55, 126–130.

Jia X, Olson DJH, Ross ARS & Wu L (2006). Structural and functional changes in human insulin induced by methylglyoxal. *FASEB J* 20, 1555–1557.

Jia Z, Misra BR, Zhu H, Li Y & Misra HP (2009b). Upregulation of cellular glutathione by 3H-1,2-dithiole-3-thione as a possible treatment strategy for protecting against acrolein-induced neurocytotoxicity. *Neurotoxicology* 30, 1–9.

Johnson JB, Summer W, Cutler RG, Martin B, Hyun D-H, Dixit VD, Pearson M, Nassar M, Telljohann R, Tellejohan R, Maudsley S, Carlson O, John S, Laub DR & Mattson MP (2007). Alternate day calorie restriction improves clinical findings and reduces markers of oxidative stress and inflammation in overweight adults with moderate asthma. *Free Radic Biol Med* 42, 665–674.

Jungas T, Motta I, Duffieux F, Fanen P, Stoven V & Ojcius DM (2002). Glutathione levels and BAX activation during apoptosis due to oxidative stress in cells expressing wild-type and mutant cystic fibrosis transmembrane conductance regulator. *J Biol Chem* 277, 27912–27918.

Kamenova P (2006). Improvement of insulin sensitivity in patients with type 2 diabetes mellitus after oral administration of alpha-lipoic acid. *Hormones (Athens)* 5, 251–258.

Kostyuk V, Potapovich A, Cesareo E, Brescia S, Guerra L, Valacchi G, Pecorelli A, Raskovic D, De Luca C, Pastore S & Korkina L (2010). Dysfunction of Glutathione S-Transferase Leads to Excess 4-Hydroxy-2-Nonenal and H2O2 and Impaired Cytokine Pattern in Cultured Keratinocytes and Blood of Vitiligo Patients. *Antioxid Redox Signal*; DOI: 10.1089/ars.2009.2976.

Laight DW, Desai KM, Gopaul NK, Anggard EE & Carrier MJ (1999a). F2-isoprostane evidence of oxidant stress in the insulin resistant, obese Zucker rat: effects of vitamin E. *Eur J Pharmacol* 377, 89–92.

Laight DW, Desai KM, Gopaul NK, Anggard EE & Carrier MJ (1999b). Pro-oxidant challenge in vivo provokes the onset of NIDDM in the insulin resistant obese Zucker rat. *Br J Pharmacol* 128, 269–271.

Lee JY, Je JH, Jung KJ, Yu BP & Chung HY (2004). Induction of endothelial iNOS by 4-hydroxyhexenal through NF-kappaB activation. *Free Radic Biol Med* 37, 539–548.

Lee JY, Jung GY, Heo HJ, Yun MR, Park JY, Bae SS, Hong KW, Lee WS & Kim CD (2006). 4-Hydroxynonenal induces vascular smooth muscle cell apoptosis through mitochondrial generation of reactive oxygen species. *Toxicol Lett* 166, 212–221.

Lee SE, Jeong SI, Kim G-D, Yang H, Park C-S, Jin Y-H & Park YS (2011). Upregulation of heme oxygenase-1 as an adaptive mechanism for protection against crotonaldehyde in human umbilical vein endothelial cells. *Toxicol Lett* 201, 240–248.

Li D, Ferrari M & Ellis EM (2011). Human aldo-keto reductase AKR7A2 protects against the cytotoxicity and mutagenicity of reactive aldehydes and lowers intracellular reactive oxygen species in hamster V79-4 cells. *Chemico-Biological Interactions*; DOI: 10.1016/j.cbi.2011.09.007.

Lieber CS, Leo MA, Cao Q, Mak KM, Ren C, Ponomarenko A, Wang X & Decarli LM (2007). The Combination of S-adenosylmethionine and Dilinoleoylphosphatidylcholine Attenuates Non-alcoholic Steatohepatitis Produced in Rats by a High-Fat Diet. *Nutr Res* 27, 565–573.

Liu W, Kato M, Akhand AA, Hayakawa A, Suzuki H, Miyata T, Kurokawa K, Hotta Y, Ishikawa N & Nakashima I (2000). 4-hydroxynonenal induces a cellular redox status-related activation of the caspase cascade for apoptotic cell death. *J Cell Sci* 113 (Pt 4), 635–641.

Liu X, Yang Z, Pan X, Zhu M & Xie J (2010). Crotonaldehyde induces oxidative stress and caspase-dependent apoptosis in human bronchial epithelial cells. *Toxicol Lett* 195, 90–98.

Long EK, Murphy TC, Leiphon LJ, Watt J, Morrow JD, Milne GL, Howard JRH & Picklo MJ (2008). Trans-4-hydroxy-2-hexenal is a neurotoxic product of docosahexaenoic (22:6; n-3) acid oxidation. *J Neurochem* 105, 714–724.

Long EK, Rosenberger TA & Picklo MJ (2010). Ethanol withdrawal increases glutathione adducts of 4-hydroxy-2-hexenal but not 4-hydroxyl-2-nonenal in the rat cerebral cortex. *Free Radic Biol Med* 48, 384–390.

Lovell MA, Ehmann WD, Mattson MP & Markesbery WR (1997). Elevated 4-hydroxynonenal in ventricular fluid in Alzheimer's disease. *Neurobiol Aging* 18, 457–461.

Lu SC (2000). S-Adenosylmethionine. *Int J Biochem Cell Biol* 32, 391–395.

Luo J, Robinson JP & Shi R (2005). Acrolein-induced cell death in PC12 cells: role of mitochondria-mediated oxidative stress. *Neurochem Int* 47, 449–457.

Lyn-Cook LE Jr, Lawton M, Tong M, Silbermann E, Longato L, Jiao P, Mark P, Wands JR, Xu H & de la Monte SM (2009). Hepatic ceramide may mediate brain insulin resistance and neurodegeneration in type 2 diabetes and non-alcoholic steatohepatitis. *J Alzheimers Dis* 16, 715–729.

Maddux BA, See W, Lawrence JC Jr, Goldfine AL, Goldfine ID & Evans JL (2001). Protection against oxidative stress-induced insulin resistance in rat L6 muscle cells by mircomolar concentrations of alpha-lipoic acid. *Diabetes* 50, 404–410.

Manini P, Andreoli R, Sforza S, Dall'Asta C, Galaverna G, Mutti A & Niessen WMA (2010). Evaluation of Alternate Isotope-Coded Derivatization Assay (AIDA) in the LC-MS/MS analysis of aldehydes in exhaled breath condensate. *J Chromatogr B Analyt Technol Biomed Life Sci* 878, 2616–2622.

Manning PJ, Sutherland WHF, Walker RJ, Williams SM, De Jong SA, Ryalls AR & Berry EA (2004). Effect of high-dose vitamin E on insulin resistance and associated parameters in overweight subjects. *Diabetes Care* 27, 2166–2171.

Martina V, Masha A, Gigliardi VR, Brocato L, Manzato E, Berchio A, Massarenti P, Settanni F, Della Casa L, Bergamini S & Iannone A (2008). Long-term N-acetylcysteine and L-arginine administration reduces endothelial activation and systolic blood pressure in hypertensive patients with type 2 diabetes. *Diabetes Care* 31, 940–944.

De Mattia G, Bravi MC, Laurenti O, Cassone-Faldetta M, Proietti A, De Luca O, Armiento A & Ferri C (1998). Reduction of oxidative stress by oral N-acetyl-L-cysteine treatment decreases plasma soluble vascular cell adhesion molecule-1 concentrations in non-obese, non-dyslipidaemic, normotensive, patients with non-insulin-dependent diabetes. *Diabetologia* 41, 1392–1396.

Maxwell SR, Thomason H, Sandler D, Leguen C, Baxter MA, Thorpe GH, Jones AF & Barnett AH (1997). Antioxidant status in patients with uncomplicated insulin-dependent and non-insulin-dependent diabetes mellitus. *Eur J Clin Invest* 27, 484–490.

McGrath LT, McGleenon BM, Brennan S, McColl D, McIlroy S & Passmore AP (2001). Increased oxidative stress in Alzheimer's disease as assessed with 4-hydroxynonenal but not malondialdehyde. *QJM* 94, 485–490.

Medina-Navarro R, Guzmán-Grenfell AM, Díaz-Flores M, Duran-Reyes G, Ortega-Camarillo C, Olivares-Corichi IM & Hicks JJ (2007). Formation of an adduct between insulin and the toxic lipoperoxidation product acrolein decreases both the hypoglycemic effect of the hormone in rat and glucose uptake in 3T3 adipocytes. *Chem Res Toxicol* 20, 1477–1481.

Menon V, Ram M, Dorn J, Armstrong D, Muti P, Freudenheim JL, Browne R, Schunemann H & Trevisan M (2004). Oxidative stress and glucose levels in a population-based sample. *Diabet Med* 21, 1346–1352.

Michiels C & Remacle J (1991). Cytotoxicity of linoleic acid peroxide, malondialdehyde and 4-hydroxynonenal towards human fibroblasts. *Toxicology* 66, 225–234.

Misonou Y, Asahi M, Yokoe S, Miyoshi E & Taniguchi N (2006). Acrolein produces nitric oxide through the elevation of intracellular calcium levels to induce apoptosis in human umbilical vein endothelial cells: implications for smoke angiopathy. *Nitric Oxide* 14, 180–187.

Miwa I, Ichimura N, Sugiura M, Hamada Y & Taniguchi S (2000). Inhibition of glucose-induced insulin secretion by 4-hydroxy-2-nonenal and other lipid peroxidation products. *Endocrinology* 141, 2767–2772.

Morris RT, Laye MJ, Lees SJ, Rector RS, Thyfault JP & Booth FW (2008). Exercise-induced attenuation of obesity, hyperinsulinemia, and skeletal muscle lipid peroxidation in the OLETF rat. *J Appl Physiol* 104, 708–715.

Münzel T, Kurz S, Rajagopalan S, Thoenes M, Berrington WR, Thompson JA, Freeman BA & Harrison DG (1996). Hydralazine prevents nitroglycerin tolerance by inhibiting activation of a membrane-bound NADH oxidase. A new action for an old drug. *J Clin Invest* 98, 1465–1470.

Nakhjavani M, Esteghamati A, Nowroozi S, Asgarani F, Rashidi A & Khalilzadeh O (2010). Type 2 diabetes mellitus duration: an independent predictor of serum malondialdehyde levels. *Singapore Med J* 51, 582–585.

Negre-Salvayre A, Coatrieux C, Ingueneau C & Salvayre R (2008). Advanced lipid peroxidation end products in oxidative damage to proteins. Potential role in diseases and therapeutic prospects for the inhibitors. *Br J Pharmacol* 153, 6–20.

Numazawa S, Takase M, Ahiko T, Ishii M, Shimizu S-I & Yoshida T (2012). Possible Involvement of Transient Receptor Potential Channels in Electrophile-Induced Insulin Secretion from RINm5F Cells. *Biol Pharm Bull* 35, 346–354.

Olusi SO (2002). Obesity is an independent risk factor for plasma lipid peroxidation and depletion of erythrocyte cytoprotectic enzymes in humans. *Int J Obes Relat Metab Disord* 26, 1159–1164.

Page S, Fischer C, Baumgartner B, Haas M, Kreusel U, Loidl G, Hayn M, Ziegler-Heitbrock HW, Neumeier D & Brand K (1999). 4-Hydroxynonenal prevents NF-kappaB activation and tumor necrosis factor expression by inhibiting IkappaB phosphorylation and subsequent proteolysis. *J Biol Chem* 274, 11611–11618.

Pan J, Keffer J, Emami A, Ma X, Lan R, Goldman R & Chung F-L (2009). Acrolein-derived DNA adduct formation in human colon cancer cells: its role in apoptosis induction by docosahexaenoic acid. *Chem Res Toxicol* 22, 798–806.

Paradis V, Kollinger M, Fabre M, Holstege A, Poynard T & Bedossa P (1997). In situ detection of lipid peroxidation by-products in chronic liver diseases. *Hepatology* 26, 135–142.

Peyroux J & Sternberg M (2006). Advanced glycation endproducts (AGEs): Pharmacological inhibition in diabetes. *Pathol Biol* 54, 405–419.

Pillon NJ, Croze ML, Vella RE, Soulère L, Lagarde M & Soulage CO (2012). The Lipid Peroxidation By-Product 4-Hydroxy-2-Nonenal (4-HNE) Induces Insulin Resistance in Skeletal Muscle through Both Carbonyl and Oxidative Stress. Endocrinology 153, 2099-2111.

Pillon NJ, Soulère L, Vella RE, Croze M, Caré BR, Soula HA, Doutheau A, Lagarde M & Soulage CO (2010). Quantitative structure-activity relationship for 4-hydroxy-2-alkenal induced cytotoxicity in L6 muscle cells. *Chem Biol Interact* 188, 171–180.

Pillon NJ, Vella RE, Soulère L, Becchi M, Lagarde M & Soulage CO (2011). Structural and functional changes in human insulin induced by the lipid peroxidation byproducts 4-hydroxy-2-nonenal and 4-hydroxy-2-hexenal. *Chem Res Toxicol* 24, 752–762.

Poirier M, Fournier M, Brousseau P & Morin A (2002). Effects of volatile aromatics, aldehydes, and phenols in tobacco smoke on viability and proliferation of mouse lymphocytes. *J Toxicol Environ Health Part A* 65, 1437–1451.

Reagan LP, Magariños AM, Yee DK, Swzeda LI, Van Bueren A, McCall AL & McEwen BS (2000). Oxidative stress and HNE conjugation of GLUT3 are increased in the hippocampus of diabetic rats subjected to stress. *Brain Res* 862, 292–300.

Riahi Y, Sin-Malia Y, Cohen G, Alpert E, Gruzman A, Eckel J, Staels B, Guichardant M & Sasson S (2010). The natural protective mechanism against hyperglycemia in vascular endothelial cells: roles of the lipid peroxidation product 4-hydroxydodecadienal and peroxisome proliferator-activated receptor delta. *Diabetes* 59, 808–818.

Riboulet-Chavey A, Pierron A, Durand I, Murdaca J, Giudicelli J & Van Obberghen E (2006). Methylglyoxal impairs the insulin signaling pathways independently of the formation of intracellular reactive oxygen species. *Diabetes* 55, 1289–1299.

Roy J, Pallepati P, Bettaieb A & Averill-Bates DA (2010). Acrolein induces apoptosis through the death receptor pathway in A549 lung cells: role of p53. *Can J Physiol Pharmacol* 88, 353–368.

Rudich A, Tirosh A, Potashnik R, Hemi R, Kanety H & Bashan N (1998). Prolonged oxidative stress impairs insulin-induced GLUT4 translocation in 3T3-L1 adipocytes. *Diabetes* 47, 1562–1569.

Rudich A, Tirosh A, Potashnik R, Khamaisi M & Bashan N (1999). Lipoic acid protects against oxidative stress induced impairment in insulin stimulation of protein kinase B and glucose transport in 3T3-L1 adipocytes. *Diabetologia* 42, 949–957.

Ruef J, Moser M, Bode C, Kubler W & Runge MS (2001). 4-hydroxynonenal induces apoptosis, NF-kappaB-activation and formation of 8-isoprostane in vascular smooth muscle cells. *Basic Res Cardiol* 96, 143–150.

Russell AP, Gastaldi G, Bobbioni-Harsch E, Arboit P, Gobelet C, Dériaz O, Golay A, Witztum JL & Giacobino J-P (2003). Lipid peroxidation in skeletal muscle of obese as compared to endurance-trained humans: a case of good vs. bad lipids? *FEBS Lett* 551, 104–106.

Sakuraba H, Mizukami H, Yagihashi N, Wada R, Hanyu C & Yagihashi S (2002). Reduced beta-cell mass and expression of oxidative stress-related DNA damage in the islet of Japanese Type II diabetic patients. *Diabetologia* 45, 85–96.

Schaur RJ (2003). Basic aspects of the biochemical reactivity of 4-hydroxynonenal. *Mol Aspects Med* 24, 149–159.

Selley ML (2004). Increased (E)-4-hydroxy-2-nonenal and asymmetric dimethylarginine concentrations and decreased nitric oxide concentrations in the plasma of patients with major depression. *J Affect Disord* 80, 249–256.

Shearn CT, Fritz KS, Reigan P & Petersen DR (2011). Modification of Akt2 by 4-hydroxynonenal inhibits insulin-dependent Akt signaling in HepG2 cells. *Biochemistry* 50, 3984–3996.

Shibata M, Hakuno F, Yamanaka D, Okajima H, Fukushima T, Hasegawa T, Ogata T, Toyoshima Y, Chida K, Kimura K, Sakoda H, Takenaka A, Asano T & Takahashi S-I (2010). Paraquat-induced oxidative stress represses phosphatidylinositol 3-kinase activities leading to impaired glucose uptake in 3T3-L1 adipocytes. *J Biol Chem* 285, 20915–20925.

Singh I, Carey AL, Watson N, Febbraio MA & Hawley JA (2008a). Oxidative stress-induced insulin resistance in skeletal muscle cells is ameliorated by gamma-tocopherol treatment. *Eur J Nutr* 47, 387–392.

Singh SP, Niemczyk M, Saini D, Awasthi YC, Zimniak L & Zimniak P (2008b). Role of the electrophilic lipid peroxidation product 4-hydroxynonenal in the development and maintenance of obesity in mice. *Biochemistry* 47, 3900–3911.

Singh SP, Niemczyk M, Zimniak L & Zimniak P (2009). Fat accumulation in Caenorhabditis elegans triggered by the electrophilic lipid peroxidation product 4-hydroxynonenal (4-HNE). *Aging (Albany NY)* 1, 68–80.

Soares AF, Guichardant M, Cozzone D, Bernoud-Hubac N, Bouzaidi-Tiali N, Lagarde M & Geloen A (2005). Effects of oxidative stress on adiponectin secretion and lactate production in 3T3-L1 adipocytes. *Free Radic Biol Med* 38, 882–889.

Srivastava S, Dixit BL, Cai J, Sharma S, Hurst HE, Bhatnagar A & Srivastava SK (2000). Metabolism of lipid peroxidation product, 4-hydroxynonenal (HNE) in rat erythrocytes: role of aldose reductase. *Free Radic Biol Med* 29, 642–651.

Syslova K, Kacer P, Kuzma M, Najmanova V, Fenclova Z, Vlckova S, Lebedova J & Pelclova D (2009). Rapid and easy method for monitoring oxidative stress markers in body fluids of patients with asbestos or silica-induced lung diseases. *J Chromatogr B Analyt Technol Biomed Life Sci*; DOI: 10.1016/j.jchromb.2009.06.008.

Tan Y, Ichikawa T, Li J, Si Q, Yang H, Chen X, Goldblatt CS, Meyer CJ, Li X, Cai L & Cui T (2011). Diabetic downregulation of Nrf2 activity via ERK contributes to oxidative stress-induced insulin resistance in cardiac cells in vitro and in vivo. *Diabetes* 60, 625–633.

Tanel A & Averill-Bates DA (2007). P38 and ERK mitogen-activated protein kinases mediate acrolein-induced apoptosis in Chinese hamster ovary cells. *Cell Signal* 19, 968–977.

Thirunavukkarasu V & Anuradha CV (2004). Influence of alpha-lipoic acid on lipid peroxidation and antioxidant defence system in blood of insulin-resistant rats. *Diabetes Obes Metab* 6, 200–207.

Toyokuni S, Yamada S, Kashima M, Ihara Y, Yamada Y, Tanaka T, Hiai H, Seino Y & Uchida K (2000). Serum 4-hydroxy-2-nonenal-modified albumin is elevated in patients with type 2 diabetes mellitus. *Antioxid Redox Signal* 2, 681–685.

Traverso N, Menini S, Cosso L, Odetti P, Albano E, Pronzato MA & Marinari UM (1998). Immunological evidence for increased oxidative stress in diabetic rats. *Diabetologia* 41, 265–270.

Traverso N, Menini S, Odetti P, Pronzato MA, Cottalasso D & Marinari UM (2002). Diabetes impairs the enzymatic disposal of 4-hydroxynonenal in rat liver. *Free Radic Biol Med* 32, 350–359.

Trevisan M, Browne R, Ram M, Muti P, Freudenheim J, Carosella AM & Armstrong D (2001). Correlates of markers of oxidative status in the general population. *Am J Epidemiol* 154, 348–356.

Uchida K, Shiraishi M, Naito Y, Torii Y, Nakamura Y & Osawa T (1999). Activation of stress signaling pathways by the end product of lipid peroxidation. 4-hydroxy-2-nonenal is a potential inducer of intracellular peroxide production. *J Biol Chem* 274, 2234–2242.

Unoki H, Bujo H, Yamagishi S, Takeuchi M, Imaizumi T & Saito Y (2007). Advanced glycation end products attenuate cellular insulin sensitivity by increasing the generation of intracellular reactive oxygen species in adipocytes. *Diabetes Res Clin Pract* 76, 236–244.

Vaillancourt F, Fahmi H, Shi Q, Lavigne P, Ranger P, Fernandes JC & Benderdour M (2008). 4-Hydroxynonenal induces apoptosis in human osteoarthritic chondrocytes: the protective role of glutathione-S-transferase. *Arthritis Res Ther* 10, R107.

Vincent HK, Bourguignon C & Vincent KR (2006). Resistance training lowers exercise-induced oxidative stress and homocysteine levels in overweight and obese older adults. *Obesity (Silver Spring)* 14, 1921–1930.

Vincent HK, Bourguignon CM, Weltman AL, Vincent KR, Barrett E, Innes KE & Taylor AG (2009). Effects of antioxidant supplementation on insulin sensitivity, endothelial adhesion molecules, and oxidative stress in normal-weight and overweight young adults. *Metab Clin Exp* 58, 254–262.

Vincent HK, Innes KE & Vincent KR (2007). Oxidative stress and potential interventions to reduce oxidative stress in overweight and obesity. *Diabetes Obes Metab* 9, 813–839.

Vindis C, Escargueil-Blanc I, Elbaz M, Marcheix B, Grazide M-H, Uchida K, Salvayre R & Nègre-Salvayre A (2006). Desensitization of platelet-derived growth factor receptor-beta by oxidized lipids in vascular cells and atherosclerotic lesions: prevention by aldehyde scavengers. *Circ Res* 98, 785–792.

Wang L, Sun Y, Asahi M & Otsu K (2011). Acrolein, an environmental toxin, induces cardiomyocyte apoptosis via elevated intracellular calcium and free radicals. *Cell Biochem Biophys* 61, 131–136.

Wang Z, Dou X, Gu D, Shen C, Yao T, Nguyen V, Braunschweig C & Song Z (2012). 4-Hydroxynonenal differentially regulates adiponectin gene expression and secretion via activating PPARγ and accelerating ubiquitin-proteasome degradation. *Mol Cell Endocrinol* 349, 222–231.

Wellen KE & Hotamisligil GS (2003). Obesity-induced inflammatory changes in adipose tissue. *J Clin Invest* 112, 1785–1788.

Wierusz-Wysocka B, Wysocki H, Byks H, Zozulinska D, Wykretowicz A & Kazmierczak M (1995). Metabolic control quality and free radical activity in diabetic patients. *Diabetes Res Clin Pract* 27, 193–197.

Yadav UCS, Ramana KV, Awasthi YC & Srivastava SK (2008). Glutathione level regulates HNE-induced genotoxicity in human erythroleukemia cells. *Toxicol Appl Pharmacol* 227, 257–264.

Yamashita C, Hayashi T, Mori T, Matsumoto C, Kitada K, Miyamura M, Sohmiya K, Ukimura A, Okada Y, Yoshioka T, Kitaura Y & Matsumura Y (2010). Efficacy of olmesartan and nifedipine on recurrent hypoxia-induced left ventricular remodeling in diabetic mice. *Life Sci* 86, 322–330.

Yoritaka A, Hattori N, Uchida K, Tanaka M, Stadtman ER & Mizuno Y (1996). Immunohistochemical detection of 4-hydroxynonenal protein adducts in Parkinson disease. *Proc Natl Acad Sci U S A* 93, 2696–2701.

Zafarullah M, Li WQ, Sylvester J & Ahmad M (2003). Molecular mechanisms of N-acetylcysteine actions. *Cell Mol Life Sci* 60, 6–20.

Zarrouki B, Soares AF, Guichardant M, Lagarde M & Geloen A (2007). The lipid peroxidation end-product 4-HNE induces COX-2 expression through p38MAPK activation in 3T3-L1 adipose cell. *FEBS Lett* 581, 2394–2400.

Zhang H & Forman HJ (2008). Acrolein induces heme oxygenase-1 through PKC-delta and PI3K in human bronchial epithelial cells. *Am J Respir Cell Mol Biol* 38, 483–490.

Zimniak P (2010). 4-Hydroxynonenal and fat storage: A paradoxical pro-obesity mechanism? *Cell Cycle* 9, 3393–3394.

Chapter 7

Lipid Peroxidation and Reperfusion Injury in Hypertrophied Hearts

Juliana C. Fantinelli, Ignacio A. Pérez Núñez,
Luisa F. González Arbeláez and Susana M. Mosca

Additional information is available at the end of the chapter

1. Introduction

Oxidative stress is characterized by an imbalance between increased exposure to reactive oxygen species (ROS), and antioxidant defenses, comprised of both small molecular weight antioxidants like glutathione, and antioxidant enzymes like superoxide dismutase. ROS cause direct damage to critical biomolecules including DNA, lipids, and proteins. Oxidative stress has been involved in the genesis of hypertension [1, 2] and implicated in the mechanisms of reversible postischemic contractile dysfunction (myocardial stunning), microvascular dysfunction, arrhythmias and cell death [3-6]. In spontaneously hypertensive rats (SHR) there are few reports showing the protective action of antioxidants against ischemia-reperfusion injury [7-9] and specifically in regard to the effects of the scavenger N-(2-mercaptopropionyl)-glycine (MPG) these have not been yet examined.

Ischemic preconditioning (IP) is acknowledged to be an endogenous mechanism of cardioprotection against ischemia and reperfusion injury [10-11]. This intervention is based in that one or more brief periods of ischemia applied previous to a prolonged ischemic period exert beneficial effects on myocardium attenuating the deleterious effects observed in the reperfused myocardium. Although there are some studies showing the beneficial effects of IP in hypertensive animals [12-15], under certain circumstances the effectiveness of that intervention is questioned [16-18]. A recent investigation performed in our laboratory shows that a single cycle of IP attenuated the myocardial stunning produced by 20-min global ischemia in SHR [19] and decreased the lipid peroxidation. Whether this protective action of IP is operating at more extended ischemic period and involves changes in oxidative stress in this rats strain is a point that needs to be clarified.

Therefore, the aim of the present study was to determine if alterations of lipid peroxidation and endogenous antioxidants are linked to myocardial and vascular postischemic damage in ischemic control, preconditioned and MPG treated hearts from SHR.

2. Material and methods

2.1. Isolated heart preparation

Experiments were performed in SHR of 5-month-old following the Guide for the Care and Use of Laboratory Animals published by the US National Institutes of Health (NIH Publication No. 85-23, revised in 1996). Beginning at 12 weeks of age, systolic blood pressure (SBP) was measured weekly in all animals by the standard tail-cuff method [20] following the modifications detailed in a recent paper by Fritz and Rinaldi [21]. Rats were anesthetized with an intraperitoneal injection of sodium pentobarbital (60 mg/kg body wt). The heart was rapidly excised and perfused by the non-recirculating Langendorff technique with Ringer's solution containing (in mmol/L): 118 NaCl, 4.7 KCl, 1.2 $MgSO_4$, 1.35 $CaCl_2$, 20 $NaCO_3H$ and 11.1dextrose. The buffer was saturated with a mixture of 95% O_2-5% CO_2, had a pH 7.4, and was maintained at 37°C. The conductive tissue in the atrial septum was damaged with a fine needle to achieve atrioventricular block, and the right ventricle was paced at 280 ± 10 beats/min. A latex balloon tied to the end of a polyethylene tube was passed into the left ventricle through the mitral valve; the opposite end of the tube was then connected to a Statham P23XL pressure transducer. The balloon was filled with water to provide an end-diastolic pressure (LVEDP) of 8-12 mmHg and this volume remained unchanged for the rest of the experiment. Coronary perfusion pressure (CPP) was monitored at the point of cannulation of the aorta and adjusted to approximately 70 mmHg. Coronary flow (CF), controlled with a peristaltic pump, was 11 ± 2 mL/min. Left ventricular pressure (LVP) and CPP data were acquired by using an analog-to-digital converter and acquisition software (Chart V4.2.3 ADInstruments).

2.2. Experimental protocols

After 10 min of stabilization, hearts from SHR were assigned to the following experimental protocols (Fig. 1):

Non-ischemic control hearts (NIC): Hearts were perfused for 3 hs without any treatment.

Ischemic control hearts (IC): Hearts were subjected to 35 min or 50 min of normothermic global ischemia followed by 2 hours of reperfusion. Global ischemia was induced by stopping the perfusate inflow line and the heart was placed in a saline bath held at 37°C.

Ischemic preconditioning (IP1): A single cycle of 5-min ischemia and 10-min reperfusion was applied previous to the 35-min and 50-min ischemic periods followed by 2-hour reperfusion.

Ischemic preconditioning (IP3): Three cycles of 2-min f ischemia and 5-min reperfusion was applied prior to the 50-min ischemic period followed by 2-hour reperfusion. Previous experiments performed by us showed that three cycles are the fewest for achieving myocardial protection of SHR when global ischemia was extended to 50 min.

MPG: Hearts were treated 10 min before ischemia and during the first 10 min of reperfusion with N-(2-mercaptopropionyl)-glycine (MPG) 2 mM. The administration time for MPG was chosen to attenuate the ROS production during ischemia and reperfusion. The dose was selected according previous experiments performed in our laboratory [22].

Additional experiments were performed (n = 6 for each protocol) to assess the biochemical parameters.

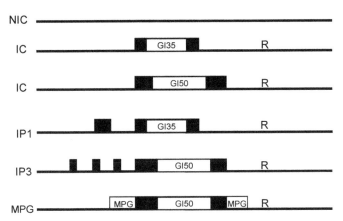

Figure 1. Scheme of the experimental protocols.

2.3. Infarct size determination

Infarct size was assessed by the widely validated triphenyltetrazolium chloride (TTC) staining technique [23]. At the end of reperfusion, atrial and right ventricular tissues were excised and left ventricle (VI) was frozen. The freeze VI was cut into six transverse slices, which were incubated for 5 minutes at 37°C in a 1% solution of triphenyltetrazolium chloride (TTC). To measure myocardial infarction, the slices were weighed and scanned. The infarcted (pale) and viable ischemic/reperfused (red) areas were measured by computed planimetry (Scion Image 1.62; Scion Corp., Frederick, Maryland, USA). Infarct weights were calculated as (A1 × W1) + (A2 × W2) + (A3 × W3) + (A4 × W4) + (A5 × W5) + (A6 × W6), where A is the infarct area for the slice and W is the weight of the respective section. Infarct size was expressed as a percentage of the total area (area at risk, AAR) [24].

2.4. Systolic and diastolic function

Myocardial contractility was assessed by the left ventricular developed pressure (LVDP), obtained by subtracting LVEDP to LVP peak, and maximal velocity of contraction (+dP/dt$_{max}$). The diastolic function was evaluated through LVEDP.

2.5. Assessment of coronary resistance (CR)

CR was calculated as a quotient between CPP and CF and expressed as difference between the values obtained at the end of reperfusion period and that observed in the preischemic period.

2.6. Preparation of tissue homogenate

At the end of reperfusion a portion of VI was homogenized in 5 volume of 25 mM PO_4KH_2 - 140 mM ClK at pH = 7.4 with a Polytron homogenizer. Aliquots of homogenate were used to assess reduced glutathione content (GSH) and thiobarbituric acid reactive substances (TBARS) as an index of lipid peroxidation. The remaining homogenate was centrifuged at 12,000 g for 5 min at 4° C and the supernatant stored at -70 $^{\circ}$C until superoxide dismutase (SOD) activity was assayed.

2.6.1. Assessment of reduced glutathione (GSH)

GSH was determined by Ellman's method [25]. This method was based on the reaction of GSH with 5, 5' dithiobis (2-nitrobenzoic acid) to give a compound that absorbs at 412 nm. GSH levels were expressed as µg/mg of protein.

2.6.2. Assessment of lipid peroxidation

TBARS concentration was determined in the supernatant following the Buege and Aust method's [26]. Absorbance at 535nm was measured and TBARS expressed in nmol/g of tissue using an extinction coefficient of 1.56×10^5 $M^{-1} cm^{-1}$.

2.6.3. Measurement of SOD cytosolic activity

SOD activity was measured by means of the nitroblue tetrazolium (NBT) method [27]. Briefly, the supernatant was added to the reaction mixture of NBT with xanthine-xanthine oxidase, and the SOD activity measured colorimetrically in the form of inhibitory activity toward blue formazan formation by SOD in the reaction mixture.

2.6.4. Protein determination

The protein concentration was evaluated by the Bradford method [28] using bovine serum albumin as a standard.

2.6.5. Correlations

The relationships between TBARS, GSH and infarct size and CR were determined by linear regression (equation $y = a + b \cdot x$).

2.7. Statistical analysis

Data are presented as mean ± SE and repeated measures of two-way analysis of variance (ANOVA) with Newman-Keuls test were used for multiple comparisons among groups. Relationships were tested for significance using the Pearson correlation coefficient (r). A P value < 0.05 was considered significant.

3. Results

Fig. 2 shows the infarct size in ischemic control and preconditioned hearts from SHR. In non-ischemic control hearts at the end of the 3-hour perfusion the infarct size was approximately 1 % of risk area. After 35-min global ischemia and 2-hour reperfusion, the infarct size was 35 ± 5 %, which was significantly decreased by one cycle of IP (IP1). When ischemia was extended to 50 min, the infarct size (58 ± 5 %) was not reduced by IP1 indicating that this preconditioning protocol is not adequate for protecting that rat strain against reperfusion injury. However, when a larger number of cycles (three in our case) were applied the hearts were protected and the infarct size diminished. A significant reduction of infarct size was also obtained when MPG was added to the perfusate during 10 min before 50-min ischemia and during the first 10 min of reperfusion.

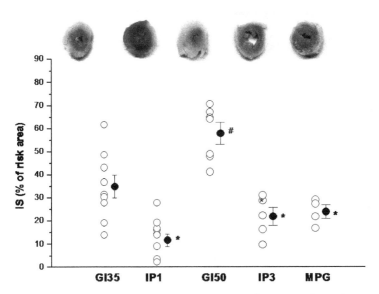

Figure 2. Infarct size (IS), expressed as percentage of risk area, in ischemic control (IC = GI35: 35-min global ischemia; GI50: 50-min global ischemia), preconditioned hearts (IP1= one cycle; IP3= three cycles) and MPG treatment. Note that hearts from SHR showed a higher IS at 50-min compared to 35-min GI. IP1 diminished the IS at 35-min ischemia but it was necessary to apply three cycles (IP3) to protect the hearts when the prolonged ischemia was extended to 50 min and that MPG decreased the IS at a similar value to IP3. * P < 0.05 with respect to GI; # P < 0.05 with respect to 35-min GI.

At the end of 3-hour non-ischemic hearts exhibited a decrease in contractility of approximately 10 %. After 35-min ischemia and 2-hour reperfusion contractility decreased approximately 90 % with respect to preischemic values. As it is depicted in Fig. 3 the recovery of systolic function was improved by both IP protocols. At the end of the reperfusion period, LVDP and +dP/dt$_{max}$ reached higher values than those obtained in ischemic control hearts. When ischemia was more prolonged (50 min) the postischemic recovery of contractility was scarce (LVDP and +dP/dt_max reached values of approximately 2 %) and it was significantly improved by IP3 and MPG treatment.

The diastolic stiffness characterized by LVEDP increased during 35-min and 50-min global ischemia and acquired greater values during reperfusion. These increases were attenuated by both IP protocols and MPG treatment. Fig. 4 shows the changes of LVEDP occurring at 50-min global ischemia in ischemic control and intervened hearts.

The increase in perfusion pressure at constant coronary flow resulted in an increase of coronary resistance. The increases (4.2 ± 0.4 and 7.0 ± 0.9 mmHg/ml x min^{-1} after 35-min and 50-min ischemia, respectively) were significantly attenuated by both IP protocols and MPG treatment (Fig. 5).

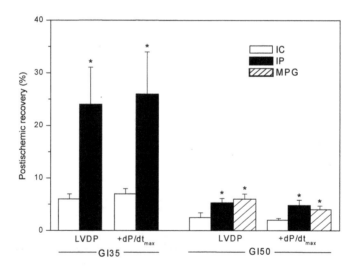

Figure 3. Values of left ventricular developed pressure (LVDP) and maximal velocity of contraction (+dP/dt_max) at the end of reperfusion period expressed as percentage of preischemic values, in ischemic control (IC = GI35: 35-min global ischemia; GI50: 50-min global ischemia), preconditioned hearts (IP) and MPG treatment. Observe that IP and MPG significantly improved the postischemic recovery of myocardial systolic function at 35-min and 50-min GI. * $P < 0.05$ with respect to IC

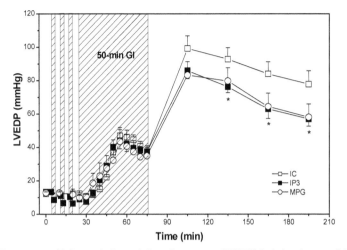

Figure 4. Time course of left ventricular end diastolic pressure (LVEDP) in ischemic control (IC = GI50: 50-min global ischemia), preconditioned hearts (IP) and MPG treatment. The three cycles of IP (IP3) and MPG attenuated in a similar manner the increase of LVEDP detected in IC hearts. * P < 0.05 with respect to IC.

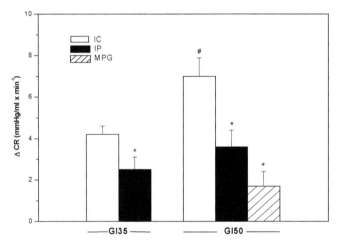

Figure 5. Changes of coronary resistance (CR) at the end of reperfusion in ischemic control (IC = GI35: 35-min global ischemia; GI50: 50-min global ischemia), preconditioned hearts (IP) and MPG treatment. The interventions attenuated the increase of CR detected in IC hearts being MPG the most effective. * P < 0.05 with respect to IC; # P < 0.05 with respect to GI35.

Given that an increase of ROS generation accompanied by a diminution of antioxidants may be responsible for myocardial reperfusion injury [29, 30], we next determined the impact of IP and MPG on myocardial GSH content, a marker of oxidative stress. Fig. 6 shows that

GSH content in non-ischemic hearts (2 ± 0.3 µg/mg prot) was significantly reduced by ischemia and reperfusion. A single or three cycles of IP and MPG treatment were able to preserve part of the GSH content.

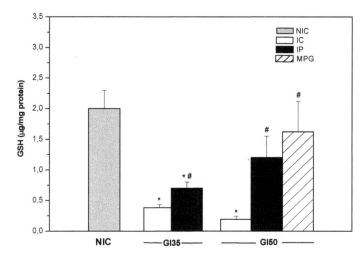

Figure 6. Myocardial reduced glutathione content (GSH, µg/mg protein) in non-ischemic control (NIC), ischemic control (IC = GI35: 35-min global ischemia; GI50: 50-min global ischemia) and preconditioned (IP) and MPG treated hearts. Observe that GSH levels decreased after ischemia and reperfusion in both ischemic periods and were partially preserved by IP and MPG. * P < 0.05 with respect to NIC; # P < 0.05 with respect to IC.

Moreover, the SOD cytosolic activity increased in ischemic controls hearts and significantly decreased in all intervened hearts (Fig. 7). Both parameters (GSH and SOD) are indicating the presence of oxidative stress caused by ischemia-reperfusion which may be attenuated by IP and MPG treatment.

Since ROS induce membrane lipid peroxidation [29], we determined TBARS content of untreated and treated ischemic-reperfused hearts. Although TBARS determination suffers from potential artifacts associated with sampling, storage and problems caused by the complexity of the biological systems, being easy and reproducible, it is one of the most widely used indexes for assessing oxidative stress. There was an increase in myocardial TBARS content in hearts submitted to ischemia and reperfusion detecting a higher value at 50-min compared to 35-min global ischemia. Preconditioned and MPG treated hearts exhibited lower TBARS levels (Fig. 8).

The analysis of data of the different interventions showed the presence of significant positive correlations TBARS vs IS (Fig. 9, A panel; r = 0,47) and TBARS vs CR (Fig. 9, B panel; n = 0,45) and negative correlations GSH vs IS (Fig. 10, A panel; n = 0,41) and GSH vs CR (Fig. 10, B panel; n = 0,40) in isolated hearts from SHR.

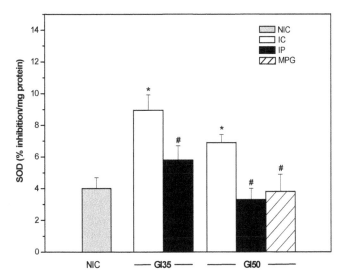

Figure 7. Myocardial SOD cytosolic activity (SOD, % inhibition/mg protein) in non-ischemic control (NIC), ischemic control (IC = GI35: 35-min global ischemia; GI50: 50-min global ischemia), preconditioned (IP) and MPG treated hearts. Note that SOD cytosolic activity increased after 35-min or 50-min GI in comparison to NIC. These increases were attenuated by both interventions (IP and MPG). * P < 0.05 with respect to NIC; # P < 0.05 with respect to IC.

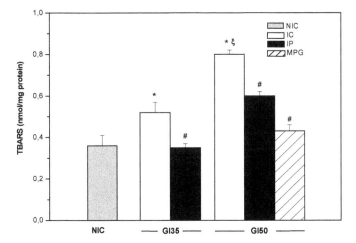

Figure 8. Myocardial thiobarbituric acid reactive substances (TBARS) concentration, expressed in nmol/mg protein in non-ischemic control (NIC), ischemic control (IC = GI35: 35-min global ischemia; GI50: 50-min global ischemia), preconditioned (IP) and MPG treated hearts. An increase of TBARS occurred at the end of reperfusion after the two ischemic periods which were attenuated by IP and MPG. * P < 0.05 with respect to NIC ; # P < 0.05 with respect to IC; ç P < 0.05 with respect to GI35.

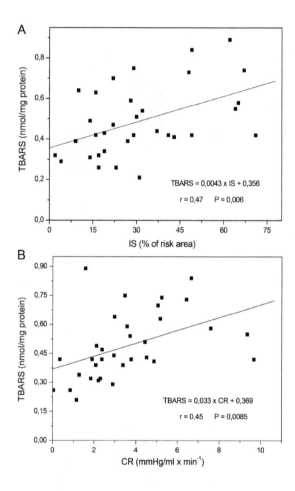

Figure 9. Relationship between TBARS and infarct size (IS, A panel) and TBARS and coronary resistance (CR, B panel) in all experimental situations. The resulting data were fitted to straight line by linear regression. Significant positive correlations between TBARS and IS and CR were found.

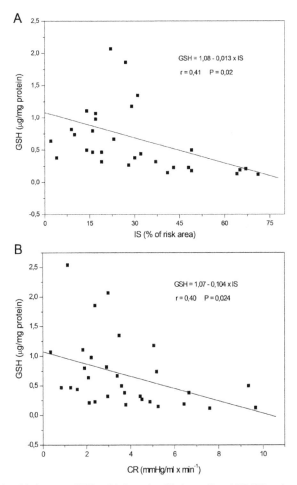

Figure 10. Relationship between GSH and infarct size (IS, A panel) and TBARS and coronary resistance (CR, B panel) in all experimental situations. The resulting data were fitted to straight line by linear regression. Significant negative correlations between GSH and IS and CR were found.

4. Discussion

To our knowledge, this is the first demonstration that the beneficial action of ischemic preconditioning and MPG against ischemia-reperfusion injury is similar in hearts from SHR and is associated with a mitigation of oxidative stress. Thus, our data show the existence of a positive correlation between TBARS concentration-used as an index of lipid peroxidation-and infarct size indicating that it will be found more infarct size when TBARS acquire higher values.

Simultaneously an inverse correlation was detected between GSH content and infarct size indicating that higher levels of GSH are associated to minor infarct size. Both variables (TBARS and GSH) suffered opposite changes due to a possible cause-effect relationship.

These results were also accompanied by changes of SOD cytosolic activity which showed lesser values in preconditioned and MPG treated hearts. Taken together, these data provide evidence to suggest that formation of lipoperoxides is a significant cause of ischemia and reperfusion injury and that the mechanism whereby IP and MPG confer cardioprotection involves, at least in part, an attenuation of those nocive products through a diminution of ROS release and/or production and an improvement of the endogenous antioxidants.

This study clearly shows that hearts from SHR suffer higher irreversible damage at 50-min compared to 35-min global ischemia accompanied with greater impairment of postichemic myocardial function. Thus, at the end of reperfusion the recovery of systolic function was scarce and diastolic stiffness significantly increased in ischemic control hearts. These alterations were attenuated by IP being one cycle of IP (IP1) effective when the ischemic period was 35 min and three cycles (IP3) when the ischemia was extended to 50 min. Thus, although the cardioprotective action of IP in hypertrophied hearts was previously reported [12-15] our study demonstrates that the optimum protocol of IP to protect SHR hearts must be selected according to the duration time of prolonged ischemia. Then, it seems to be possible that the number of IP cycles appears as other key factor for determining the efficacy of IP. Moreover MPG treated hearts in the same way that the preconditioned showed lesser infarct size and improved postischemic recovery of myocardial function in comparison to ischemic control hearts.

Hypertension is associated with an elevation of ROS and frequently with an impairment of endogenous antioxidant mechanisms [30]. These alterations have also been described during ischemia and reperfusion [3, 4, 31-33]. In this study, at the end of reperfusion after ischemic period cardiac tissue showed lesser GSH content, higher TBARS concentration and SOD cytosolic activity in comparison to non-ischemic control hearts. Major changes of GSH and TBARS were detected at 50-min compared to 35-min global ischemia. However, SOD cytosolic activity showed higher increase at 35 min of ischemia. This result may explain the lesser lipid peroxidation found in this experimental group. All these changes were partially reversed by both IP protocols and MPG treatment. Thus, GSH content was higher and SOD cytosolic activity was lower than the values observed in untreated hearts. The favorable changes in GSH and SOD cytosolic activity were reflected in the lower lipid peroxidation (decreased TBARS concentration) observed in preconditioned hearts and in those treated with MPG in comparison to ischemic control hearts. In other words the improvement of the antioxidant systems (SOD and GSH) by IP and MPG treatment were enough to attenuate the oxidative damage detected in untreated hearts. These results suggest that changes of lipid peroxidation and antioxidant systems would be sufficient to promote differences in the cell death and the attenuation of oxidative stress would be considered as a factor contributing to the cardioprotection by IP and MPG treatment in hearts from SHR.

On the other hand, a balance between the production of nitric oxide (NO) and ROS controls the endothelial function [34, 35]. When the NO production is normal its bioavailability may be reduced because of the oxidative inactivation by an excessive production of superoxide (O_2^-) in the vascular wall. The available data on the NO system in SHR are limited and apparently contradictory. Increased ROS in SHR have been demonstrated to enhance NO inactivation and reduce NO bioavailability [36], which contributes to the maintenance of hypertension. According to a previous study the peroxynitrite- product of NO and O_2^- combination- may also be involved in maintenance of the high levels of blood pressure in SHR [37. Furthermore in this rats strain was reported that the activity and/or expression of the different nitric oxide synthase (NOS) isoforms would be altered [38-40] which might act as a compensatory mechanism to maintain the production of bioactive NO in the face of increased oxidant stress [41]. In our study, ischemic control hearts showed an increase of coronary resistance at the end of reperfusion compared to pre-ischemic period which was greater after 50-min than 35-min global ischemia. These increases were attenuated by IP and MPG treatment being this last intervention the most effective. Thus, the beneficial effect of IP and MPG on coronary resistance would be attributed to a greater NO availability mediated by an attenuation of oxidative stress. This mechanism could explain the significant correlations between TBARS, GSH and coronary resistance found in this study and reinforces the idea that changes of oxidative stress constitute the basis of myocardial and vascular postischemic alterations.

5. Conclusions

We can conclude that the level of lipid peroxidation and antioxidant defenses are linked to reperfusion injury in hypertrophied hearts from SHR. The finding that IP and MPG reduce the postischemic myocardial and vascular injury as well as levels of TBARS and improve the endogenous antioxidants suggest that the decrease in ROS levels would be the common mechanism of cardioprotection of both interventions.

Author details

Juliana C. Fantinelli, Ignacio A. Pérez Núñez,
Luisa F. González Arbeláez and Susana M. Mosca
Centro de Investigaciones Cardiovasculares, Facultad de Ciencias Médicas,
Universidad Nacional de La Plata, La Plata, Buenos Aires, Argentina

Acknowledgement

This work was supported in part by the grant PICT 1046 from Agencia Nacional de Promoción Científica y Técnica of Argentina to Dr Susana M Mosca.

6. References

[1] Newaz MA, Nawal NN (1999) Effect of gamma-tocotrienol on blood pressure, lipid peroxidation and total antioxidant status in spontaneously hypertensive rats (SHR). Clin. Exp. Hypertens. 21: 1297-1313.

[2] Vaziri ND, Sica DA (2004) Lead-induced hypertension: role of oxidative stress. Curr. Hypertens. Rep. 6: 314–320.

[3] Griendling KK, FitzGerald GA (2003) Oxidative stress and cardiovascular injury: Part II: animal and human studies. Circulation 108: 2034-2040.

[4] Papaharalambus CA, Griendling KK (2007) Basic mechanisms of oxidative stress and reactive oxygen species in cardiovascular injury. Trends Cardiovasc. Med. 17: 48-54.

[5] Ravingerová T, Slezák J, Tribulová J, Džurba A, Uhrík B, Ziegelhöffer A (1999) Reactive oxygen species contribute to high incidence of reperfusion-induced arrhythmias in isolated rat heart. Life Sci. 65: 1927-1930.

[6] Kalaycioglu S, Sinci V, Imren Y, Oz E (1999) Metoprolol prevents ischemia–reperfusion injury by reducing lipid peroxidation. Jpn. Circ. J. 63:718–721.

[7] Abebe W, Liu JY, Wimborne H, Mozaffari MS (2010) Effects of chromium picolinate on vascular reactivity and cardiac ischemia-reperfusion injury in spontaneously hypertensive rats. Pharmacol. Rep. 62: 674-682.

[8] Potenza MA, Gagliardi S, De Benedictis L, Zigrino A, Tiravanti E, Colantuono G, Federici A, Lorusso L, Benagiano V, Quon MJ, Montagnani M (2009) Treatment of spontaneously hypertensive rats with rosiglitazone ameliorates cardiovascular pathophysiology via antioxidant mechanisms in the vasculature. Am. J. Physiol. Endocrinol. Metab. 297(3): 685-694.

[9] Sahna E, Deniz E, Bay-Karabulut A, Burma O (2008) Melatonin protects myocardium from ischemia-reperfusion injury in hypertensive rats: role of myeloperoxidase activity. Clin. Exp. Hyperten. 30: 673-681.

[10] Mosca SM, Gelpi RJ, Milei J, Fernandez Alonso G, Cingolani HE (1998) Is stunning prevented by ischemic preconditioning? Mol . Cell. Biochem. 186: 123-129.

[11] Yellon DM, Downey JM (2003) Preconditioning the myocardium: from cellular physiology to clinical cardiology. Physiol. Rev. 83: 1113–1151.

[12] Speechly-Dick ME, Baxter GF, Yellon DM (1994) Ischaemic preconditioning protects hypertrophied myocardium. Cardiovasc. Res. 28(7):1025-1029.

[13] Boutros A, Wang J (1995) Ischaemic preconditioning, adenosine and bethanechol protect spontaneously hypertensive isolated rat hearts. J. Pharmacol. Exp. Ther. 275: 1148-1156.

[14] Pantos CI, Davos CH, Carageorgiou HC, Varonos DV, Cokkinos DV (1996) Ischaemic preconditioning protects against myocardial dysfunction caused by ischaemia in isolated hypertrophied rat hearts. Basic Res. Cardiol. 91: 444-449.

[15] Nakagawa C, Asayama J, Katamura M, Matoba S, Keira N, Kawahara A, Tsuruyama K, Tanaka T, Kobara M, Akashi K, Ohta B, Tatsumi T, Nakagawa M (1997) Myocardial strech induced by increased left ventricular diastolic pressure preconditions isolated perfused hearts of normotensive and spontaneously hypertensive rats. Basic Res. Cardiol. 92: 410-416.

[16] Ebrahim Z, Yellon DM, Baxter GF (2007) Ischemic preconditioning is lost in aging hypertensive rat heart: independent effects of aging and longstanding hypertension. Exp. Gerontol. 42(8): 807-814.

[17] Ebrahim Z, Yellon DM, Baxter GF (2007) Attenuated cardioprotective response to bradykinin, but not classical ischaemic preconditioning, in DOCA-salt hypertensive left ventricular hypertrophy. Pharmacol. Res. 55(1): 42-48.

[18] Dai W, Simkhovich BZ, Kloner RA (2009) Ischemic preconditioning maintains cardioprotection in aging normotensive and spontaneously hypertensive rats. Exp. Gerontol. 44(5): 344-349.

[19] Fantinelli JC, Mosca SM (2007) Comparative effects of ischemic pre and postconditioning on ischemia-reperfusion injury in spontaneously hypertensive rats (SHR). Mol. Cell . Biochem. 296: 45-51.

[20] Buñag RA (1973) Validation in awake rats of tail cuff method for measuring systolic pressure. J.Appl. Physiol. 34: 279-282.

[21] Fritz M, Rinaldi G (2008) Blood pressure measurement with the tail-cuff method in Wistar and spontaneously hypertensive rats: Influence of adrenergic- and nitric oxide-mediated vasomotion. J. Pharmacol. Toxicol. Methods 58(3): 215-221.

[22] Fantinelli JC, Cingolani HE, Mosca SM (2006) Na$^+$/H$^+$ exchanger inhibition at the onset of reperfusion decreases the myocardial infarct size: role of ROS. Cardiovasc. Pathol. 15/4: 179-184.

[23] Fishbein MC, Meerbaum S, Rit J, Lando U, Kanmatsuse K, Mercier JC, Corday E, Ganz W (1981) Early phase acute myocardial infarct size quantification: validation of the triphenyl tetrazolium chloride tissue enzyme staining technique. Am. Heart. J. 101(5): 593– 600.

[24] Suzuki M, Sasaki N, Miki T, Sakamoto N, Ohmoto-Sekine Y, Tamagawa M, Seino S, Marbán E, Nakaya H (2001) Role of sarcolemmal KATP channels in cardioprotection against ischemia/reperfusion injury in mice. J. Clin.Invest. 109: 509–516.

[25] Sedlak J, Lindsay RH (1968) Estimation of total, protein-bound, and nonprotein sulfhydryl groups in tissue with Ellman's reagent. Anal. Biochem. 25:192-205.

[26] Buege JA, Aust SD (1978) Microsomal lipid peroxidation. Methods Enzymol. 52: 302-310.

[27] Beauchamp Ch, Fridovich I (1971) Superoxide dismutase: improved assays and an assay applicable to acrylamide gels. Anal. Biochem. 44: 276-287.

[28] Bradford MM (1976) A rapid and sensitive method for the quantitation of microgram quantitites of protein utilizing the principle of protein-dye binding. Anal. Biochem. 72: 248-254.

[29] Ambrosio G, Flaherty JT, Duilio C, Tritto I, Santoro G, Elia PP, Condorelli M, Chiariello M (1991) Oxygen radicals generated at reflow induce peroxidation of membrane lipids in reperfused hearts. J. Clin. Invest. 87: 2056-2066.

[30] Lassègue B, Griendling KK (2010) NADPH oxidases: functions and pathologies in the vasculature. Arterioscler. Thromb. Vasc. Biol. 30: 653-661.

[31] Bolli R, Jeroudi MO, Patel BS, DuBose CM, Lai EK, Roberts R, McCay PB (1989) Direct evidence that oxygen-derived free radicals contribute to postischemic myocardial dysfunction in the intact dog. Proc. Natl. Acad. Sci. USA 86: 4695-4699.

[32] Leichtweis S, Ji LL (2001) Glutathione deficiency intensifies ischaemia-reperfusion induced cardiac dysfunction and oxidative stress. Acta Physiol. Scand. 172(1): 1-10.

[33] Haramaki N, Stewart DB, Aggarwal S, Ikeda H, Reznick AZ, Packer L (1998) Networking antioxidants in the isolated rat heart are selectively depleted by ischemia-reperfusion. Free Rad. Biol. Med. 25(3): 329-339.

[34] Kodja G, Harrison D (1999) Interactions between NO and reactive oxygen species: pathophysiological importance in atherosclerosis, hypertension, diabetes and heart failure. Cardiovasc. Res. 43: 562-571.

[35] Cai , Harrison DG (2000) Endothelial dysfunction in cardiovascular diseases: the role of oxidant stress. Circ. Res. 87: 840-844.

[36] Shah AM, MacCarthy PA (2000) Paracrine and autocrine effects of nitric oxide in myocardial function. Pharmacol. Ther. 86: 49-86. [37] Kagota S, Tada Y, Kubota Y, Nejime N, Yamaguchi Y, Nakamura K, Kunitomo M, Shinozuka K (2007) Peroxynitrite is involved in the dysfunction of vasorelaxation in SHR/NDmcr-cp rats, spontaneously hypertensive rats. J. Cardiovasc. Pharmacol. 50: 677-685.

[38] Piech A, Dessy C, Havaux X, Feron O, Balligand JL (2003) Differential regulation of nitric oxide synthases and their allosteric regulators in heart and vessels of hypertensive rats. Cardiovasc. Res. 57: 456-467.

[39] Vaziri ND, Ni Z, Oveisi F (1998) Upregulation of renal and vascular nitric oxide synthase in young spontaneously hypertensive rats. Hypertension 31: 1248-1254.

[40] Nava E, Noll G, Luscher TF (1995) Increased activity of constitutive nitric oxide synthase in cardiac endothelium in spontaneous hypertension. Circulation 91: 2310-2313.

[41] Vapaatalo H, Mervaala E, Nurminen ML (2000) Role of endothelium and nitric oxide in experimental hypertension. Physiol. Res. 49: 1–10.

Chapter 8

Lipid Peroxidation in Hepatic Fibrosis

Ichiro Shimizu, Noriko Shimamoto,
Katsumi Saiki, Mai Furujo and Keiko Osawa

Additional information is available at the end of the chapter

1. Introduction

Hepatic fibrosis is a complex dynamic process which is mediated by death of hepatocytes and activation of hepatic stellate cells (HSCs). Lipid peroxidation including the generation of reactive oxygen species (ROS), transforming growth factor-β, and tumor necrosis factor-α can be implicated as a cause of hepatic fibrosis.

Damage of any etiology, such as infection with hepatitis C virus (HCV) or hepatitis B virus (HBV), heavy alcohol intake, and iron overload, to hepatocytes can produce oxygen-derived free radicals and other ROS derived from lipid peroxidative processes. Persistent production of ROS constitutes a general feature of a sustained inflammatory response and liver injury, once antioxidant mechanisms have been depleted. The major source of ROS production in hepatocytes is NADH and NADPH oxidases localized in mitochondria (Figure 1). NADH and NADPH oxidases leak ROS as part of its operation. Kupffer cells (hepatic resident macrophages), infiltrating inflammatory cells such as macrophages and neutrophils, and HSCs also produce ROS in the injured liver.

2. Oxidative stress in liver injury

ROS include the free radicals superoxide (O_2^-) and hydroxyl radical (HO^-) and non-radicals such as hydrogen peroxide (H_2O_2). A number of reactive nitrogen species including nitric oxide (NO) and peroxynitrite ($ONOO^-$) are also ROS. Superoxide production is mediated mainly by NADH oxidase. Hydrogen peroxide is more stable and membrane permeable in comparison to other ROS. Thus, hydrogen peroxide plays an important role in the intracellular signaling under physiological conditions. With respect to pathological actions, ROS participate in the development of liver disease. In this regard, hydrogen peroxide is converted into the hydroxyl radical, a harmful and highly reactive ROS, in the presence of transition metals such as iron (Figure 1). The hydroxyl radical is able to induce not only lipid peroxidation in the structure of membrane phospholipids, which results in irreversible modifications of cell membrane structure and function (membrane injury), but DNA cleavage (DNA injury) as well. Such a chain of events due to increased ROS production exceeding cellular antioxidant defense systems are called oxidative stress, inducing cell death.

Malondialdehyde (MDA) and 4-hydroxynonenal (HNE) (Figure 1), end products of lipid peroxidation, are discharged from destroyed hepatocytes into the space of Disse (Figure 2). Cells are well equipped to neutralize the effects of ROS by virtue of a series of the antioxidant protective systems, including superoxide dismutase (SOD), glutathione peroxidase, glutathione (GSH), and thioredoxin. Upon oxidation, GSH forms glutathione disulfide (GSSG).

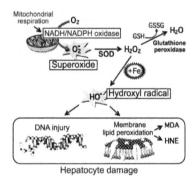

Figure 1. Oxidative stress and hepatocyte damage (Shimizu et al., 2012). A primary source of reactive oxygen species (ROS) production is mitochondrial NADPH/NADH oxidase. Hydrogen peroxide (H_2O_2) is converted to a highly reactive ROS, the hydroxyl radical, in the presence of transition metals such as iron (+Fe) and copper. The hydroxyl radical induces DNA cleavage and lipid peroxidation in the structure of membrane phospholipids, leading to cell death and discharge of products of lipid peroxidation, malondialdehyde (MDA) and 4-hydroxynonenal (HNE) into the space of Disse. Cells have comprehensive antioxidant protective systems, including SOD, glutathione peroxidase and glutathione (GSH). Upon oxidation, GSH forms glutathione disulfide (GSSG).

A single liver injury eventually results in an almost complete resolution, but the persistence of the original insult causes a prolonged activation of tissue repair mechanisms, thereby leading to hepatic fibrosis rather than to effective tissue repair. Hepatic fibrosis, or the excessive collagen deposition in the liver (see next section), is associated with oxidative stimuli and cell death. Cell death is a consequence of severe liver damage that occurs in many patients with chronic liver disease, regardless of the etiology such as HCV/HBV infection, heavy alcohol intake, and iron overload.

3. What is hepatic fibrosis?

At the cellular levels, origin of hepatic fibrosis is initiated by the damage of hepatocytes, followed by the accumulation of neutrophils and macrophages including Kupffer cells on the sites of injury and inflammation in the liver. When hepatocytes are continuously damaged, leading to cell death, production of extracellular matrix proteins such as collagens predominates over hepatocellular regeneration. Overproduced collagens are deposited in injured areas instead of destroyed hepatocytes. In other words, hepatic fibrosis is fibrous scarring of the liver in which excessive collagens build up along with the duration and extent of persistence of liver injury. Hepatic fibrosis itself causes no symptoms but can lead to the end-stage cirrhosis. In cirrhosis the failure to properly replace destroyed hepatocytes and the excessive collagen deposition to distort blood flow through the liver (portal hypertension) result in severe liver dysfunction. Cirrhosis is an important host-related risk factor for the development of hepatocellular carcinoma (HCC) in chronic hepatitis C and B, as well as a major factor predicting a poor response to interferon-based antiviral therapy in chronic hepatitis C. Staging of chronic liver disease by assessment of hepatic fibrosis always is a major function of prognostic interpretation of individual data including liver biopsy. Of the commonly used staging systems, the METAVIR fibrosis score has been widely used (Huwart et al., 2008). The stages are determined by both the quantity and location of the fibrosis. With this score, F0 represents no fibrosis; F1 (mild fibrosis), portal fibrosis without septa; F2 (moderate fibrosis), portal fibrous and few septa; F3 (severe fibrosis), numerous septa without cirrhosis; and F4, cirrhosis (Figure 3), the tissue is eventually composed of nodules surrounded completely by fibrosis.

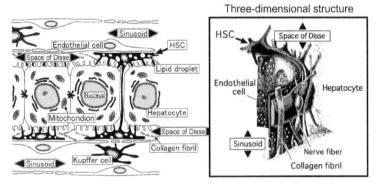

Figure 2. Schema of the sinusoidal wall of the liver (Shimizu et al., 2012). Schematic representation of hepatic stellate cells (HSCs) was based on the studies by Wake (Wake, 1999). Kupffer cells (hepatic resident macrophages) rest on fenestrated endothelial cells. HSCs are located in the space of Disse in close contact with endothelial cells and hepatocytes, functioning as the primary retinoid storage area. Collagen fibrils course through the space of Disse between endothelial cells and the cords of hepatocytes.

4. Activation of HSCs

Normal liver has a connective tissue matrix which includes collagen type IV (non-fibrillary), glycoproteins such as fibronectin and laminin, and proteoglycans such as heparan sulphate. These comprise the low density basement membrane in the space of Disse. Following liver injury there is a 3- to 8-fold increase in the extracellular matrix which is of a high density interstitial type, containing fibril-forming collagens (types 1 and III) as well as fibronectin, hyaluronic acid and proteoglycans. Collagen types 1 and III are major components of the extracellular matrix, which is principally produced by cells known as HSCs. HSCs are located in the space of Disse in close contact with hepatocytes and sinusoidal endothelial cells (Figure 2). Their three-dimensional structure consists of the cell body and several long and branching cytoplasmic processes (Wake, 1999). In the resting liver, HSCs have intracellular droplets containing retinoids. Retinoids refer to a group of chemical compound associated with vitamin A. HSCs contain approximately 50-80% of the whole body stores of retinoids (Blomhoff et al., 1990). In contrast, in the injured liver, HSCs are regarded as the primary target cells for inflammatory and oxidative stimuli, and they are proliferated, enlarged and transformed into myofibroblast-like cells. These HSCs are referred to as activated cells and are responsible for the overproduction of collagens during hepatic fibrosis to cirrhosis. This activation is accompanied by a loss of cellular retinoids, and the synthesis of α-smooth muscle actin (α-SMA), and large quantities of the major components of the extracellular matrix including collagen types I, III, and IV, fibronectin, laminin and proteoglycans. α-SMA is produced by activated HSCs (myofibroblast-like cells) but not by resting (quiescent) HSCs, thereby a marker of HSC activation. Moreover, activated HSCs produce ROS and transforming growth factor-β (TGF-β) (Figure 4). TGF-β is a major fibrogenic cytokine, regulating the production, degradation and accumulation of the extracellular matrix in hepatic fibrosis. TGF-β expression correlates with the extent of hepatic fibrosis (Castilla et al., 1991). This cytokine induces its own expression in activated HSCs, thereby creating a self-perpetuating cycle of events, referred to as an autocrine loop. TGF-β is also released in a paracrine manner from Kupffer cells, endothelial cells, and infiltrating inflammatory cells following liver injury. Similarly, ROS are produced by activated HSCs in response to ROS released from adjacent cells such as destroyed hepatocytes and activated Kupffer cells.

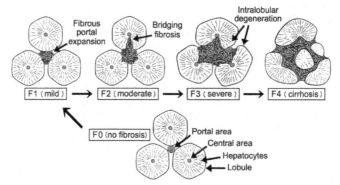

Figure 3. Stages of hepatic fibrosis in chronic hepatitis according to the five stages (0-4) of the METAVIR scoring system (1994). With this score, F0 represents no fibrosis; F1 (mild fibrosis), fibrous expansion of portal areas without septa; F2 (moderate fibrosis), fibrous septa extend to form bridges between adjacent vascular structures, both portal to portal and portal to central, occasional bridges; F3 (severe fibrosis), numerous bridges or septa without cirrhosis; and F4 (cirrhosis), the tissue is eventually composed of nodules surrounded completely by fibrosis.

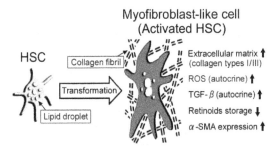

Figure 4. During liver injury, HSCs are proliferated, enlarged and transformed into myofibroblast-like cells (Shimizu, 2001). These activated HSCs produce large quantities of collagens, α-smooth muscle actin (α-SMA), ROS, and transforming growth factor-β (TGF-β), and lose cellular retinoids.

HSCs are activated mainly by ROS, products of lipid peroxidation (MDA and HNE) (Lee et al., 1995; Parola et al., 1993), and TGF-β, which are released from destroyed hepatocytes, activated Kupffer cells and infiltrating macrophages and neutrophils in the injured liver (Figure 5). In addition to ROS, exogenous TGF-β increases the production of ROS, particularly hydrogen peroxide, by HSCs, whereas the addition of hydrogen peroxide induces ROS and TGF-β production and secretion by HSCs (De Bleser et al., 1999). This so-called autocrine loop of ROS by HSCs is regarded as mechanism corresponding to the autocrine loop of TGF-β which HSCs produce in response to this cytokine with an increased collagen expression in the injured liver (Itagaki et al., 2005).

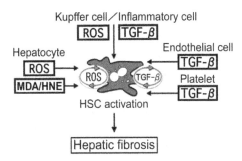

Figure 5. Activation of HSCs. HSCs are activated by such factors as ROS, lipid peroxidation products (MDA and HNE), and TGF-β released when adjacent cells including hepatocyte, Kupffer cells, and endothelial cells are injured. ROS and TGF-β are also produced by HSCs in response to exogenous ROS and TGF-β in an autocrine manner.

Other important factors for HSC activation are platelet-derived growth factor (PDGF) released from platelets, and endothelin-1 from endothelial cells. PDGF is the most potent mitogen. HSCs congregate in the area of injury, through proliferation and migration from elsewhere, in response to the release of PDGF and monocyte chemotactic peptide-1 (MCP-1). MCP-1 is produced by activated Kupffer cells and infiltrating macrophages and neutrophils. The number of activated HSCs also increases after liver injury (Enzan et al., 1994).

At the molecular levels, HSCs express the genes which encode for enzymes such as matrix metalloproteinase (MMP)-1 (interstitial collagenase) (Casini et al., 1994), which digests native fibrillar collagen types I and III, and MMP-2 (Milani et al., 1994), which digests denatured collagen types I and III and native collagen type IV, as well as tissue inhibitors of MMPs (TIMP)-1 and TIMP-2 (Iredale et al., 1992). Imbalance between matrix synthesis and degradation plays a major role in hepatic fibrosis (Shimizu, 2001). Matrix degradation depends upon the balance between MMPs, TIMPs and converting enzymes (MT1-MMP and stromelysin) (Li and Friedman, 1999). Collagen types I and III constitute the main framework of the so-called "fibrillar matrix". The space of Disse is a virtual space constituted by an extracellular matrix network composed of collagen type IV and non-collagenous components such as laminin. The large majority of collagen types III and IV, and laminin are synthesized by HSCs and endothelial cells, whereas all cell types synthesize small amounts of collagen type I. During hepatic fibrosis, however, HSCs become the major extracellular matrix producing cell type, with a predominant production of collagen type I (Maher and McGuire, 1990). In the resting liver, a balance between matrix synthesis and degradation exists, whereas, in the injured liver, the balance is disrupted. The net result of the changes during hepatocyte damage is increased degradation of the normal basement membrane collagen, and reduced degradation of interstitial-type collagens. The latter can be explained by increased TIMP-1 and TIMP-2 expressions relative to MMP-1. The degradative portion of the remodeling process is coordinated by MMPs and TIMPs.

5. Oxidative stress and intracellular pathway

Origin of hepatic fibrosis is initiated by the damage of hepatocytes, resulting in the recruitment of inflammatory cells and platelets, and activation of kupffer cells, with subsequent release of cytokines and growth factors. HSCs are the primary target cells for these inflammatory and oxidative stimuli, because during hepatic fibrosis, HSCs undergo an activation process to a myofibroblast-like cell, which represents the major matrix-producing cell. In the injured liver, hydrogen peroxide seems to act as a second messenger to regulate signaling events including mitogen activated protein kinase (MAPK) activation. The MAPK family includes three major subgroups, extracellular signal regulated kinase (ERK), p38 MAPK (p38), and c-Jun N-terminal kinase/stress activated protein kinase (JNK). MAPK participates in the intracellular signaling to: (1) induce the gene expression of redox sensitive transcription factors, such as activator protein-1 (AP-1) and nuclear factor κB (NF-κB) (Pinkus et al., 1996), (2) stimulate apoptosis (Clement and Pervaiz, 1999), and (3) modulate cell proliferation (Lundberg et al., 2000). ERK and JNK lie upstream of AP-1. JNK and p38 activation are more important in stress responses such as inflammation, which can also activate NF-κB. AP-1 and NF-κB induce the expression of multiple genes involved in inflammation and oxidative stress response, cell death and fibrosis, including proinflammatory cytokines such as tumor necrosis factor-α (TNF-α), interleukin-1 and interleukin-6 and growth factors such as PDGF and TGF-β. TGF-β is a major fibrogenic cytokine, acting as a paracrine and autocrine (from HSCs) mediator as already noted. TGF-β

triggers and activates the proliferation, enlargement and transformation of HSCs, but it exerts its inhibitory effect on hepatocyte proliferation (Nakamura et al., 1985).

Since many cytokines exert growth factor like activity, in addition to their specific proinflammatory effects, the distinction between cytokines and growth factors is somewhat artificial. No growth factor or cytokine acts independently. The injured liver, predominantly Kupffer cells and infiltrating macrophages and neutrophils, produces TNF-α, interleukin-1 and interleukin-6. These proinflammatory cytokines may also also inhibit hepatic regeneration. In particular, TNF-α plays a dichotomous role in the liver, where it not only induces hepatocyte proliferation and liver regeneration but also acts as a mediator of cell death (Schwabe and Brenner, 2006). During TNF-α-induced apoptosis in hepatocytes, hydrogen peroxide is an important mediator of cell death (Bohler et al., 2000).

In liver injury of hepatitis virus infection, transgenic mice expressing HBsAg exhibit the generation of oxidative stress and DNA damage, leading to the progression of hepatic fibrosis and carcinogenesis (Hagen et al., 1994; Nakamoto et al., 2004). In addition, HBV X protein changes the mitochondrial transmembrane potential and increases ROS production in the liver (Waris et al., 2001). Moreover, structural and non-structural (NS) proteins of HCV are involved in the production of ROS in an infected liver. HCV core protein is associated with increased ROS, decreased intracellular and/or mitochondrial glutathione content, and increased levels of lipid peroxidation products (Moriya et al., 2001). Glutathione is an antioxidant. NS3 protein of HCV activates NADPH oxidase in Kupffer cells to increase production of ROS, which can exert oxidative stress on nearby cells (Thoren et al., 2004).

6. Conclusion

Hepatic fibrosis is a complex dynamic process which is mediated by death of hepatocytes and activation of HSCs. Lipid peroxidation including the generation of ROS, TGF-β, and TNF-α can be implicated as a cause of hepatic fibrosis. HSCs are regarded as the primary target cells for inflammatory stimuli, and produce extracellular matrix components. HSCs are activated by such factors as ROS, lipid peroxidation products (MDA and HNE), and TGF-β released when adjacent cells including hepatocyte, Kupffer cells, and endothelial cells are injured. ROS and TGF-β are also produced by HSCs in response to exogenous ROS and TGF-β in an autocrine manner. During TNF-α-induced death in hepatocytes, ROS is an important mediator of cell death. The most common cause of hepatic fibrosis is currently chronic HCV/HBV infection.

Understanding the basic mechanisms underlying the ROS-mediated fibrogenesis provides valuable information on the search for effective antifibrogenic therapies.

Author details

Ichiro Shimizu*, Noriko Shimamoto, Katsumi Saiki, Mai Furujo and Keiko Osawa
Showa Clinic, Shin Yokohama, Kohoku-ku, Yokohama, Kanagawa, Japan

* Corresponding Author

7. References

The French METAVIR Cooperative Study Group (1994). Intraobserver and interobserver variations in liver biopsy interpretation in patients with chronic hepatitis C. Hepatology 20:15-20.

Blomhoff R, Green MH, Berg T & Norum KR . (1990). Transport and storage of vitamin A. Science 250:399-404.

Bohler T, Waiser J, Hepburn H, Gaedeke J, Lehmann C, Hambach P, Budde K & Neumayer HH . (2000). TNF-alpha and IL-1alpha induce apoptosis in subconfluent rat mesangial cells. Evidence for the involvement of hydrogen peroxide and lipid peroxidation as second messengers. Cytokine 12:986-991.

Casini A, Ceni E, Salzano R, Milani S, Schuppan D & Surrenti C . (1994). Acetaldehyde regulates the gene expression of matrix-metalloproteinase- 1 and -2 in human fat-storing cells. Life Sci 55:1311-1316.

Castilla A, Prieto J & Fausto N . (1991). Transforming growth factors beta 1 and alpha in chronic liver disease. Effects of interferon alfa therapy. N Engl J Med 324:933-940.

Clement MV & Pervaiz S . (1999). Reactive oxygen intermediates regulate cellular response to apoptotic stimuli: an hypothesis. Free Radic Res 30:247-252.

De Bleser PJ, Xu G, Rombouts K, Rogiers V & Geerts A . (1999). Glutathione levels discriminate between oxidative stress and transforming growth factor-beta signaling in activated rat hepatic stellate cells. J Biol Chem 274:33881-33887.

Enzan H, Himeno H, Iwamura S, Saibara T, Onishi S, Yamamoto Y & Hara H . (1994). Immunohistochemical identification of Ito cells and their myofibroblastic transformation in adult human liver. Virchows Arch 424:249-256.

Hagen TM, Huang S, Curnutte J, Fowler P, Martinez V, Wehr CM, Ames BN & Chisari FV . (1994). Extensive oxidative DNA damage in hepatocytes of transgenic mice with chronic active hepatitis destined to develop hepatocellular carcinoma. Proc Natl Acad Sci U S A 91:12808-12812.

Huwart L, Sempoux C, Vicaut E, Salameh N, Annet L, Danse E, Peeters F, ter Beek LC, Rahier J, Sinkus R, Horsmans Y & Van Beers BE . (2008). Magnetic resonance elastography for the noninvasive staging of liver fibrosis. Gastroenterology 135:32-40.

Iredale JP, Murphy G, Hembry RM, Friedman SL & Arthur MJ . (1992). Human hepatic lipocytes synthesize tissue inhibitor of metalloproteinases-1. Implications for regulation of matrix degradation in liver. J Clin Invest 90:282-287.

Itagaki T, Shimizu I, Cheng X, Yuan Y, Oshio A, Tamaki K, Fukuno H, Honda H, Okamura Y & Ito S . (2005). Opposing effects of oestradiol and progesterone on intracellular pathways and activation processes in the oxidative stress induced activation of cultured rat hepatic stellate cells. Gut 54:1782-1789.

Lee KS, Buck M, Houglum K & Chojkier M . (1995). Activation of hepatic stellate cells by TGF alpha and collagen type I is mediated by oxidative stress through c-myb expression. J Clin Invest 96:2461-2468.

Li D & Friedman SL . (1999). Liver fibrogenesis and the role of hepatic stellate cells: new insights and prospects for therapy. J Gastroenterol Hepatol 14:618-633.

Lundberg AS, Hahn WC, Gupta P & Weinberg RA . (2000). Genes involved in senescence and immortalization. Curr Opin Cell Biol 12:705-709.

Maher JJ & McGuire RF . (1990). Extracellular matrix gene expression increases preferentially in rat lipocytes and sinusoidal endothelial cells during hepatic fibrosis in vivo. J Clin Invest 86:1641-1648.

Milani S, Herbst H, Schuppan D, Grappone C, Pellegrini G, Pinzani M, Casini A, Calabro A, Ciancio G & Stefanini F . (1994). Differential expression of matrix-metalloproteinase-1 and -2 genes in normal and fibrotic human liver. Am J Pathol 144:528-537.

Moriya K, Nakagawa K, Santa T, Shintani Y, Fujie H, Miyoshi H, Tsutsumi T, Miyazawa T, Ishibashi K, Horie T, Imai K, Todoroki T, Kimura S & Koike K . (2001). Oxidative stress in the absence of inflammation in a mouse model for hepatitis C virus-associated hepatocarcinogenesis. Cancer Res 61:4365-4370.

Nakamoto Y, Suda T, Momoi T & Kaneko S . (2004). Different procarcinogenic potentials of lymphocyte subsets in a transgenic mouse model of chronic hepatitis B. Cancer Res 64:3326-3333.

Nakamura T, Tomita Y, Hirai R, Yamaoka K, Kaji K & Ichihara A . (1985). Inhibitory effect of transforming growth factor-beta on DNA synthesis of adult rat hepatocytes in primary culture. Biochem Biophys Res Commun 133:1042-1050.

Parola M, Pinzani M, Casini A, Albano E, Poli G, Gentilini P & Dianzani MU . (1993). Stimulation of lipid peroxidation or 4-hydroxynonenal treatment increases procollagen (I) gene expression in human liver fat-storing cells. Biochem Biophys Res Commun 194:1044-1050.

Pinkus R, Weiner LM & Daniel V . (1996). Role of oxidants and antioxidants in the induction of AP-1, NF-kappaB, and glutathione S-transferase gene expression. J Biol Chem 271:13422-13429.

Schwabe RF & Brenner DA . (2006). Mechanisms of Liver Injury. I. TNF-alpha-induced liver injury: role of IKK, JNK, and ROS pathways. Am J Physiol Gastrointest Liver Physiol 290:G583-G589.

Shimizu I . (2001). Antifibrogenic therapies in chronic HCV infection. Curr Drug Targets Infect Disord 1:227-240.

Shimizu I, Kamochi M, Yoshikawa H & Nakayama Y. (2012). Gender difference in alcoholic liver disease. In: Shimizu I (ed) Trends in alcoholic liver disease research: Clinical and scientific aspects. InTech-Open Access Publisher, Rijeka, Croatia, pp 23-40.

Thoren F, Romero A, Lindh M, Dahlgren C & Hellstrand K . (2004). A hepatitis C virus-encoded, nonstructural protein (NS3) triggers dysfunction and apoptosis in lymphocytes: role of NADPH oxidase-derived oxygen radicals. J Leukoc Biol 76:1180-1186.

Wake K . (1999). Cell-cell organization and functions of 'sinusoids' in liver microcirculation system. J Electron Microsc 48:89-98.

Waris G, Huh KW & Siddiqui A . (2001). Mitochondrially associated hepatitis B virus X
    protein constitutively activates transcription factors STAT-3 and NF-kappa B via
    oxidative stress. Mol Cell Biol 21:7721-7730.

Chapter 9

Role of Lipid Peroxidation in the Pathogenesis of Age-Related Cataract

Bojana Kisic, Dijana Miric, Lepsa Zoric and Aleksandra Ilic

Additional information is available at the end of the chapter

1. Introduction

The occurrence and development of cataract affect the decline of visual, working and living comfort. Cataract is the leading cause of blindness, accounting for 50% of blindness worldwide [1]. Cataract is progressive lens opacity in humans of 45 years or more, occurring without any known cause such as trauma, inflammation, hypocalcemia, medications or congenital factors. Risk factors for the occurrence of cataract are numerous: aging, diabetes mellitus, UV radiation, malnutrition, smoking, hypertension, renal disease, and others. Free oxygen radicals and oxidative stress are considered to be an important factor contributing to age-related cataract [1,2]. Oxidative stress has been shown to cause cataract in in vitro models [3]. This hypothesis is supported by studies that examined the anticatarogenic effect of different nutritional and physiological antioxidants [4].

Oxygen does not manifest toxic effects on cells of aerobic organisms in molecular form, but in the form of free oxygen radicals. Free radicals occur in univalent transfer of electrons to molecular oxygen. Due to its biochemical nature, and the low activation energy, they are able to react with biomolecules of all cellular structures, thereby carrying out their chemical and physiological modification. Under physiological conditions, the level of free radicals is controlled by mechanism of antioxidant protection. The balance between the production and catabolism of oxidants by cells and tissue is essential for maintenance of the biologic integrity of the tissue. Ocular tissues contain antioxidants that prevent damage from excessive oxygen metabolites: antioxidant enzymes, proteins, ascorbic acid, glutathione, amino acids cysteine and tyrosine, and other.

Changes in the oxidation of biomolecules can be found in many human diseases of the body, but the cataract is one of the most common diseases, where oxidative modifications of proteins [5] and lipids [1,2] is a dominant metabolic substrate of pathological disorders. Oxidative modification of lens proteins, loss of protein function and the creation of protein aggregates of high molecular mass, which increases the scattering of light, are the main features of age-related cataract [6]. These protein modifications might be caused by oxidative stress resulting in higher levels of reactive oxygen radicals.

2. Sources of reactive oxygen species (ROS) in the lens

2.1. Reactive oxygen species generated in the lens by the UV irradiation

Human lens has several systems of defense from ROS and oxidative stress, which are together responsible for the maintenance of lens transparency and prevention of cataract. But during the life the lens is exposed to multiple sources of oxidative stress, endogenous (altered mitochondrial respiration, respiratory burst of phagocytes, viral infection) and exogenous (UV light, metals,

drugs, cigarette smoke), which can lead to production of reactive oxygen species: superoxide anion ($O_2^{\bullet-}$), hydrogen peroxide (H_2O_2), hydroxyl radical ($HO^{\bullet}$) and others. The lens has a protective role for the other eye structures also, because light and oxygen are synergistically involved in the pathogenesis of cataract. By absorbing the part of the ultraviolet spectrum the lens protects deeper structures of the eye from the harmful effects of the solar radiation, whereby it is only subject to photooxidative damage. Photooxidative stress and the formation of reactive oxygen species by photosensitizing mechanisms are due to absorption of light by the biomolecules of the lens. Specifically, UV irradiation can mediate damage of lens structures, due to: direct absorption of the incident light by the cellular components, resulting in excited state formation and subsequent chemical reaction, and photosensitization mechanisms, where the light is absorbed by endogenous photosensitizers that are excited to their triplet states [6]. The excited photosensitizers can induce cellular damage by electron transfer and hydrogen abstraction processes to yield free radicals or energy transfer with O_2 to yield the reactive excited state, singlet oxygen.

Experiments in organ culture have shown that cataract can be caused by photochemical production of superoxide radicals, hydroxyl radicals and H_2O_2 [7]. Other researchers [8,9] indicate that photochemical generation of reactive species of oxygen in the lens and aqueous and consequent damage to the tissue has been implicated in the genesis of age-related cataract. The fact that the incidence of cataract is higher in the population that is more exposed to sunlight [10] imposes the assumption that photocatalytic conversion of molecular oxygen from ground state to excitatory states, which are highly reactive ($O_2^{\bullet-}$, H_2O_2, $HO^{\bullet}$ and others) occurs. High concentration of ascorbate in the aqueous humor is assumed to represent a kind of filter that prevents the penetration of UV light in the lens and thus protects tissue from oxidative damage, particularly photoinduced damage [11].

Photosensible substance, that absorbs certain wavelengths of light, activates and subtracts hydrogen or electrons from the substrate by converting them into free radicals. In the presence of O_2 , the energy is transferred from excitatory substance and produce $^1O_2$, which can initiate the process of lipid peroxidation. In the ocular tissues numerous substances can initiate photodynamic reactions. These substances are riboflavin, heme derivatives, tryptophan and its oxidation product N-formylkynurenine, lipofuscin, visible pigments (retinol), and photosensible substances of exogenous origin, such as drugs [12]. Key link between photo-oxidation and cataract is that photo-oxidation of thiol groups on lens crystallyne produces disulfide bridges between molecules and, the build-up of these will lead to protein aggregation and hence cataract.

3. Mitochondria as a source of reactive oxygen species in the lens

In the ocular tissues, including the lens, as in other organ systems, ROS are formed in the mitochondria via the electron transport chain where inefficient electron coupling leads to the formation of superoxide anion. Molecular oxygen is tightly bound to the enzyme complex cytochrome C oxidase. However, the bond on the vectors of electrons in the respiratory chain in front of the system cytochrome C oxidase, on the level of NADH-coenzyme-Q reductase and the reduced forms of coenzyme Q, is not that strong and some of transferred electrons can "leak" from the system on molecular oxygen, forming $O_2^{\bullet-}$. Superoxide production is significantly increased during reperfusion of tissues, when the availability of oxygen is increased.

The human lens consists of three metabolically different zones: the epithelium, the cortex and lens nucleus. Epithelial cells and superficial cortical fibers are metabolically most active, and the greatest part of mitochondrion respiration and aerobic glycolysis in the lens occurs in them [13]. One third of total energy produced (ATP) in the lens is produced in epithelial cells under aerobic conditions, while the metabolic activity of nuclear part of the lens is at much lower level. Intense metabolic activity makes epithelial cells susceptible to oxidative damage, especially their membrane pump systems and DNA. Oxidation of unsaturated lipids in epithelial cells could be the initial step that leads to generation of oxidation products. If reactive oxygen species or secondary products of lipid oxidation from the epithelium were to migrate to the fiber cells, it is possible that prolonged accumulation of lipid oxidative products could eventually lead to alterations in fiber cell structure and increased opacity, which leads to the development of cataract.

Thiol (-SH) groups of membrane proteins, the lens epithelial cells, which are significant for regulation of ion transport, are very susceptible to oxidative attack, especially when the concentration if intracellular GSH is reduced. The optimal membrane function of lens epithelial cells depends on reduced state of protein-SH groups. The oxidation of membrane thiol (-SH) groups of the lens cells leads to breakdown of active transport through the membrane, to the increase of membrane permeability and consequently intracellular alternations, which is involved in the development of cataract. The consequence of impaired active transport is also the reduced level of ascorbic acid in the lens. Studies have confirmed that ascorbic acid (AA) levels in human lenses with the development of cataract are reduced [34], and concentration of dehydroascorbic acid (DHA) is increased [14,35]. Timely removal of dehydroascorbic acid from the lens is important because of its potential toxicity as oxidant. Increase of the current concentration of DHA/AA redox balance can be an indicator of oxidative stress in the lens [35].

4. The importance of ascorbic acid in lens

The role of ascorbic acid is important, as a strong reductant and effective scavenger of hydroxyl and superoxide anion radical. Vitamin C has antioxidant, but also prooxidant properties. In which direction will vitamin C work, depends on the concentration of vitamin C, oxygen and the presence of metal ions. Oxidation is the cause of modification of lens proteins which accumulate over a lifetime. Some believe that ascorbate can contribute to protein modifications, react as prooxidant and participate in reactions that generate radicals [14].

These reactions may be caused by light or metal-catalyzed oxidation of endogenous ascorbic acid. It is known that copper and iron are present in micromolar concentrations and that autooxidative processes can occur in the lens. Fenton-type reactions, where H_2O_2 reacts with free metal ions, iron (Fe^{2+}) or copper (Cu^{2+}) to produce the $HO^{\bullet}$ radical, are a major source of oxidative stress initiated by transition metals [15] and are thought to be involved in the formation of cataract [16]. In the presence of metals, especially iron and copper, and oxygen, ascorbic acid is oxidized to dehydroascorbate, which produces hydrogen peroxide and metal is reduced. Hydrogen peroxide can react with reduced metal, generating hydroxyl radical and other reactive oxygen radicals [17]. When copper and protein-bound iron is included in this reaction, the radicals cause oxidative modification of amino-acids that are near the metal. In this way ascorbate can actually become a prooxidant and lead to protein damage via both H_2O_2 and Fenton production of $HO^{\bullet}$. These reactions become important when cells lose their ability to remove metals, making it available for reaction and/or when cells lose their ability to maintain their vitamin C in a reduced form. It is noted that during the aging of lens, as in cataract lenses the concentration of copper and iron increases [18,19]. The data that confirm the level of iron and copper ions is lower in non-cataract lenses and study that compared cortical nuclear and mature cataracts found higher iron levels in the mature cataract [20] suggest that metal ions that mediate the production of $HO^{\bullet}$, may be important in the development of age-related cataract [16].

Experiments on isolated proteins showed that oxidation products of ascorbate (dehydroascorbic acid) can form cross-link with crystalline lens, producing molecules of high molecular weight, which cause light scattering typical for cataract [21]. It is assumed that similar modification of lens proteins occurs in vivo during the development of age-related cataract [22].

5. Lipid peroxidation in the lens

In physiological/controlled conditions the process of lipid peroxidation affects the permeability of cell membranes, the metabolism of membrane lipids and proteins, provides control of cell proliferation, but the adverse effects of this process occurring under conditions of oxidation stress, ie. in conditions of impaired balance of prooxidative and antioxidative factors of the cell. Lipid peroxidation (LPO) is considered a pathogenetic factor of cataractogenesis [1,2,23,24,25]. LPO in the lens may be induced by endogenous or exogenous factors: enzymes, reactive oxygen species, metal ions, UV irradiations, heat, radical-initiating chemicals, drugs. Cell membrane lipids (phospholipids, glycolipids) are the most common substrates of oxidative attacks, and since the cell membranes have lipoprotein structure, the structure of membrane proteins is disturbed at the same time. That causes the disturbance of cell membrane barrier function, leading to a larger entry of calcium and other ions [26]. Structural changes of the cell membrane and its increased permeability change the cell volume and the configuration of the lens, leading to refractory changes that are associated with the early cataract.

Cell membranes are very sensitive to the effects of oxygen radicals, due to the presence of polyunsaturated fatty acids in lipids. Fatty acids in lipids of cell membranes contain a different number of carbon atoms (14 to 24), and the present double bonds are in cis configuration. The presence of double bond in the neighborhood destabilize the bond between the carbon and the hydrogen of methylene group in the chain of polyunsaturated fatty acid, and by subtracting hydrogen from such a methylene group by reactive oxidant, begins the process of oxidative modification of fatty acids - lipid peroxidation [27]. Non-enzymatic peroxidation of polyunsaturated fatty acids is a process that takes place in three stages: initiation, propagation and termination. The intensity of this process, as well as the ability of partial or complete repair of damage, depends on pro/anti-oxidative environment in which this process occurs.

Free radicals formed during lipid peroxidation have a local effect due to short life, but the degradation products of lipid peroxidation can be second messengers of oxidative stress, because of their longer half-life and ability to diffuse from the place of formation. These degradation products, mainly aldehydes, such as malonaldehyde, hexanal, 4-hydroxynonenal or acrolein, have biological roles in cellular signaling, in normal and pathological conditions and in regulation of cell cycle [28]. Due to their chemical reactivity these products can covalently modify macromolecules as nucleic acids, proteins and lipids and consequently exhibit different biological effects. They are also biomarkers of lipid peroxidation and oxidative stress.

During the development of cataract, non-enzymatic lipid peroxidation occurs. It is a nonspecific and uncontrolled process in which the resulting reactive oxygen species readily react with surrounding molecules, leading to intensification of the process, and the damage of the cell membrane. The process LPO in the lens can be initiated by hydroxyl radical, singlet oxygen, peroxyl radical. These radicals seize from the unsaturated fatty acids H^+ from the methyl group ($-CH_2$) in the α- position of the double bond, to form unsaturated fatty acid lipid radical ($L^\bullet$).

By intramolecular rearrangement of double bonds in lipid radicals, conjugated dienes are formed. By adding the molecular oxygen to conjugated dienes, lipoperoxyl radical ($LOO^\bullet$) is formed. Lipoperoxyl radicals have significant oxidative potential, they can further initiate the seizure of hydrogen from the neighbouring unsaturated fatty acid, by which lipid peroxidation enters the phase of propagation and autooxidaton, resulting in formation of lipid hydroperoxyde ($LOOH$) and new lipid radicals ($L^\bullet$).

Lipid hidroperoxydes (LOOH), as primary molecular products of lipid peroxidation process, are well-soluble, have ability to migrate from the site of development and are important potential sources for the formation of reactive hydroxyl radical (HO•). In the presence of metals with variable valence they may reopen a cascade of lipid peroxidation [27]. In the presence of Fe^{+2} in the classical Fenton-type reactions, by decomposing of LOOH, HO• and alkoxyl radical (LO•) occur. In the presence of Fe^{+3}, LOOH is decomposed to peroxyl radical (LOO•). Alkoxyl and peroxyl radical are responsible for initiation and propagation of LPO processes.

By removing hydroperoxides (LOOH), which occur during the first phase of lipid peroxidation, by activity of glutathione-dependent peroxidase which catalyzes reduction of hydroperoxide into the corresponding alcohol, the reactions of lipid peroxidation propagation can be prevented.

Compounds that react with reactive oxygen species produced during the chain reaction, such as peroxyl and alkoxyl radicals (ROO• and RO•), and lead to the formation of the species that are unable to remove the hydrogen atom from unsaturated fatty acids, are considered to be antioxidant switches of chain reactions, of which the most important is liposoluble α-tocopherol. In series of complex degradation reactions of hydro- and dihydro-peroxides of polyunsaturated fatty acids, many other aldehydes are produced: 4-hydroxy-alkenals, 4-hydroperoxy-alkenals, 4-oxo-alkenals and bis-aldehyde malonaldehyde (MDA) [28]. These aldehydes are highly reactive compounds because they have electrophilic properties, they are less volatile than hydroperoxides, and can diffuse from the place of origin and express their reactivity through biotransformation and adduction to biomolecules (proteins, DNA).

Consequences of peroxidative damage of lens cell membranes are multiple. On the one hand, a number of membrane functions may be altered due to a direct attack by reactive oxygen species on the membrane components responsible for these functions. Indirect consequences of peroxidative damage of membranes are important. Lipid peroxidation modifies the environment of not only membrane proteins and may in this way influence their functional efficiency. The consequences of adduct formation at the protein level is associated with numerous cytotoxic consequences including the disruption of cell signalling, altered gene regulation, inhibition of enzyme activity, mitochondrial dysfunction, impaired energy metabolism. Two different ways of oxidative modifications of cellular constituents have to be considered in cataractogenesis: the direct modification by reactive oxygen species and the indirect modification via reactive products of lipid peroxidation.

Lipid peroxidation products can diffuse across membranes, allowing the reactive aldehyde to covalently modify proteins localized throughout the cell and relatively far away from the initial site of reactive oxygen species (ROS) formation. LPO is implicated in human cataractogenesis because the toxic peroxidation products induce fragmentation of soluble lens proteins and damage vital membrane structures, correlating with an increase in lens opacity and changes in the refractive properties of the lens [23]. The results obtained by the authors, that caused posterior subcapsular cataract in the rabbit, by application of peroxidative products in the vitreous, were published [29].

The resulting lipid peroxidation products, such as, 4-hydroxynonenal (HNE), can form protein-HNE adducts that may result in altered protein functions and can mediate oxidative stress-induced cell death in the lens epithelial cells [30]. Also, the resulting MDA, can react with amino-groups of proteins, forming intra-molecular cross bonds and bind two distinct proteins by forming the inter-molecular bonds [31] and thus affect the structural and functional properties of proteins. MDA is linked to the Lys residues of proteins and enzymes, and forms Schiff bases, for phospholipids, nucleic acids (shown mutagenic properties), and also MDA inhibits a number of thiol dependent enzymes: glucose-6-phosphatase, Na^+-K^+-ATP-ase, Ca^{++}-ATP-ase.

At the level of cell membranes in the lens lipid hydroperoxides induce changes of the lipoprotein structure and permeability [13], and oxidation inhibition of membrane enzymes Na^+/K^+-ATP-ase and Ca^{++}-ATP-ase, which are responsible for osmotic regulation and transport of metabolites. Disturbance of ion transport through the cell membrane is associated with ATP hydrolysis, which increases membrane permeability to protons, as the result of LPO. The Na^+/K^+-ATP- and Ca^{++}-ATP-ases would be directly affected because of their regulation by proton concentration. Other transport proteins would be affected in a secondary sense because they are coupled to ions whose intracellular concentrations are regulated by the ATP-ase. For example, glucose is the main source of energy for lens cells and is typically co-transported with Na^+, and similar transport are known for the import of amino acids across cell membranes. Formation of the lens opacity follows from either osmotic imbalance or a cytoplasmic imbalance of specific cations, in particular Ca^{+2} [32]. Disturbance of function Ca^{++}-ATP-ase and the consequent increase of Ca^{+2} in the lens leads to electrostatic changes in the crystalline, which can disrupt the protein conformation and interaction. Maintenance of calcium homeostasis is critical to lens clarity and cataractous lens has elevated calcium levels. An in vitro binding study indicates that human lens lipids have the capacity to bind nearly all the calcium present in the human lens and that age and cataract diminished the capacity of lens lipids to bind calcium. It is possible that the increased concentration of intracellular Ca^{+2} and reduced ability of lens lipids to bind calcium, to initiate further disturbances that lead to an increase of light scattering from proteins and lipids [33]. Disturbance of the lens cell membrane permeability during the development of cataract, and reduced ability of active transport of substances against the concentration gradient lead to changes in concentration of intracellular compounds and metabolic changes within the cells. This is manifested by lowering the content of GSH, ATP, and other intracellular compounds and electrolytes in lens cells.

Lipid peroxides as potential causes of cataracts, lead to the changes of not only the lens cell membrane, but also of the cytosol, because it serves to reduce concentration of glutathione and cause the change of redox relationship GSH/GSSG [1].

6. Defense against lipid peroxidation in the lens

Low molecular mass compounds which act primarily against peroxyl radicals involved in radical propagation, provide first line of defense against lipid peroxidation in the lens.

These compounds (GSH, ascorbic acid, α- tocopherol) can terminate the propagation of free radical mediated reactions and interrupt the autocatalytic chain reaction of lipid peroxidation. GSH is a major antioxidant in the lens, and helps to reduce proteins, contains a side chain of sulfhydryl (-SH) residue that enables it to protect cells against oxidants. GSH can directly scavenge ROS or enzymatically via two major antioxidant enzyme systems, glutathione peroxidases (GPx) and glutathione S-transferases (GST). Enzymes such as superoxide dismuatase (SOD), catalase (CAT) and GPx can decompose ROS and prevent the damage to cellular constituents and initiation of lipid peroxidation. In the event of ROS induced lipid peroxidation, secondary defense enzymes are involved in the removal of LOOH to terminate the autocatalytic chain of lipid peroxidation and protect membranes. GPx and GST which catalyze GSH-dependent reduction of LOOH through their peroxidase activity are the major secondary defenses in the lens against ROS induced lipid peroxidation. Resynthesis of GSH from the oxidized form is catalyzed by glutathione reductase, where the necessary NADPH$^+$+H$^+$ is produced in pentose pathway of carbohydrates [36]. Glutathione reductase (GR) is a control enzyme glutathione-redox cycle and by maintaining of intracellular levels of GSH can affect cation transport systems, lens hydration, sulfhydryl groups of proteins, and membrane integrity. Detoxifying role of GST is reflected in its ability to catalyze reactions of conjugation of reduced glutathione with endogenous electrophiles, mostly by the products of oxidative stress, lipid hydroperoxides and final products of lipid peroxidation [37]. Class μ and π GST isoenzymes are expressed in the human lens, with the π isoenzyme predominating. The highest GST activity occurs in the peripheral and the equatorial cortexes, with the lowest activity in the nucleus [38]. Glutathione peroxidase activity is shown by the enzymes that catalyze the reduction of hydrogen peroxide, organic hydroperoxide and phospholipid hydroperoxide using GSH as a hydrogen donor. Superoxide dismutase (SOD) catalyzes the reaction of dismutation of superoxide anion radicals ($O_2^-\bullet$) in the presence of hydrogen donor to hydrogen peroxide and molecular oxygen. By removing the $O_2^-\bullet$, SOD prevents the formation of $^1O_2$ which can initiate the process of lipid peroxidation.

Aim. Our studies have focused on measuring the products of lipid peroxidation in corticonuclear lens blocks, with different type and different degrees of maturity of age-related cataract. In addition to measuring products of lipid peroxidation in cataract lenses, our study included the determination of activities and ability of lens glutathione peroxidase and glutathione S-transferase to remove hydroperoxides, which are probably involved in the early stages of cataractogenesis and development of mature cataract through oxidative stress.

7. Material and methods – patients

Clinical and biochemical researches were carried out in 101 patients with age-related cataract, 46 women and 55 men. The average age of the group was 72.5 (SD $\pm$ 7.9). According to the cataract maturity degree the patients were classified into two groups as follows: age-related cataract incipient (N=41) and matura (N=60). In the group age-related cataract incipient there were 23 patients with posterior subcapsular (PS), 9 patients with nuclear subcapsular (NP) and 9 patients with cortical nuclear (CN) cataract. In the group age-related catataract matura there were 19 patients with cataract which started as a posterior subcapsular, 15 patients which started as a nuclear subcapsular, 16 patients with matura, which started as cortical nuclear and 10 patients diagnosed with matura, which started as cortical cataract.

Samples corticonuclear blocks/parts of lens (without epithelial cells) were obtained from patients undergoing extracapsular extraction of cataracts and used as the test material. Types of cataract were estimated during ophthalmologic examination and confirmed during its extraction. Immediately after sample acquisition, samples were closed in individual capsules and deeply frozen. This research has been conducted following the tenets of the Declaration of Helsinki and approved by the ethics committee of Medical Faculty. Informed consent was provided from all patients after a careful explanation of the aims of the study.

Homogenate of lenses from each group was prepared in 0.2 mol/L potassium phosphate buffer (pH 7.2). For analysis we used supernatant obtained by centrifugation of homogenates at 5000 rpm for 15 min at 4°C.

The concentration of conjugated dienes was measured spectrophotometrically at 233 nm [39].

Lens MDA concentrations were measured as the product of the reaction with thiobarbituric acid (TBA) using a modification of the method Ledwozyw et al [40].

Fluorescent products (lipid- and water-soluble) of lipid peroxidation were determined by spectrofluorimetric analysis at 360/430 (excitation/emission) nm [41].

The concentration of GSH in the sample was determined in the reaction 5,5'-dithiobis-2-nitrobenzoic acid (DTNB) (Ellman's reagent), after removal of proteins by perchloric acid [42].

8. Enzyme assays

The activity of glutathione peroxidase was determined at 412 nm, by the method Chin et al. [43]. The conjugation of GSH with 1-chloro,2-4 dinitrobenzene (CDNB), a hydrophilic substrate, was examined spectrophotometrically at 340 nm to measure glutathione S-transferase activity [44]. One unit of GST was defined as the amount of enzyme required to conjugate 1μmol of CDNB with GSH/min. The activity of glutathione reductase was assayed by the procedure of Glatzle et al. [45]. Superoxide dismutase (SOD) activity was determined by the method Misra and Fridovich [47], based on the inhibition of the adrenochrome during the spontaneous oxidation of adrenaline in basic conditions. The change in absorbance was read at 480 nm on a spectrophotometer. The SOD activity was expressed as kU/g protein (one unit was considered to be the amount of enzyme that inhibited adrenaline auto-oxidation by 50%).

To calculate the specific enzyme activity, protein in each sample was estimated by the method of Lowry et al [46].

9. Results and discussion

Senile cataract is manifested in the later years of life, so it is estimated that the costs of operation would be reduced by 45% if the incidence of age-related cataract could be delayed for ten years [48]. By studying the oxidation changes of lens structures during the development of cataract, we attempted to contribute to clearing up the process of cataractogenesis, of which even today there are many unknowns.

Lipid peroxidation is one of the possible mechanisms of cataractogenesis, caused by excessive production of reactive oxygen species in aqueous environment and reduced antioxidant defense of the lens. By studying the corticonuclear lens block of the patients with age-related cataract were detected increased concentrations in primary molecular products LPO (diene conjugates and lipid hydroperoxides) and end fluorescent LPO products (table 1).

	Age-related cataract incipient (n=41)	Age-related cataract matura (n=60)
Conjugated diens (nmol/g weight of lens)	2.48 ± 0.84*	1.57 ± 0.49
Fluorescent products/ g protein (lipid soluble)	46.59 ± 14.40	70.94 ± 13.21*
Fluorescent products/ g protein (water soluble)	57.53 ± 18.23	103.08 ± 27.81*
MDA (nmol/g weight of lens)	1.81 ± 0.67	3.17± 0.78*

Data is presented as means ± SD *p<0.001

Table 1. Lipid peroxidation products in cataractous lenses

In the group of patients with the incipient cataract, we obtained significantly higher concentration of diene conjugates in the lenses compared to matura cataract (p<0.001) (table 1). This can be explained by the fact that at the early stages of the development of cataract the most intense is the proces of lipid peroxidation, which is either the initiator of cataractogenesis proces or initiated by creation of reactive oxigen types, and continues to affect changes in the lens by its propagation.

In the lenses with the incipient cataract, the concentration of conjugated diens is significantly higher in cortical nuclear cataract (CN) compared with the lenses with posterior subcapsular cataract (PS) (p=0.001) (table 2). Also, in matura cataract the concentration of conjugated diens is the highest in the lenses diagnosed with CN cataract (p<0.05) (table 3).

Cataract incipient (type)	PS (n=23)	NP (n=9)	CN (n=9)
Conjugated diens (nmol/g lens)	2.12 ± 0.55	2.63 ± 0.97	3.24 ± 0.84*
Fluorescent products/g protein (lipid soluble)	40.09 ± 10.18	52.34 ± 13.92‡	57.47 ± 16.33†
Fluorescent products/g protein (water soluble)	50.17 ± 12.88	61.99 ± 14.58	71.90 ± 24.12†
MDA (nmol/g weight of lens)	1.60 ± 0.56	1.74 ± 0.73‡	2.41 ± 0.55*

Data is presented as means ± SD *$p<0.001$, ‡$p<0.05$, †$p<0.01$.
PS - posterior subcapsular, NP - nuclear subcapsular, CN - cortical nuclear

Table 2. Products LPO in lenses with cataract incipient

Cataract matura	Cataract matura began as PS (N=19)	Cataract matura began as NP (N=15)	Cataract matura began as CN (N=16)
Conjugated diens (nmol/g lens)	1.41 ± 0.30	1.46 ± 0.67	1.91 ± 0.39†
Fluorescent products/ g protein (lipid soluble)	67.32 ± 15.45	76.69 ± 14.65‡	74.74 ± 7.49
Fluorescent products/ g protein (water soluble)	94.83 ± 25.43	105.29 ± 24.72	120.68 ± 27.08†
MDA (nmol/g lens)	3.32 ± 1.07	3.23 ± 0.80	3.06 ± 0.48

Data is presented as means ± SD †$p<0.01$, ‡$p<0.05$.

Table 3. Products LPO in lenses with cataract matura

In the presence of free metal ions with variable valence (Fe^{+2} or Cu^{2+}) hydrogen peroxide in Fenton's reaction is translated into highly reactive hydroxyl radical, while the lipid peroxides are translated into peroxyl and alkoxyl radicals. Because of the longer half-life time compared to the alkoxyl radical, peroxyl radical is ideal for propagation of oxidative chain reactions, while the oxidation of alkoxyl radicals produce dihydroperoxide, which is degraded to toxic aldehydes, such as 4-hydroxy-2,3-trans-nonenal, 4-hydroxy-pentanal, short chain malondialdehyde. Such conditions may exist in humane senile lens [49]. MDA and 4-hydroxy-2,3-trans-nonenal by forming of Schiff's bases with amino groups amino acid residues of protein contribute to increase of carbonyl groups content and produce fluorescent products of lipid peroxidation. We have measured significantly higher concentration of these fluorescent products of LPO in the lenses with mature cataract (table

1), and the highest concentration was measured in the lenses with the NP and CN cataract in relation to posterior subcapsular cataract (PS) ($p<0.05$) (table 2,3). Some authors have identified fluorescent Schiff bases in higher concentration in human cataract lenses compared to healthy lenses, resulting from the interaction of reactive carbonyl groups MDA with amino groups of lens membrane phospholipids [50].

By measuring the concentration of MDA in the homogenate of cataract lenses, we obtained significantly higher concentration in the group of patients with matura compared to incipient cataract ($p<0.001$) (table 1). This can be explained by the fact that the malondialdehyde is one of the final products of LPO, which accumulates in the lens during the process of lipid peroxidation and the development of cataract. Author's results with the experimentally induced cataract that have measured significantly higher concentration of MDA in cataract lenses compared to the control group were published [51], as well as author's works that obtained significantly higher concentration of MDA in cataract lenses of diabetics [2], myopic lenses and senile cataract [52]. Reduced activity of glutathione peroxidase enzyme and glutathione S-transferase, that are important for the removal of malondialdehyde, contribute to the increase of its concentration in the lens.

The reasons that cause significantly higher concentration of lipid peroxidation products in lenses with cortical nuclear (CN) and nuclear subcapsular (NP) cataract, in relation to the posterior subcupsular (PS) are numerous. The results of other researchers [49] indicate that during the life time in the lens some kind of "internal" lens barrier is developed between nuclear and corticular parts, which hinders the diffusion of molecules to the nucleus. This barrier prevents the diffusion of antioxidative molecules to the nuclear part, which increases the sensitivity of central part of the lens to oxidative damage. Also, it is possible that unstable prooxidant molecules have longer residency in the central part of the lens. The endogenous lens hromofore, tryptophan metabolites (kynurenine, 3-hydroxykynurenine, 3-hydroxykynurenine glucoside), which are relatively inert photochemically, during oxidative stress and/or aging are formed photochemically active tryptophan metabolite (N-formyl-kynurenine, xanthurenic acid) that have photochemical properties, and also act as an endogenous photosensitizers in the lens [53].

Through photosensible reactions tryptophan products transfer absorbed energy to oxygen, which further leads to a series of cellular changes through the oxidation. With age, the level of free components of UV filters ie. tryptophan derivatives in the lens are reduced, and their binding to lens proteins increases [54]. Tryptophan products are subject non-enzyme deamination, at physiological pH, resulting in α,β-unsaturated ketone, reactive intermediates [55], which can covalently bind to amino acids, usually His, Cys, or Lys residues in proteins of human lenses, or react with the Cys residue of glutathione (GSH) to form GSH-3OHKynG [54]. GSH that is present in the lens in relatively high concentrations may compete with the amino acid residues for the unsaturated ketone derivative of kynurenine, thereby protecting the crystalline from modification. This covalent modification is particularly expressed in the nucleus of the lens containing the older proteins, causing altered transport/diffusion of small molecules in the lens. Specifically, it develops a barrier in to the movement of molecules between metabolically active cortex and inert nucleus. The barrier also restricts the flow of GSH from the cortex, which reduces the concentration of GSH in the nucleus of the lens, so the response to oxidative damages in this part of the lens is also reduced [49]. During the life time, lens fibers are very compactly arranged in nuclear part, with minimal presence of extracellular space. Nuclear plasma of the membrane undergo oxidation damage, whereas the phospholipid molecules modified by oxygen accumulate in the lipid layer, leading to changes in the structure and violate lipid-lipid and protein-lipid interactions in membranes of the lens fibers. This probably contributes to concentration of the lipid peroxidation products to be the highest in the lenses with early nuclear cataract.

Reduced glutathione (GSH) and GSH-dependent enzymes, glutathione peroxidase (GPx) and glutathione S-transferase (GST), are very important in defending the lens structures of the products of lipid peroxidation. One reason for increased production and accumulation of lipid hydroperoxides in the lens with cataract may also be reduced activity of GPx and GST.

The primary biological role of superoxide dismutase (SOD) is to catalyze reaction of dismutations of superoxide anion radicals ($O_2^-\bullet$) in the presence of hydrogen donor to hydrogen peroxide and molecular oxygen. Superoxides can first be degraded into H_2O_2 by SOD, and subsequently, catalyzed into ground-state oxygen and water by catalase and enzymes of the glutathione redox cycle, including glutathione reductase and glutathione peroxidase. At low levels of hydrogen peroxide, the glutathione redox cycle is responsible for protecting against H_2O_2-induced damage and maintaining high levels of GSH in the lens, whereas, at a higher concentration, the principal mechanism for the removal of hydrogen peroxide is catalase. The study of the lens cell culture where the expression of DNK for superoxide dismutase was performed on intact lens cells, showed that the cells with higher activity of SOD resistant to oxidative damage, caused by hydrogen peroxide, superoxide anion radical and UV radiation. Expression of the superoxide dismutase enzyme prevented the beginning of the cataract in lens cells [56].

GSH and other sulfhydryls are particularly important in the protection of thiol (-SH) groups of crystalline and prevent the formation of aggregates which reduce transparency of the lens [57]. The oxidized glutathione (GSSG) is reduced back to GSH by a NADPH-dependent glutathione reductase, which in physiological conditions maintains a high ratio of GSH/GSSG in the lens and other ocular tissues. Oxidative stress, induced by accumulation of LPO products in the lens during the development of cataract, causes consumption of GSH and disruption of redox balance in the lens, so the age-related cataract is associated with progressive reduction of GSH concentration in the lens. Probably with the progression of the cataract than the consumption of GSH against toxic compounds, the synthesis of GSH is reduced, as the result of deficient availability of substrates and reduced activity of enzymes for its synthesis (γ-glutamyl-cysteine syntethase). This reduces the amount of available GSH for optimal function GPx and GST, which causes the peroxide metabolism disorder.

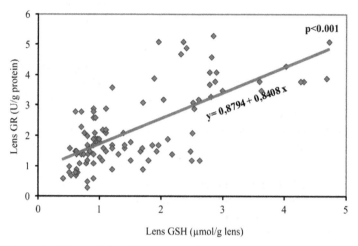

Figure 1. Lens glutathione and glutathione reductase in cataract.

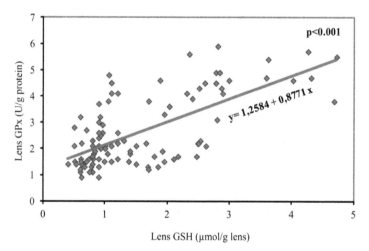

Figure 2. Lens glutathione and glutathione peroxidase in cataract.

The significance of GSH, as an important compound for the function of GPx and GST in the tested lenses, is confirmed also by the positive correlations between the concentration of GSH and glutathione reductase ($r= 0.7011$, $p< 0.001$) (figure 1), between the concentration of GSH and glutathione peroxidase activity ($r= 0.6749$, $p< 0.001$) (figure 2), and between activities of glutathione S-transferase and the concentration of GSH ($r= 0.6379$, $p< 0.001$) (figure 3).

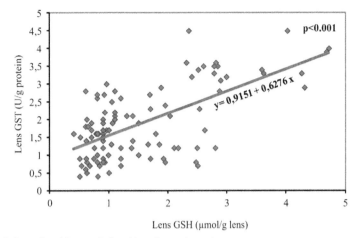

Figure 3. Lens glutathione and glutathione S-transferase in cataract.

Experiments showed that applying the injections of buthionine-sulphoximine as inhibitors of reduced glutathione synthesis, on the newborn rats, causes the development of cataract [58].

	Cataract incipient (n=41)	Cataract matura (n=60)
GSH	2.55 ± 0.9	$0.88 \pm 0.26^*$
GPx	3.40 ± 1.47	$2.09 \pm 0.90^*$
GR	3.03 ± 1.29	$1.61 \pm 0.71^*$
GST	2.46 ± 1.14	$1.50 \pm 0.67^*$
SOD	4.13 ± 2.14	$2.14 \pm 0.91^*$

Data is presented as means $\pm$ SD $^*p<0.001$
GSH (μmol/g weight of lens), GPx, GR, GST (U/g protein), SOD (kU/g protein).

Table 4. Antioxidative defense factors in cataractous lenses

In tested corticonuclear lens blocks with mature cataract, we measured lower activity of GR, GPx and GST enzymes in relation to the incipient ($p<0.001$) (table 4).

Cataract incipient (type)	PS (n=23)	NP (n=9)	CN (n=9)
GSH	2.99 ± 0.90	$2.16 \pm 0.55\dagger$	$1.81 \pm 0.42^*$
GPx	4.58 ± 0.70	$2.17 \pm 0.53^*$	$1.64 \pm 0.17^*$
GR	4.03 ± 0.66	$1.91 \pm 0.73^*$	$1.61 \pm 0.27^*$
GST	3.31 ± 0.60	$1.61 \pm 0.75^*$	$1.14 \pm 0.29^*$
SOD	5.77 ± 1.25	$2.34 \pm 0.59^*$	$1.71 \pm 0.58^*$

Data is presented as means $\pm$ SD $^*p<0.001$, $\dagger p<0.01$.
GSH (μmol/g weight of lens), GPx, GR, GST (U/g protein), SOD (kU/g protein)

Table 5. Antioxidative defense factors in lenses with cataract incipient

The lowest activity of these enzymes was measured in the lenses with the CN and NP in relation to the PS cataract (table 5). Reduced activity of GR, GPx and GST enzymes is followed by a significant decrease in the concentration of GSH in lens homogenates with matura (0.88 ± 0.26 μmo/g weight), compared to the initial cataract (2.55 ± 0.9 μmo/g weight) ($p<0.001$) (table 4), with lower concentrations of GSH measured in CN and NP compared to the PS cataract (table 5,6).

Cataract matura	Cataract matura began as PS (N=19)	Cataract matura began as NP (N=15)	Cataract matura began as CN (N=16)
GSH	0.98 ± 0.19	0.78 ± 0.20	0.81 ± 0.35
GPx	2.96 ± 0.99	$1.89 \pm 0.57^*$	$1.44 \pm 0.31^*$
GR	2.26 ± 0.60	$1.31 \pm 0.49^*$	$1.06 \pm 0.47^*$
GST	2.13 ± 0.50	$1.13 \pm 0.43^*$	$1.07 \pm 0.49^*$
SOD	2.96 ± 0.65	$1.77 \pm 0.52^*$	$1.25 \pm 0.36^*$

Data is presented as means ± SD *$p<0.001$.
GSH (μmol/g weight of lens), GPx, GR, GST (U/g protein), SOD (kU/g protein)

Table 6. Antioxidative defense factors in lenses with cataract matura

Also the results of other researchers show that the level of reduced glutathione in the lens decreases with the development of the cataract [24]. Such changes are probably a reflection of oxidative processes, increased by the formation of toxic products of lipid peroxidation, compared to a weakened antioxidative capacity of cataract lenses. In cataract lenses the concentration of GSH is being reduced, since, as a main representative of non-protein thiols, it is included and consumed in oxido-reduction processes in terms of excess oxidized substrates. A possible reason for the consumption of GSH during oxidative stress is its conversion into oxidized form, which can be conjugated with protein thiol groups to form mixed disulphides (PSSG), via a process called protein-S-thiolation [59].

Results of other researchers also show that the human lens with age-related cataract glutathione peroxidase activity was significantly reduced compared with normal lenses. The kinetic study of GPx showed that lipid hydroperoxides achieve saturation of enzymes at a concentration that is approximately 1 mmol ie. that Km GPx is achieved at a concentration of lipid hydroperoxides of 0.434 mmol [60]. Because of these kinetic properties, GPx activity was probably inhibited in age-related cataract by products of lipid peroxidation, using non-competitive inhibition principle. GPx activity, aside from availability of GSH, is affected by other factors such as glutathione reductase activity, the amount of produced $NADPH^+ + H^+$ in the pentose pathway and availability of selenium. The authors who have examined the activity of GPx in the lenses of experimental mice and compared the degree of lens blur, and age of mice with specific activity of GPx, showed significant correlation between decreased activity of GPx and the level of lens blur, as well as the age of mice [61].

Considering the function of GST to catalyze reactions of conjugation of lipid peroxidation products with GSH [62], thereby reducing the toxicity of electrophylic compounds and their reactivity towards nucleophilic groups in biomolecules, it is logical that the activity of the GST measured in incipient cataract is higher (2.46 ± 1.14 U/g protein), because at the beginning of the development of the cataract the most intense is the process of lipid peroxidation. With advancing of the process of cataractogenesis, the amount of GSH is reduced for glutathione-S-transferase, and the enzyme activity is significantly decreased in the lenses with mature cataract (1.5 ± 0.67 U/g protein) (p<0.001) (table 4).

In the study of GST activity in epithelial cells of operated cataract lenses, the group of authors showed that lens epithelial cells with cortical nuclear and cortical cataract show complete loss of activity of glutathione-S-transferase [63].

By analyzing the activity of glutathione reductase, significantly higher activity was found in the lenses with the incipient cataract (3.03 ± 1.29 U/g protein) compared to the lenses with mature cataract (1.61 ± 0.71 U/g protein) (p<0.001) (table 4). Other researchers have obtained similar results after comparing the activities of GR in cataract lenses and intact lenses of older persons [64]. Glutathione reductase plays a key role in maintaining thiol (-SH) groups in the lens, and this is probably the most important role of this enzyme in maintaining lens transparency. It is possible that lower GR activity in comparison to normal activities, can be one of the causes of lens blur. In addition to its predominantly cortical distribution within lens fiber cells, high susceptibility of GR to post-translational modifications [65] could also be of critical importance for the early dysfunction of GPx and GST under oxidative stress.

Based on the fact that the activity of GPx and GST is focused on degradation of lipid peroxidation products, we tested the relationship between the activity of the antioxidant defense enzyme and products of LPO in cataract lenses.

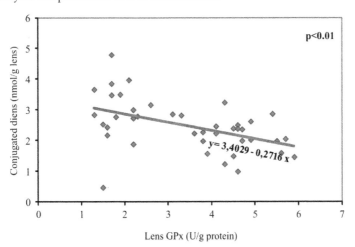

Figure 4. Lens conjugated diens and glutathione peroxidase relationship in cataract.

Reduced specific GPx activity (U/g protein) in the lenses shows significant correlation with the increased concentration of conjugated diens (r= -0.476, p<0.01) (figure 4), lipid-soluble fluorescent products (r= -0.429, p<0.01) (figure 5), water-soluble fluorescent products (r= -0.367, p<0.05) (figure 6) and MDA (r= -0.328, p<0.05) (figure 7).

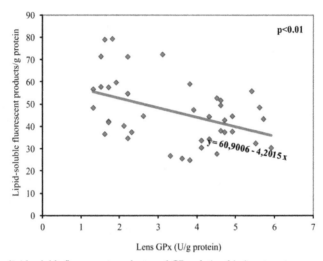

Figure 5. Lens lipid-soluble fluorescent products and GPx relationship in cataract.

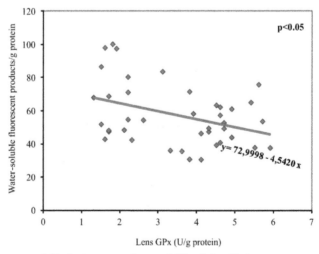

Figure 6. Lens water-soluble fluorescent products and GPx relationship in cataract.

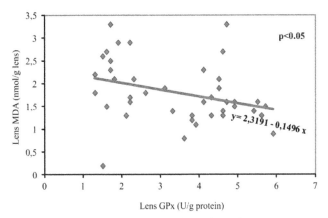

Figure 7. Lens malondialdehide and glutathione peroxidase relationship in cataract.

Also, the reduced GST activity (U/g protein) shows the significant correlation with the increased concentration of total hydroperoxides (r= -0.313, p<0.05), conjugated diens (r= -0.465, p<0.01) (figure 8), lipid-soluble fluorescent products (r= -0.398, p=0.01) (figure 9), water-soluble fluorescent products (r= -0.347, p<0.05) (figure 10) and MDA (r= -0.345, p<0.05) (figure 11) in homogenates of the lenses with the incipient cataract.

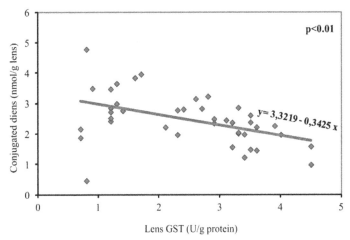

Figure 8. Lens conjugated diens and glutathione S-transferase relationship in cataract.

Probably, the consumption of GSH and other antioxidants in reactions of degradation of lipid peroxidation products, affects the decrease of GPx and GST enzymes activity, and as the process of cataractogenesis progresses towards mature cataract, all lens structures become affected by the changes, which probably leads to the change of enzyme molecules themselves.

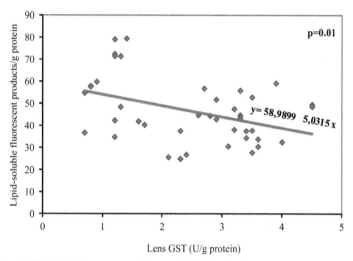

Figure 9. Lens lipid-soluble fluorescent products and GST relationship in cataract.

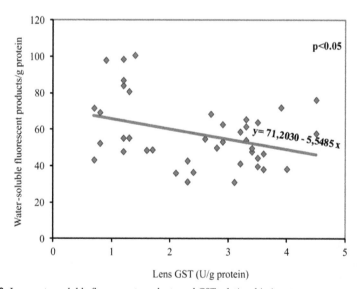

Figure 10. Lens water-soluble fluorescent products and GST relationship in cataract.

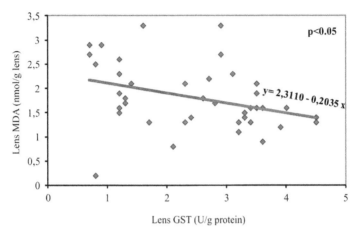

Figure 11. Lens malondialdehide and glutathione S-transferase in cataract.

We measured higher activity of SOD (4.13 ± 2.14 kU/g protein) in lenses with incipient cataract in comparison to mature cataract (2.14 ± 0.91 kU/g protein) ($p<0.001$) (table 2). Such changes are probably a reflection of oxidative processes, increased by formation of toxic products of lipid peroxidation in relation to the weakened antioxidant capacity of cataract lenses. Also, other researchers have measured decreased SOD activity in the lenses of patients with senile and diabetic cataract [66]. The decreased activity of superoxide dismutase in the lenses with age-related cataract may be due to denaturation of enzyme molecules, and/or slow enzyme synthesis. Reduced SOD activity and consequently, the increase of concentration of H_2O_2 lead to the formation of hydroxyl radical from Fenton's type reaction. Subsequently, hydroxyl radical induces the formation of superoxide radical, which may initiate the process of lipid peroxidation in the lens.

10. Conclusion

Based on the results of measuring products of lipid peroxidation and antioxidant enzyme activity in corticonuclear lens blocks with age-related cataract, we can say that the lens structure changes induced by lipid peroxidation may with other risk factors present, affect the beginning or the development of cataract. The changes in redox system are particularly pronounced in cortical nuclear cataract, but are reflected in all parts of the lens, regardless of the initial localization of the lens blur. The lowest level of oxidative stress was detected in posterior subcapsular cataract, so it is possible that it has less importance in the development of PS cataract. The development of cataract can probably be prevented/slowed down by preventing the accumulation of products of lipid peroxidation in the lens and maintaining adequate level of GSH and function of GSH-dependent antioxidant enzymes.

Author details

Bojana Kisic and Dijana Miric
Institute of Biochemistry, Serbia

Lepsa Zoric
Clinic for Eye Diseases, Serbia

Aleksandra Ilic
Institute of Preventive Medicine, Faculty of Medicine, Kosovska Mitrovica, Serbia

11. References

[1] Babizhayev MA (2012) Biomarkers and special features of oxidative stress in the anterior segment of the eye linked to lens cataract and the trabecular meshwork injury in primary open-angle glaucoma: challenges of dual combination therapy with N-acetylcarnosine lubricant eye drops and oral formulation of nonhydrolyzed carnosine. Fundam Clin Pharmacol. 26(1):86-117.

[2] Donma O, Yorulmaz E, Pekel H, Suyugul N (2002) Blood and lens lipid peroxidation and antioxidant status in normal individuals, senile and dibetic cataractous patients. Curr Eye Res. 25(1): 9-16.

[3] Spector A, Wang GM, Wang RR, Garner WH, Moll H (1993) The prevention of cataract caused by oxidative stress in cultured rat lenses. I. H_2O_2 and photochemically induced cataract. Curr Eye Res. 12 (2):163–179.

[4] Yagci R, Aydin B, Erdurmus M, Karadag R, Gurel A, Durmus M, Yigitogly R (2006) Use of melatonin to prevent selenite-induced catarct formation in rat eyes. Curr Eye Res. 31(10): 845-850.

[5] Boscia F, Grattagliano I, Vendemiale G, Micelli-Ferrari T, Altomare E (2001) Protein oxidation and lens opacity in humans. Invest Ophthalmol Vis Sci. 41(9):2461-2465.

[6] Davies MJ, Truscott RJ (2001) Photo-oxidation of proteins and its role in cataractogenesis. J Photochem Photobiol B. 63(1-3):114-125.

[7] Spector A (2000) Oxidative stress and disease. J Ocular Pharmacol. 16:193-201.

[8] McCarty CA, Taylor HR (2002) A review of the epidemilogic evidence linking ultraviolet radiation and cataracts. Dev Ophthalmol. 35: 21-31.

[9] Abraham AG, Cox C, West S (2010) The differential effect of ultraviolet light exposure on cataract rate across regions of the lens. Invest Ophthalmol Vis Sci. 51(8):3919-3923.

[10] Delcourt C, Carrière I, Ponton-Sanchez A, Lacroux A, Covacho MJ, Papoz L and the POLA Study Group (2000) Light Exposure and the Risk of Cortical, Nuclear, and Posterior Subcapsular Cataracts: The Pathologies Oculaires Liées à l'Age (POLA) Study. Arch of Ophthalmol.118(3):385–392.

[11] Kannan R, Stolz A, Ji Q, Prasad PD, Ganapathy V (2001) Vitamin C Transport in Human Lens Epithelial Cells: Evidence for the Presence of SVCT2. Exp Eye Res. 73(2):159-165.

[12] Varma SD, Hegde KR (2010) Kynurenine-induced photo oxidative damage to lens in vitro: protective effect of caffeine. Mol Cell Biochem. 340(1-2):49-54.

[13] Huang L, Tang D, Yappert MC, Borchman D (2006) Oxidation-induced changes in human lens epithelial cells 2. Mitochondria and the generation of reactive oxygen species. Free Radic Biol Med. 41(6):926-936.

[14] Linetsky M, Shipova E, Cheng R, Ortwerth BJ (2008) Glycation by ascorbic acid oxidation products leads to the aggregation of lens proteins. Biochim et Bioph Acta: Molec Basis of Disease. 1782(1):22-34.

[15] Fridovich I (1997) Superoxide anion radical (O_2^{-}), superoxide dismutases, and related matters. J Biol Chem. 272(30):18515-18517.

[16] Garner B, Davies M, Truscott RJ (2000) Formation of hydroxyl radicals in the human lens is related to the severity of nuclear cataract. Exp Eye Res. 70:81-88.

[17] Garland DL (1991) Ascorbic acid and the eye. Am J of Clin Nutrit. 54(6):1198S-1202S

[18] Balaji M, Sasikala K, Ravindran T. (1992) Copper levels in human mixed, nuclear brunescence, and posterior subcapsular cataract. British J of Ophthalmol. 76(11):668-669.

[19] Garner B, Roberg K, Qian M, Eaton JW, Truscott RJ (2000) Distribution of ferritin and redox-active transition metals in normal and cataractous human lenses. Exp Eye Res. 71(6):599-607.

[20] Dawczynski J, Blum M, Winnefeld K, Strobel J (2002) Increased content of zinc and iron in human cataractous lenses. Biol Trace Elem Res. 90(1-3):15-23.

[21] Nagaraj RH, Monnier VM (1995) Protein modification by the degradation products of ascorbate: formation of a novel pyrrole from the Maillard reaction of L-threose with proteins. Biochim et Biophys Acta. 1253(1):75-84

[22] Fan X, Reneker LW, Obrenovich ME, Strauch C, Cheng R, Jarvis SM, Ortwerth BJ, Monnier VM (2006) Vitamin C mediates aging of lens crystallins by the Maillard reaction in a humanized mouse model. Proc of the Nat Acad of Sci of the United States. 103(42):16912-16917.

[23] Ansari NH, Wang L, Srivastava SK (1996) Role of lipid aldehydes in cataractogenesis: 4-hydroxynonenal-induced cataract. Biochem Mol Med. 58(1):25-30;

[24] Zoric L, Elek-Vlajic S, Jovanovic M, Kisic B, Djokic O, Canadanovic V, Cosic V, Jaksic V (2008) Oxidative stress intensity in lens and aqueous depending on age-related cataract type and brunescense. Eur J Ophthalmol. 18(5):669-674.

[25] Kisic B, Miric D, Zoric L, Dragojevic I, Stolic A (2009) Role of lipid peroxidation in pathogenesis of senile cataract. Vojnosanit Pregl. 66(5):371-375.

[26] Stark G (2005) Functional consequences of oxidative membrane damage. J Memb Biol. 205(1):1-16

[27] Halliwell B, Chirico S (1993) Lipid peroxidation: its mechanism, measurement, and significance. Am J Clin Nutr. 57: 715S-725S.

[28] Guéraud F, Atalay M, Bresgen N, Cipak A, Eckl PM, Huc L, Jouanin I, Siems W, Uchida K (2010) Chemistry and biochemistry of lipid peroxidation products. Free Rad Res. 44(10):1098–1124.

[29] Goosey JD, Tuan WM, Garcia CH (1984) A lipid peroxidative mechanism for posterior subcapsular cataract formation in the rabbit. A possible model for cataract formation in tapetoretinal diseases. Invest Ophthalmol Vis Sci. 25:608-612.

[30] Choudhary S, Srivastava S, Xiao T, Andley UP, Srivastava SK, Ansari NH (2003) Metabolism of lipid derived aldehyde, 4-hydroxynonenal in human lens epithelial cells and rat lens. Invest Ophthalmol Vis Sci. 44(6):2675-2682.

[31] Berlett BS, Stadtman ER (1997) Protein oxidation in aging, disease, and oxidative stress. J Biol Chem. 272(33):20313-20316.

[32] Biju PG, Rooban BN, Lija Y, Devi VG, Sahasranamam V, Abraham A (2007) Drevogenin D prevents selenite-induced oxidative stress and calpain activation in cultured rat lens. Mol Vis. 13:1121-1129.

[33] Tang D, Borchman D, Yappert MC, Vrensen GF, Rasi V (2003) Influence of age, diabetes, and cataract on calcium, lipid-calcium, and protein-calcium relationships in human lenses. Invest Ophthalmol Vis Sci. 44(5):2059-2066.

[34] Huang W, Koralewska-Makar A, Bauer B, Akesson B (1997) Extracellular glutathione peroxidase and ascorbic acid in aqueous humor and serum of patients operated on for cataract. Clin Chim Acta. 261(2):117-130.

[35] Kisic B, Miric D, Zoric L, Ilic A, Dragojevic I (2012) Antioxidant Capacity of Lenses with Age-Related Cataract. Oxidative Medicine and Cellular Longevity. Available: http://www.hindawi.com/journals/oximed/2012/467130/.

[36] Linetsky M, Chemoganskiy VG, Hu F, Ortwerth BJ (2003) Effect of UVA Light on the Activity of Several Aged Human Lens Enzymes. Invest Ophthalmol Vis Sci 44(1): 264-74.

[37] Singhal SS, Awasthi S, Srivastava SK, Zimniak P, Ansari NH, Awasthi YC (1995) Novel human ocular glutathione S-transferases with high activity toward 4-hydroxynonenal. Invest Ophthalmol Vis Sci. 36(1):142-150.

[38] Alberti G, Oguni M, Podgor M, Sperduto RD, Tomarev S, Grassi C, Williams S, Kaiser-Kupfer M, Maraini G, Hejtmancik JF (1996) Glutathione S-transferase M1 genotype and age-related cataracts. Lack of association in an Italian population. Invest Ophthalmol Vis Sci. 37(6):1167-1173.

[39] Recknagel RO, Glende EA JR (1984) Spectrophotometric detection of lipid conjugated dienes. Methods Enzymol.105:331-337.

[40] Ledwozyw A, Michalak B, Stepien A, Kadziolka A (1986) The relationship between plasma triglicerides, cholesterol, total lipids and lipid peroxidation products during human atherosclerosis. Clin Chim Acta. 155(3):275-284.

[41] Shimasaki H (1994) Assay of fluorescent lipid peroxidation products. Methods Enzymol. 233:338-346

[42] Beutler E, Duron O, Kelly BM (1963) Improved method for the determination of blood glutathione. J Lab Clin Med. 61:882-888.

[43] Chin PTY, Stults FH, Tapell AL (1976) Purification of rat lung soluble glutathione peroxidase. Biochem Biophys Acta. 445 (3):558-566.

[44] Habig WH, Pabst MJ, Jakoby WB (1974) Glutatione-S-transferases. J Biol Chem. 249:7130-7134.

[45] Glatzle D, Vuillenmir JP, Weber F, Decker K (1974) Glutatione reductase test with whole blood, a convenient procedure for the assesment of the riboflavine status in humans. Experimentia. 30(6):665-667.

[46] Lowry OH, Rosebrough NJ, Farr AL, Randall RJ (1951) Protein measurement with the Folin phenol reagent. J Biol Chem.193(1):265–275.

[47] Misra HP, Fridovich I (1972) The role of superoxide anion in the autoxidation of epinephrine and a simple assay for superoxide dismutase. J Biol Chem. 247(10):3170-3175.

[48] Taylor A, Jacques PF, Epstein EM (1995) Relations among aging, antioxidant status, and cataract. Am J Clin Nutr. 62(6):1439-1447.

[49] Ttuscott RJW (2005) Age-related nuclear cataract – oxidation is the key. Exp Eye Res. 80:251-259.

[50] Bhuyan KD, Master RWP, Bhuyan KC (1996) Crosslinking of aminophospholipids in cellular membranes of lens by oxidative stress in vitro. Biochim Biophys Acta. 1285 (1):21-28.

[51] Gupta SK, Trivedi D, Srivastava S, Joshi S, Halder N, Verma SD (2003) Lycopene attenuates oxidative Stress Induced experimental Cataract Development: An In Vitro and In Vivo Study. Nutrition.19(9):794-799.

[52] Micelli-Ferrari T, Vendemiale G, Grattagliano I, Boscia F, Arnese L, Altomare E, Cardia L (1996) Role of lipid peroxidation in the pathogenesis of myopic and senile cataract. Br J Ophthalmol. 80(9): 840-843.

[53] Roberts JE, Finley EL, Patat SA, Schey KL (2001) Photooxidation of lens proteins with xanthurenic acid: a putative chromophore for cataractogenesis. Photochem Photobiol. 74(5):740-744.

[54] Vazquez S, Aquilina JA, Jamie JF, Sheil MM, Truscott RJ (2002) Novel protein modification by kynurenine in human lenses. J Biol Chem. 277: 4867-4873.

[55] Taylor LM, Aquilina JA, Jamie JF, Truscott RJ (2002) UV filter instability: consequences for the human lens. Exp Eye Res. 75:165-175.

[56] Lin D, Barnett M, Grauer L, Robben J, Jewell A, Takemoto L, Takemoto DJ (2005) Expression of superoxide dismutase in whole lens prevents cataract formation. Molecular Vision 11:853-858.

[57] Ganea E, Harding JJ (2006) Glutathione-related enzymes and the eye. Curr Eye Res. 31(1):1-11.

[58] Li W, Calvin HI, David LL, Wu K, McCormack AL, Zhu GP, Fu SC (2002) Altered patterns of phosphorylation in cultured mouse lenses during development of buthionine sulfoximine cataracts. Exp Eye Res. 75(3):335-346.

[59] Lou MF (2000) Thiol Regulation in the lens. Journ of Ocular Pharmacol and Therapeut. 16 (2):137-148.

[60] Babizhayev MA (1996) Failure to withstand oxidative stress induced by phospholipid hydroperoxides as a possible cause of the lens opacities in systemic diseases and ageing. Biochim Biophys Acta.1315(2):87-99.

[61] Rieger G, Winkler R (1994) Changes of Glutathione Peroxidase Activity in Eye Tissues of Emory Mice in Relation to Cataract Status and Age. Ophthalmologica. 208(1):5-9.

[62] Srivastava SK, Singhal SS, Awasthi S, Pikula S, Ansari NH, Awasthi YC (1996) A glutathione S-transferases isozyme (bGST 5.8) involved in the metabolism of 4-hydroxy-2-trans-nonenal is localized in bovine lens epithelium. Exp Eye Res. 63(3):329-337.

[63] Huang QL, Lou MF, Straatsma BR, Horwitz J (1993) Distribution and activity of glutathione-S-transferse in normal human lenses and in cataractous human epithelia. Curr Eye Res. 12(5): 433-437.

[64] Yan H, Harding JJ, Xing K, Lou MF (2007) Revival of Glutathione in Human Cataractous and Clear Lens Extracts by Thioredoxin and Thioredoxin Reductase, in Conjunction with α-Crystallin or Thioltransferase. Curr Eye Res. 32(5): 455-463.

[65] Linetsky M, Chemoganskiy VG, Hu F, Ortwerth BJ (2003) Effect of UVA light on the activity of several aged human lens enzymes. Invest Ophthalmol Vis Sci. 44(1):264-74.

[66] Ozmen B, Ozmen D, Erkin E, Habif S, Bayindir O (2002) Lens speroxide dismutase and catalase activities in diabetic cataract. Clin Biochem. 35(1):69-72.

Chapter 10

Region Specific Vulnerability to Lipid Peroxidation in the Human Central Nervous System

Alba Naudí, Mariona Jové, Victòria Ayala, Omar Ramírez, Rosanna Cabré, Joan Prat, Manuel Portero-Otin, Isidre Ferrer and Reinald Pamplona

Additional information is available at the end of the chapter

1. Introduction

Around 100 billion neurons in the human nervous system orchestrate an exceptionally wide range of motor, sensory, regulatory, behavioural, and executive functions. Such diverse functional output is the product of different molecular events occurring in nervous cells and particularly, neurons. Morphologically, central nervous system (CNS) neurons differ in size, number and complexity of dendrites, number of synaptic connections, length of axons and distance across which synaptic connections are established, extent of axonal myelination, and other cellular characteristics. Neuronal diversity is also amplified by the inclusion of chemical specificity on the basis of the neurotransmitters, which they use for chemical transmission or neuromodulation. This great diversity among neuronal populations is a strong indication that although all neurons contain the same genetic code in their genome, each neuronal population has their own gene expression profile. While the diversity of neuronal structures and functions are well documented, what is less appreciated is the diverse response of neurons to stresses and adverse factors during aging or as a result of neurodegenerative diseases. Furthermore, to add complexity to an already heterogeneous landscape, non-neuronal populations, often described as a 'supporting matrix' are recently being recognized as active, information-rich, cellular counterpart in CNS function.

In this scenario of cellular diversity emerges the concept of selective neuronal vulnerability (SNV). SNV is described as the differential sensitivity of neuronal populations in the nervous system to stresses that cause cell damage or death and can lead to neurodegeneration [1,2]. The fact that specific regions of the nervous system exhibit differential vulnerabilities to aging and various neurodegenerative diseases is a reflection of both the specificity in the aetiology of each disease and of the heterogeneity in neuronal (and non-neuronal) responses to cell-damaging processes associated with each of the diseases [3]. The appearance of SNV is not limited to cross-regional differences in the nervous system, as within a single e.g. brain region, such as the hippocampus or the entorhinal cortex, where SNV is manifested as internal, sub-regional differences in relative sensitivities to stress and disease [2,3]. Among these cell-damaging processes one could count inflammatory, proteotoxicity, vascular and many other pathophysiological processes, including an excess of lipid peroxidation (see later).

Oxidative stress (OS) is involved in the basic mechanisms of nervous system aging; whilst an excessive oxidation has been invoked as an etiopathogenic or physiopathologic mechanism for neurodegeneration. Oxidative stress, the result of an imbalance between production of free radicals and the enzymatic or non-enzymatic detoxification of these highly reactive species, is detrimental to cells because free radicals chemically modify lipids, proteins, and nucleic acids. So, it is very important to define the vulnerability of the different neuronal populations in terms of susceptibility to oxidative stress in physiological conditions in order to extent this knowledge to improve our understanding of how this particular form of cell vulnerability causes selective neuronal losses in nervous system, as well as reveal potential molecular and cellular mechanisms that bring about relative resistance or sensitivity of neurons to stresses.

2. Mitochondrial free radical generation and membrane fatty acid composition in the CNS

2.1. Mitochondrial free radical production

Chemical reactions in cells of the nervous system are under strict enzyme control and conform to a tightly regulated metabolic program in order to minimize unnecessary side reactions. Nevertheless, apparently uncontrolled and potentially deleterious reactions occur, even under physiological conditions. Reactive oxygen species (ROS) express a variety of molecules and free radicals (chemical species with one unpaired electron) physiologically generated from the metabolism of molecular oxygen [4]. They are extremely reactive and have damaging effects. Superoxide anion, the product of a one-electron reduction of oxygen, is the precursor of most ROS and a mediator in oxidative chain reactions [4] (**Figure 1**). The character of radical is not circumscribed to oxygen containing species, as nitrogen, chloride and sulphide containing molecules could also play a significant role. Globally, in cells of the CNS the major sites of physiological ROS generation are the complex I and III of the mitochondrial electron transport chain, which contains several redox centers (flavins, iron–sulphur clusters, and ubisemiquinone) capable of transferring one electron to oxygen to form superoxide anion [5,6]. Oxidative damage is a broad term used to cover the attack upon biological molecules by free radicals. ROS attack/damage all cellular constituents [6], but especially biological membranes.

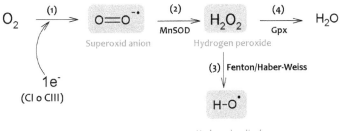

Figure 1. The three main physiological ROS. CI, mitochondrial complex I; CIII, mitochondrial complex III; MnSOD, Manganese superoxide dismutase; Gpx, glutathione peroxidase. MnSOD and Gpx are antioxidants enzymes.

2.2. Membrane fatty acid composition in neural cells

All living organisms have lipid membranes. Biological membranes are dynamic structures that generally consist of bilayers of amphipathic molecules hold together by non-covalent bonds [7]. Phospholipids, consisting of a hydrophilic head group with attached hydrophobic acyl chains, are the predominant membrane lipids and are, from a topographic point of view, asymmetrically distributed across the bilayer. The variation in head groups and aliphatic chains allows the existence of a huge range of different phospholipid species [8,9]. The acyl chains are either saturated, monounsaturated or polyunsaturated hydrocarbon chains that normally vary from 14 to 22 carbons in length (**Figure 2**), with an average chain length strictly maintained around 18 carbon atoms, and a relative distribution between saturated and unsaturated fatty acids of 40:60 (SFA and UFA, respectively) [10]. Polyunsaturated fatty acids (PUFAs) are essential components of cellular membranes that strongly affect their fluidity, flexibility and selective permeability, as well as many cellular and physiological processes [11].

Long-chain polyunsaturated fatty acids are highly enriched in the nervous system. Docosahexaenoic acid (DHA; 22:6n-3; 4,7,10,13,16,19-C22:6), in particular, is the most abundant PUFA in the brain and is concentrated in aminophospholipids of cell membranes. Numerous studies have indicated that this concentration of DHA in the nervous system is essential for optimal neuronal functions. Although the underlying mechanisms of its essential function are still not clearly understood, emerging evidence suggests that unique metabolism of DHA in relation to its incorporation into neuronal membrane phospholipids plays an important role.

Accretion of DHA in the CNS actively occurs during the developmental period, primarily relying on circulating plasma DHA derived from diet or from biosynthesis in the liver [12]. However, local biosynthesis of DHA also occurs in the brain, providing an alternative source of DHA for its accumulation in the brain [13]. It is well established that DHA can be biosynthesized from α-linolenic acid (18:3n-3; 9,12,15-C18:3), a shorter chain n-3 fatty acid precursor, through chain elongation and desaturation processes [14] (**Figure 3**). Linolenic

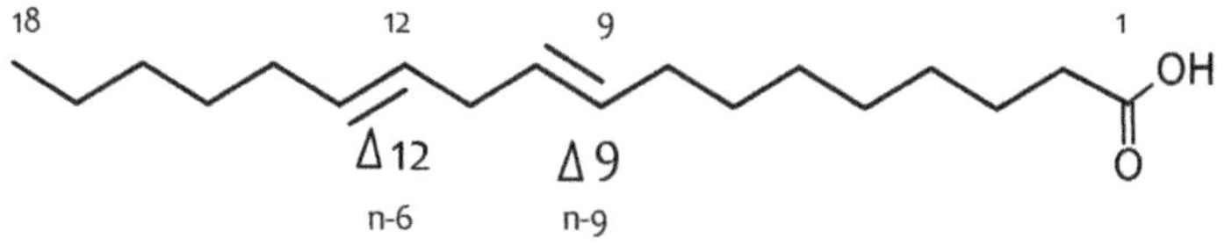

n= chain length= 18 carbons

x= 2 double bounds   Cn:x(n-y) $\longrightarrow$ C18:2(n-6)

x-y=n-6

position of the first double bound from the methyl end

Figure 2. Fatty acid nomenclature. As an example, the structure of linoleic acid (C18:2n-6) is given.

acid is desaturated to 18:4n-3 (6,9,12,15-C18:4) by Δ6-desaturase, chain-elongated to 20:4n-3 (8,11,14,17-C20:4), and subsequently converted to eicosapentaenoic acid (20:5n-3; 5,8,11,14,17-C20:5) by Δ5-desaturase in the endoplasmic reticulum (ER). Mammalian Δ5- and Δ6-desaturases have been identified and cloned [15]. However, Δ4-desaturase, responsible for making 22:6n-3 directly from 22:5n-3, an elongation product of 20:5n-3, has been identified only in microalgae [16]. In mammals, 22:5n-3 is further elongated to 24:5n-3 (9,12,15,18,21-C24:5) followed by desaturation by Δ6-desaturase to 24:6n-3 (6,9,12,15,18,21-C24:6). Subsequently, 24:6n-3 is transferred to peroxisomes and converted to 22:6n-3 by removing two carbon chains by β-oxidation. DHA thus formed is transferred back to the ER and quickly incorporated into membrane phospholipids by esterification during de novo synthesis or by a deacylation-reacylation reaction. Because biosynthesis of both fatty acids and phospholipids occurs in ER, a particular fatty acid intermediate can be either incorporated into phospholipids or further chain-elongated/desaturated, although the regulation of these processes is still poorly understood. Long-chain n-6 fatty acids are biosynthesized from linoleic acid (18:2n-6; 9,12-C18:2) using the analogous pathway and the same enzyme system (**Figure 3**). In most tissues, the commonly observed long-chain n-6 fatty acid is arachidonic acid (AA; 20:4n-6; 5,8,11,14-C20:4). Docosapentaenoic acid (DPAn-6; 22:5n-6; 4,7,10,13,16-C22:5) produced by further elongation and desaturation of AA and subsequent peroxisomal β-oxidation, is rather a minor component, and yet it accumulates in the brain in place of DHA when the DHA supply is inadequate, especially during developmental periods. The distinctive fatty acid profile in the brain enriched with DHA or DPAn-6 may reflect the brain-specific uptake and/or regulation of fatty acid synthesis and esterification into membrane phospholipids. The liver is considered to be the primary site for biosynthesis of DHA, which becomes available to brain uptake through subsequent secretion into the circulating blood stream. Among neural cells, consisting of neurons, astrocytes, microglia, and oligodendrocytes, the capacity to synthesize DHA has been demonstrated only in astrocytes [13]. Despite the fact that neurons are major targets for DHA accumulation, they cannot produce DHA because of lack of desaturase activity. Cerebromicrovascular endothelia can also elongate and desaturate shorter carbon chain fatty acids. However, they cannot perform the final desaturation step to produce either 22:5n-6 or 22:6n-3 [17].

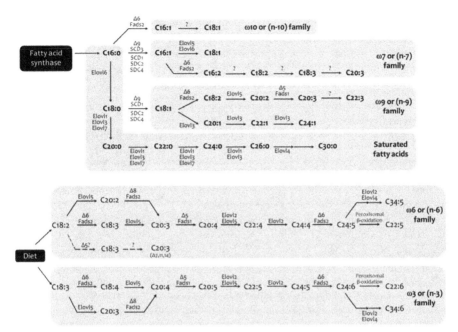

Figure 3. Long chain and very long-chain fatty acid biosynthesis in mammals. The long chain saturated fatty acids and unsaturated fatty acids of the n-10, n-7 and n -9 series can be synthesized from palmitic acid (C16:0) produced by the fatty acid synthase (FAS). Long-chain fatty acids of the n-6 and n-3 series can only be synthesized from precursors obtained from dietary precursors (DIET). Elovl, elongation of very long chain fatty acids (fatty acid elongase); Fads, fatty acid desaturases.

DHA synthesis in astrocytes is negatively influenced by the availability of preformed DHA [18] and thus may represent a quantitatively minor source for the neural DHA accretion when the circulating DHA supply is adequate. Incorporation of circulating DHA across the blood brain barrier appears to be an important route for maintaining adequate levels of DHA in the brain. In agreement with this notion, it has been shown that constant basal turnover of esterified DHA in the brain with unesterified DHA in plasma occurs at an estimated rate of 2–8% per day in adult rats [19]. Generally, it is difficult to deplete DHA from the neural membranes of adult mammals even with a DHA low diet, presumably because of preferential uptake of DHA into the brain to support the basal turnover. In the case of insufficient supply of n-3 fatty acids during development, the loss of DHA does occur but is compensated with DPAn-6 through reciprocal replacement, suggesting a requirement of very long-chain, highly unsaturated fatty acids in neural membranes.

Whether brain DHA is derived from the circulating plasma pool or biosynthesized locally, in the astrocytes, which are situated in close contact with neurons, appear to play an important role in supplying DHA to neurons. DHA can be released readily from astroglial membranes under basal and stimulated conditions, and supplied to neurons. Despite its high abundance in neuronal membranes, DHA is not easily released but is tenaciously retained in the neuronal membranes under the conditions in which AA can be released. Considering the fact that astroglia support neurons by providing neurotrophic factors, DHA supplied by astroglia may also be trophic. Indeed, DHA has been shown to promote neuronal survival [20] and differentiation [21] in both transformed and primary neuronal cells in culture.

3. Membrane unsaturation and lipoxidation-derived molecular damage

The susceptibility of membrane phospholipids to oxidative demise is related to two inherent traits, the physico-chemical properties of the membrane bilayer and the intrinsic chemical reactivity of the fatty acids composing the membrane [10,22]. The first property is related to the fact that oxygen and free radicals are more soluble in the fluid lipid bilayer than in the aqueous solution. Thus, membranes contain an interior organic phase in which the oxygen may tend to concentrate. Therefore, these differences in solubility are important when considering the availability of oxygen/free radicals for chemical reactions inside living systems: organic regions may contain more free radicals than aqueous regions [8] and, consequently, membrane lipids become primary targets of oxidative damage. The second property is related to the fact that PUFA residues of phospholipids are extremely sensitive to oxidation. Every membrane phospholipid contains an unsaturated fatty acid residue esterified to the 2-hydroxyl group of its glycerol moiety. Many of these are polyunsaturated and the presence of a methylene group between two double bonds renders the fatty acid more sensitive to ROS-induced damage. Therefore, the sensitivity of these molecules to oxidation increase exponentially as a function of the number of double bonds per fatty acid molecule [22,23]. Consequently, the high concentration of PUFAs in phospholipids not only makes them prime targets for reaction with oxidizing agents but also enables them to participate in long free radical chain reactions. Reactive free radicals can pull off hydrogen atoms from PUFA side chains. A hydrogen atom (H•) has only one electron. This hydrogen is bonded to a carbon in the fatty acid backbone by a covalent bond. Hence, the carbon from which H• is abstracted now has an unpaired electron (i.e it is a free radical). PUFA side chains (two or more double bonds) are more sensitive to attack by radicals than are SFAs (no double bonds) or monounsaturated fatty acids (MUFA,one double bond) side chains. When C• radicals are generated in the hydrophobic interior of membranes, their most likely fate is combination with oxygen dissolved in the membrane. The resulting peroxyl radical is highly reactive: it can attack membrane proteins and oxidize adjacent PUFA side chains. So, the reaction is repeated and the whole process continues in a free radical chain reaction, generating lipid hydroperoxides [4]. Lipid hydroperoxides are more hydrophilic than unperoxidized fatty acid side chains. They try to migrate to the membrane surface to interact with water, thus disrupting the membrane structure, altering fluidity and other functional properties and making the membrane leaky.

Lipid peroxidation generates hydroperoxides as well as endoperoxides, which undergo fragmentation to produce a broad range of reactive intermediates called reactive carbonyl species (RCS) (**Figure 4**) with three to nine carbons in length, the most reactive being α,β-

unsaturated aldehydes [4-hydroxy-trans-2-nonenal (HNE) and acrolein], di-aldehydes [malondialdehyde (MDA) and glyoxal], and keto-aldehydes [4-oxo-trans-2-nonenal (ONE) and isoketals] [24]. 2-Hydroxyheptanal (2-HH) and 4-hydroxyhexenal (4-HHE) are aldehydic product of lipid peroxidation of PUFAn-6. Additionally, a number of other short chain aldehydes are produced during lipid peroxidation through poorly understood mechanisms. These carbonyl compounds, ubiquitously generated in biological systems, have unique properties contrasted with free radicals. Thus, compared with reactive oxygen and nitrogen species, reactive aldehydes have a much longer half-life (i.e., minutes to hours instead of microseconds to nanoseconds for most free radicals). Further, the non-charged structure of aldehydes allows them to migrate relatively ease through hydrophobic membranes and hydrophilic cytosolic media, thereby extending the migration distance far from the production site. Based on these features alone, these carbonyl compounds can be more destructive than ROS and may have far-reaching damaging effects on target sites within or outside membranes. For the same reason, their long half-life allows them to subserve as second-messenger, signalling for important cellular responses in a mostly unknown fashion.

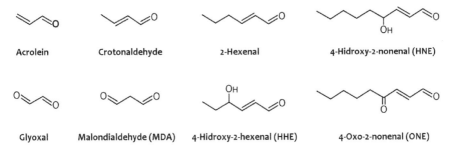

Figure 4. General structures of principal lipoxidative reactive carbonyl species detected in biological systems.

Carbonyl compounds react with nucleophilic groups in macromolecules like proteins, DNA, and aminophospholipids, among other, resulting in their chemical, nonenzymatic, and irreversible modification and formation of a variety of adducts and crosslinks collectively named Advanced Lipoxidation Endproducts (ALEs) [10,25,26] (**Figure 5**) . Thus, by reacting with nucleophilic sites in proteins (belonging basically to Cys, Lys, Arg, and His residues), carbonyl compounds generate ALE adducts such as MDA-Lys, HNE-Lys, FDP-Lys, carboxymethyl-lysine (CML) and S-carboxymethyl-cysteine; and the crosslinks glyoxal-lysine dimer (GOLD), and methylglyoxal-lysine dimer (MOLD), among several others. The accumulation of MDA adducts on proteins is also involved in the formation of lipofuscin (a nondegradable intralysosomal fluorescent pigment formed through lipoxidative reactions). Lipid peroxidation-derived endproducts can also react at the exocyclic amino groups of deoxyguanosine, deoxyadenosine, and deoxycytosine to form various alkylated products. Guanine is, however, the most commonly modified DNA base because of its high nucleophilicity. Some common enals that cause DNA damage, analogously to proteins, are

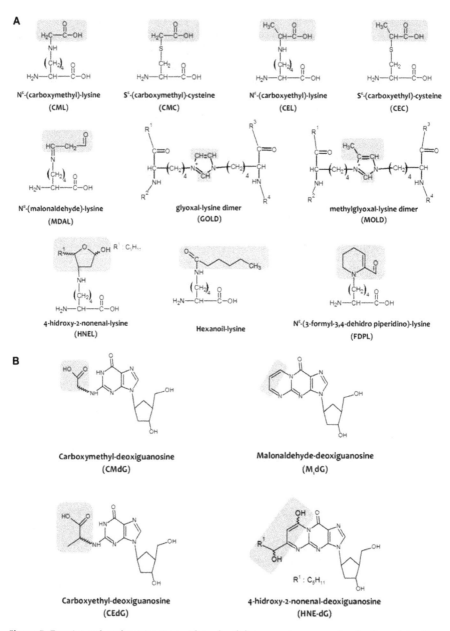

Figure 5. Reactive carbonyl species react with nucleophilic groups in macromolecules (A, proteins; B, DNA) resulting in their chemical, nonenzymatic, and irreversible modification and formation of a variety of adducts and crosslinks collectively named Advanced Lipoxidation Endproducts (ALEs).\

MDA, HNE, and acrolein, among others. Thus, the most common adducts arising from enals are exocyclic adducts such as etheno adducts, and MDA-deoxyguanosine (M1dG). Finally, the amino group of aminophospholipids can also react with carbonyl compounds and initiate some of the reactions ocurring in proteins and DNA, leading to the formation of adducts like MDA-phosphatidylethanolamine, and carboxymethyl-phosphatidylethanolamine [10].

4. ALEs: Molecular and cellular effects

Reactive carbonyl species (RCS) generated during lipid peroxidation reactions exhibit a wide range of molecular and biological effects, ranging from protein, DNA, and phospholipid damage to signaling pathway activation and/or alteration. The detailed mechanisms of 'toxicity' are, however, mostly unknown.

4.1. Molecular damage

Lipoxidation reactions lead to structural and functional changes in proteins [25-27] such as i) alterations in physico-chemical properties (conformation, charge, hydrophobicity, elasticity, solubility, and electrophoretic mobility, among others); ii) decrease/inhibition in enzyme activity and growth factors; iii) alteration of protein degradation; iv) alteration in traffic and processing of proteins; and v) formation of intra- and inter-molecular protein cross-links and aggregates.

DNA lipoxidative damage is present in the genome of healthy humans and other animal species at biologically significant levels similar or even higher that oxidation markers *sensu stricto*. DNA damage is mutagenic, carcinogenic, and have powerful effects on signal transduction pathways [28].

Finally, the amino group of aminophospholipids can also react with carbonyl compounds and initiate some of the reactions ocurring in proteins [29]. Biological processes involving aminophospholipids could be potentially affected by this process. Among these processes, it may be highlighted i) asymmetrical distribution of aminophospholpids in cellular and different subcellular membranes; ii) translocation between and lateral diffusion in the membrane; iii) membrane physical properties; iv) biosynthesis and turnover of membrane phospholipids; and v) activity of membrane-bound proteins that require aminophospholipids for their function.

4.2. Cellular adaptive responses

The peroxidation of the PUFA chains of phospholipids generates a complex mixture of carbonyl compounds. Initially, these aldehydes were believed to produce only "cytotoxic" effects associated with oxidative stress (by inducing molecular damage and damaging cellular responses based on inflammatory responses, changes in gene expression and apoptosis) [10], but as depicted above, evidence is increasing in the sense that these compounds can also have specific signaling roles inducing adaptive responses driven to decrease oxidative damage and improve antioxidant defences.

Thus, available studies support the notion that superoxide radical produced by the mitochondrial electron transport chain can cause mild uncoupling of mitochondria by activating the membrane proton conductance by uncoupling proteins (UCPs). Insight into the molecular mechanism by which superoxide radical activates UCPs comes from the finding that the lipid peroxidation product 4-HNE and its homologs induce uncoupling of mitochondria through UCPs and also through the adenine nucleotide translocase [30]. This and other observations support a model in which endogenous superoxide production generates carbon-centred radicals that initiate lipid peroxidation, producing alkenals like 4-HNE that activate UCPs and adenine nucleotide translocase. So, UCPs respond to overproduction of matrix superoxide by catalyzing mild uncoupling, which lower proton motive force and would decrease superoxide production by the electron transport chain. This negative feedback loop will protect cells from ROS-induced damage and might represent the ancestral function of all UCPs.

In addition, RCS can also activate the 'antioxidant response' likely to prevent their accumulation to toxic levels [31]. This signaling cascade culminates in the nuclear translocation of and transactivation by the transcription factor Nrf2, the master regulator of the response [32]. Nrf2 activity is repressed by an inhibitory binding protein, Keap1. Keap1 retains Nrf2 in the cytosol, closely associated with the actin cytoskeleton, and promotes proteasomal degradation of Nrf2 through Cullin3-dependent polyubiquitination. Following exposure to RCS, Keap1 can be directly modified on several cysteine residues, and this modification can promote release of Nrf2. Nrf2 contains a C-terminal basic leucine zipper structure that facilitates dimerization and DNA binding, specifically to the antioxidant response element (ARE). The binding of Nrf2 to the ARE stimulates transcription of downstream cytoprotective genes [32].

5. The selective neuronal vulnerability

The idea that oxygen radicals, especially those of mitochondrial origin, are causally related to the basic aging process is increasingly receiving support from several independent sources [reviewed in 6,33]. Accordingly, the mitochondrial oxygen radical theory of aging apparently fulfils the main characteristics of this natural process: reactive oxygen species (ROS) are endogenously produced at mitochondria under normal physiological conditions, they are produced continuously throughout life (and can thus lead to progressive aging changes), and their deleterious effects on macromolecules may inflict irreversible damage during aging in post-mitotic tissues. The detrimental effects of aging are best observed in postmitotic tissues because cells that are irreversibly damaged or lost cannot be replaced by mitosis of intact ones. Nervous system is considered a postmitotic tissue, and therefore highly susceptible to aging. As this process is involved as a risk factor in most neurodegenerative diseases, and oxidative modifications play a key role in aging, it is often accepted that these diseases should have increased oxidative damage.

In this context, and from an inter-organ comparative approach, two main properties emerge as characteristics that render nervous system as especially sensible to oxidative modification: i) the % free radical leak, and ii) the membrane unsaturation. The % free radical leak (%FRL) refers the fraction (%) of electrons out of sequence which reduce oxygen to oxygen radicals (instead of reducing oxygen to water at cytochrome oxidase) in the mitochondrial respiratory chain. Since two electrons are needed to reduce one molecule of oxygen to H_2O_2, whereas four electrons are needed to reduce one molecule of oxygen to water, the free radical leak is easily calculated by dividing the rate of ROS production by 2 times the rate of oxygen consumption, the result being multiplied by 100. Results show that the higher %FRL corresponds to brain, suggesting that mitochondria are more inefficient in this tissue than in other organs (**Tables 1** and **2**, and **Figure 6**).

		Brain	Heart	Kidney	Liver	Skeletal muscle
Pyruvate/malate	State 4	11.2 ± 1.7	28.4 ± 4.7	-	7.8 ± 1.0	22 ± 3
	State 3	23.1 ± 3.0	72.2 ± 5.4	-	19.5 ± 2.4	127 ± 18
Glutamate/malate	State 4	11.5 ± 2.1	-	23.2 ± 4.2	9.4 ± 0.8	-
	State 3	23.6 ± 4.1	-	88.1 ± 13.4	72.1 ± 5.3	-
Succinate/rotenone	State 4	15.8 ± 1.6	85.8 ± 9.4	55.6 ± 11.6	26.2 ± 2.0	97 ± 12
	State 3	24.0 ± 2.0	112. ± 5.3	148.1 ± 27.0	100.9 ± 9.1	239 ± 34

Values are mean ± SEM from 8 different animals. State 4, oxygen consumption in the absence of ADP; State 3, oxygen consumption in the presence of ADP. Data from references: 34-36 and unpublished results.

Table 1. Rates of mitochondrial oxygen consumption (nmoles of O_2/ min mg protein) of different organs from male adult rats.

Substrate	Brain	Heart	Kidney	Liver	Skeletal muscle
Pyr/mal	0.2 ± 0.04	0.24 ± 0.04	-	0.06 ± 0.01	0.084 ± 0.020
Pyr/mal + Rot	1.05 ± 0.13	1.49 ± 0.12	-	0.38 ± 0.04	0.73 ± 0.14
Glut/mal	0.17 ± 0.04	-	0.075 ± 0.05	0.16 ± 0.02	-
Glut/mal + Rot	0.72 ± 0.10	-	0.66 ± 0.12	0.46 ± 0.04	-
Succ + Rot	0.26 ± 0.05	0.51 ± 0.05	0.10 ± 0.04	0.32 ± 0.04	0.31 ± 0.06
Succ	1.72 ± 0.17	0.71 ± 0.12	0.52 ± 0.08	1.00 ± 0.18	1.57±0.31
Succ + AA	2.67 ± 0.31	-	5.97 ± 0.45	3.00 ± 0.41	2.57± 0,30

Values are means ± SEM from 8 different animals. Pyr/mal = pyruvate/malate; Glu/mal = glutamate/malate; Succ, succinate; Rot, rotenone; AA, antimycin A. Data from references: 34-36 and unpublished results.

Table 2. Rates of mitochondrial H_2O_2 production (nanomoles H_2O_2/min mg protein) of different organs from male adult rats.

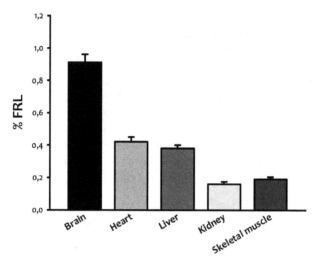

Figure 6. % Free radical leak (% FRL) of different organs from male adult rats. Data from references: 34-36 and unpublished results.

As mentioned above, the susceptibility of biological membranes to oxidative alterations is related to two inherent traits, the physico-chemical properties of the lipid bilayer and the chemical reactivity of the fatty acids which make up the membrane. **Table 3** shows the mitochondrial fatty acid composition (mol%) of different organs. Data clearly indicate that average chain length is maintained around 18 carbon atoms and that the % saturated:unsaturated follows the ratio around 40:60. By contrast, the more relevant

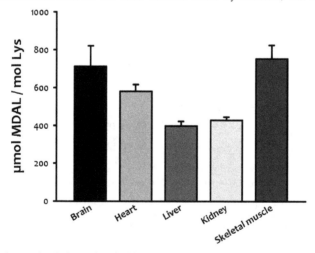

Figure 7. Steady-state level of mitochondrial lipoxidative-derived protein damage in different organs from male adult rats. MDAL, malondialdehyde-lysine. Units: μmol/mol lysine.

differences are droved to the distribution of the different types of PUFAs. Thus, brain is characterized by the presence of the higher content of monounsaturated fatty acids, as well as PUFAn-3, and particularly, the docosahexaenoic acid (22:6n-3) compared to the other postmitotic organs (**Table 3**). This higher PUFAs content which are more susceptible to oxidative damage leads to a higher steady-state level of lipoxidation-derived molecular damage at least at mitochondrial level (**Figure 7**).

	Brain	Heart	Kidney	Liver	Skeletal muscle
14:0	0.14±0.03	1.17±0.03	0.36±0.03	0.18±0.01	0.82±0.14
16:0	10.55±0.46	22.68±0.60	10.00±0.38	17.31±0.44	20.39±0.45
16:1n-7	0.28±0.02	0.91±0.04	0.48±0.09	1.11±0.07	0.67±0.05
18:0	18.12±0.39	26.49±0.34	12.67±0.56	17.19±0.24	18.34±0.73
18:1n-9	24.78±0.31	13.23±0.39	8.97±0.44	9.09±0.17	8.44±0.98
18:2n-6	1.55±0.07	9.27±0.42	14.10±0.42	19.27±0.66	17.73±1.23
18:3n-3	0.12±0.03	0.38±0.01	0.47±0.11	0.18±0.04	0.80±0.12
18:4n-6	2.38±0.08	0.11±0.01	4.67±0.40		
20:0	1.05±0.07	0.20±0.01	0.51±0.07		
20:1n-9	4.12±0.18	0.19±0.02	0.69±0.09		
20:2n-6	0.52±0.05	0.97±0.14	0.92±0.06	0.35±0.01	
20:3n-6	0.10±0.01	0.79±0.06	0.38±0.03	0.27±0.02	
20:4n-6	12.74±0.13	15.69±0.45	36.32±1.21	26.61±0.44	13.96±0.79
20:5n-3	0.14±0.01	0.15±0.01	0.40±0.07	0.35±0.02	1.66±0.44
22:0	1.57±0.26	0.21±0.02	2.24±0.16		
22:4n-6	4.19±0.16	0.73±0.06	0.46±0.02	0.16±0.008	1.54±0.35
22:5n-6	0.68±0.03	1.40±0.22	0.46±0.04	0.28±0.04	1.99±0.23
22:5n-3	2.00±0.16	0.84±0.04	1.56±0.11	0.79±0.04	2.95±0.23
22:6n-3	12.81±0.33	4.28±0.26	1.81±0.19	6.79±0.44	10.65±0.96
24:0	0.14±0.04	0.25±0.04	0.27±0.04		
24:5n-3	0.26±0.07		0.93±0.22		
24:6n-3	1.65±0.16		1.24±0.28		
ACL	19.12±0.05	18.15±0.03	18.97±0.03	18.49±0.01	18.54±0.02
SFA	31.61±0.52	51.02±0.71	26.07±0.62	34.69±0.50	39.56±1.04
UFA	68.38±0.52	48.97±0.71	73.92±0.62	65.30±0.50	60.43±1.04
MUFA	29.19±0.32	14.34±0.44	10.15±0.51	10.20±0.18	9.11±1.01
PUFA	39.19±0.73	34.63±0.94	63.77±1.07	55.09±0.56	51.32±1.41
PUFAn-6	22.18±0.20	28.97±0.92	57.33±1.16	46.97±0.95	36.91±1.18
PUFAn-3	17.00±0.57	5.66±0.25	6.43±0.32	8.12±0.41	14.41±0.88

Values: mean±SEM. N x group: 8. ACL, average chain length; SFA, saturated faaty acids; UFA, unsaturated fatty acids; PUFA n-6/n-3, polyunsaturated fatty acids n-6 or n-3 series; MUFA, monounsaturated fatty acids. Data from references: 34-36 and unpublished results.

Table 3. Mitochondrial fatty acid composition (mol%) of different organs from male adult rats.

Compared to other postmitotic organs, brain seems to be the tissue more susceptible to oxidative damage. However, are there cross-regional differences in the nervous system? Available data seems to indicate that this is the case. Thus, **Table 4** reflects the fatty acid composition in seven different regions from human nervous system samples. The data are conclusive: despite to maintain a stable average chain length (around 18 carbon atoms) and a ratio saurated:unsaturated practically identical (40:60), there is a cross-regional difference with respect to the type of PUFA distribution affecting very specially to both monounsaturated and polyunsaturated fatty acids that seems to be inversely related. In other words, the higher the presence of monounsaturated fatty acids for a given region, the lower the PUFA content. The meaning of this differential distribution remains to be elucidated, but it is evident that determine a differential susceptibility to oxidative damage.

	Frontal Cortex	Occipital Cortex	Hippocampus	Amygdala	Substantia Nigra	Medulla Oblongata	Spinal Cord
14:0	0.51±0.04	0.54±0.05	0.42±0.05	0.56±0.05	0.54±0.04	0.50±0.04	0.48±0.05
16:0	21.13±0.51	20.30±1.70	19.54±0.27	20.69±0.67	12.68±0.39	13.57±0.38	14.75±0.25
16:1n-7	0.98±0.12	0.76±0.19	0.89±0.10	1.48±0.12	1.68±0.18	1.21±0.03	0.94±0.06
18:0	21.35±1.23	19.23±0.41	21.89±0.20	20.19±0.67	24.61±0.27	25.64±0.61	22.07±1.72
18:1n-9	23.78±1.51	28.57±1.16	26.12±0.50	28.40±0.77	33.57±0.19	30.13±0.87	33.51±1.43
18:2n-6	0.77±0.12	0.57±0.07	0.59±0.04	0.53±0.09	0.33±0.06	0.56±0.01	0.96±0.61
18:3n-3	0.14±0.01	0.21±0.04	0.23±0.02	0.08±0.006	0.17±0.003	0.23±0.01	0.34±0.05
20:0	1.25±0.23	2.40±0.52	1.44±0.09	1.71±0.12	4.95±0.12	5.58±0.23	7.62±0.53
20:1	0.19±0.04	0.32±0.05	0.37±0.01	0.28±0.02	0.22±0.02	0.45±0.02	0.75±0.09
20:2n-6	0.23±0.01	0.22±0.01	0.23±0.02	0.20±0.02	0.23±0.03	0.31±0.06	0.28±0.01
20:3n-6	0.63±0.07	0.64±0.04	0.90±0.10	0.60±0.08	0.61±0.02	0.99±0.03	0.87±0.16
20:4n-6	8.28±0.19	6.07±0.35	8.05±0.24	7.42±0.42	3.70±0.29	3.95±0.33	3.13±0.09
22:4n-6	4.77±0.14	5.06±0.69	6.24±0.20	4.79±0.27	4.32±0.15	4.07±0.08	3.16±0.17
22:5n-6	0.67±0.09	0.59±0.09	0.92±0.08	0.94±0.11	0.12±0.007	0.40±0.02	0.24±0.04
22:5n-3	0.26±0.03	0.19±0.02	0.33±0.02	0.13±0.04	0.06±0.006	0.09±0.01	0.38±0.14
22:6n-3	13.68±0.17	13.71±0.71	10.97±0.31	9.99±0.58	9.74±0.28	9.28±0.45	8.84±0.71
24:0	0.38±0.06	0.31±0.05	0.45±0.04	0.47±0.03	0.64±0.06	1.19±0.22	1.00±0.08
24:1n-9	0.93±0.23	0.21±0.03	0.33±0.04	1.46±0.32	1.73±0.18	1.75±0.11	0.59±0.10
ACL	18.60±0.01	18.56±0.04	18.58±0.008	18.48±0.04	18.59±0.005	18.64±0.03	18.52±0.01
SFA	44.64±1.33	42.80±0.74	43.77±0.37	43.65±0.94	43.43±0.34	46.51±0.47	45.95±1.81
UFA	55.35±1.33	57.19±0.74	56.22±0.37	56.34±0.94	56.56±0.34	53.48±0.47	54.04±1.81
MUFA	25.90±1.70	29.88±1.11	27.74±0.52	31.62±1.05	37.22±0.24	33.56±0.92	35.80±1.53
PUFA	29.45±0.46	27.31±0.38	28.48±0.25	24.71±1.02	19.33±0.42	19.92±0.45	18.24±0.43
PUFAn-6	15.36±0.36	13.18±0.25	16.95±0.36	14.49±0.66	9.34±0.26	10.30±0.30	8.67±0.81
PUFAn-3	14.08±0.18	14.12±0.63	11.53±0.34	10.21±0.59	9.99±0.28	9.61±0.46	9.57±0.57

Values: mean±SEM. N x group: 3-9. ACL, average chain length; SFA, saturated faaty acids; UFA, unsaturated fatty acids; PUFA n-6/n-3, polyunsaturated fatty acids n-6 or n-3 series; MUFA, monounsaturated fatty acids.
(#) Unpublished results.

Table 4. Fatty acid composition (mol%) in the human nervous system: a cross-regional comparative approach [#].

This means that saturated and monounsaturated fatty acyl chains (SFA and MUFA) are essentially resistant to peroxidation while polyunsaturates (PUFA) are damaged. Furthermore the greater the degree of polyunsaturation of PUFA the more prone it is to peroxidative damage. Indeed Holman [23] empirically determined (by measurement of oxygen consumption) the relative susceptibilities the different acyl chains (see **Figure 8**): Docosahexaenoic acid (DHA), the highly polyunsaturated omega-3 PUFA with six double bonds is extremely susceptible to peroxidative attack and is eight-times more prone to peroxidation than linoleic acid (LA) which has only two double bonds. DHA is 320-times more susceptible to peroxidation than the monounsaturated oleic acid (OA) [22,23].

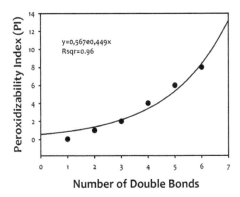

Number of Double Bonds

Figure 8. The relative susceptibilities of selected unsaturated fatty acids to peroxidation. Data are from [23], and all were empirically determined as rates of oxygen consumption. They are expressed relative to the rate for linoleic acid (18:2n-6) which is arbitrarily given a value of 1.

Combining the relative susceptibilities of different fatty acids with the fatty acid composition of membrane lipids it is possible to calculate a peroxidizability index[1] (a measure of the susceptibility to peroxidation) for any particular membrane. The peroxidation index of a membrane is not the same as its unsaturation index (sometimes also called its "double bond index") which is a measure of the density of double bonds in the membrane. For example, a membrane bilayer consisting solely of MUFA will have an unsaturation index of 100 and a peroxidation index of 2.5, while a membrane bilayer consisting of 95% SFA and 5% DHA will have an unsaturation index of 30 and a peroxidation index of 40. This means that although the 5% DHA-containing membrane has only 30% the density of double bonds of the monounsaturated bilayer, it is 16-times more susceptible to peroxidative damage. In this context, data clearly show the existence of very important cross-regional differences in peroxidizability index in human central nervous system (**Figure 9**).

---

[1] *[Peroxidizability Index (PI) = 0.025 x (% monoenoics) + 1 x (% dienoics) + 2 x (% trienoics) + 4 x (% tetraenoics) + 6 x (% pentaenoics) + 8 x (% hexaenoics), while Unsaturation index (UI) = 1 x (% monoenoics) + 2 x (% dienoics) + 3 x (% trienoics) + 4 x (% tetraenoics) + 5 x (% pentaenoics) + 6 x (% hexaenoics)]*

The different cross-regionally PIs observed are due to changes in the type of unsaturated fatty acid that participates in membrane composition. So, there is a systematic redistribution between the types of PUFAs present from highly unsaturated fatty acids to the less unsaturated that is region-specific. Surprisingly, the change shows a gradient that follows the cranio-caudal axis considering the structural organization of the CNS. The mechanism(s) responsible for the cross-regional-related differences in fatty acid profile can be related, in principle, to the fatty acid desaturation pathway, and the deacylation-reacylation cycle. The available delta-5 and delta-6 estimated desaturase activities indicate that they are several folds higher in frontal cortex than in spinal cord (**Figure 10**). The mechanism underlying to the membrane unsaturation regulation could explain the differences in membrane fatty acid composition and, in turn, the peroxidizability index, and suggest a regulatory mechanism region-specific that is expressed differentially in a cranio-caudal axis likely associated to the development process and even the evolution of the central nervous system.

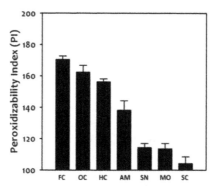

Figure 9. Cross-regional differences in the peroxidizability index in the human central nervous system. FC, frontal cortex; OC, occipital cortex; HC, hippocampus; AM, amygdala; SN, substantia nigra; MO, medulla oblongata; SC, spinal cord.

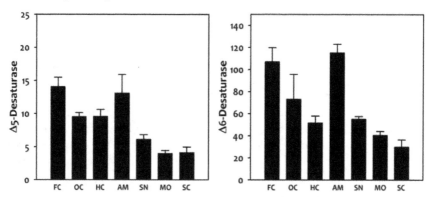

Figure 10. Delta-5 and delta-6 desaturase activities in different regions of the human central nervous system. FC, frontal cortex; OC, occipital cortex; HC, hippocampus; AM, amygdala; SN, substantia nigra; MO, medulla oblongata; SC, spinal cord.

In summary, membrane unsaturation is a key characteristic able to define the selective neuronal vulnerability. In this context, it is plausible to postulate that membrane unsaturation could be a main determinant factor in determining differences in the rate of aging for different regions of the CNS, and in the occurrence of neurodegenerative disorders [e.g., Alzheimer's disease (AD), Parkinson's disease (PD), or amyotrophic lateral sclerosis (ALS), among others] during the sixth, seventh and eighth decades of life. Interestingly, this property is also causally related to the aging process and the lifespan of animal species [6].

6. Conclusions and perspectives

A major goal of research into aging is to extend 'healthspan' by identifying approaches for delaying or preventing age-related diseases. The fact that many individuals maintain a well-functioning nervous system and continue productive lives through their seventies, eighties and even nineties is encouraging. The implication is that if the cellular and molecular mechanisms that determine whether nervous systems adapt positively or develop a disease during aging can be identified, then disease processes can be averted. In this regard, oxidative and metabolic stress and impaired cellular stress adaptation, are mechanisms of aging that render neurons vulnerable to degeneration. On this background of age-related endangerment, genetic and environmental factors likely determine whether a disease process develops. These include causal mutations, more subtle genetic risk factors and environmental factors, including aspects of diet and lifestyle. Because of the cellular and molecular complexity of the nervous system, and the signalling mechanisms that influence neuronal plasticity and survival, the basis of SNV remains elusive. However, available evidence from both an inter-organ comparative approach and cross-regional differences seems to confirm this idea, highlighting membrane unsaturation as a key trait associated with selective neuronal vulnerability. Interestingly, this property is also causally related to the aging process and the lifespan of animal species [6,10], and it is apparently operative in multiple neurodegenerative disorders [3,37]. Currently, most efforts to prevent and treat neurodegenerative disorders are focusing on diet, on lifestyle modification, and on drugs that target disease processes [3,6,33]. Although data on humans is still limited, the emerging evidence that dietary restriction (along with exercise and cognitive stimulation) can bolster neuroprotective mechanisms suggests that diet and lifestyle changes could reduce the risk of neurodegenerative disorders [6,26,33]. Therefore, it seems likely that extension of neural healthspan is possible for most individuals.

Author details

Alba Naudí, Mariona Jové, Victòria Ayala, Omar Ramírez, Rosanna Cabré, Joan Prat, Manuel Portero-Otin, Isidre Ferrer and Reinald Pamplona

Department of Experimental Medicine, University of Lleida-Biomedical Research Institute of Lleida, Lleida, Spain

Acknowledgement

Work carried out at the Department of Experimental Medicine was supported in part by R+D grants from the Spanish Ministry of Science and Innovation [BFU2009-11879/BFI], and the Autonomous Government of Catalonia [2009SGR735]. The authors declare no conflict of interest.

7. References

[1] Mattson MP, Magnus T (2006) Ageing and neuronal vulnerability. Nat. Rev. Neurosci. 7(4):278-94.

[2] Wang X, Michaelis EK (2010) Selective neuronal vulnerability to oxidative stress in the brain. Front. Aging Neurosci. 2:12.

[3] Martínez A, Portero-Otin M, Pamplona R, Ferrer I (2010) Protein targets of oxidative damage in human neurodegenerative diseases with abnormal protein aggregates. Brain Pathol. 20(2):281-97.

[4] Halliwell B, Gutteridge JMC (2007) Free radicals in biology and medicine. Oxford: Oxford University Press. 851 p.

[5] Barja G. (2007) Mitochondrial oxygen consumption and reactive oxygen species production are independently modulated: implications for aging studies. Rej. Res. 10(2):215-24.

[6] Pamplona R, Barja G (2011) An evolutionary comparative scan for longevity-related oxidative stress resistance mechanisms in homeotherms. Biogerontology, 12(5):409-35.

[7] Vereb G, Szollosi J, Matko J, Nagy P, Farkas T, Vigh L, Matyus L, Waldmann TA, Damjanovich S (2003) Dynamic, yet structured: The cell membrane three decades after the Singer-Nicolson model. Proc. Natl. Acad. Sci. U.S.A. 100(14):8053-8.

[8] Van Meer G, Voelker DR, Feigenson GW (2008) Membrane lipids: where they are and how they behave. Nature Rev. Mol. Cell. Biol. 9(2):112-24.

[9] Dowhan W (1997) Molecular basis for membrane phospholipid diversity: why are there so many lipids? Annu. Rev. Biochem. 66:199-232.

[10] Pamplona R (2008) Membrane phospholipids, lipoxidative damage and molecular integrity: a causal role in aging and longevity. Biochim. Biophys. Acta. 1777(10):1249-62.

[11] Wallis JG, Watts JL, Browse J (2002) Polyunsaturated fatty acid synthesis: what will they think of next? Trends Biochem. Sci. 27(9):467.

[12] Scott BL, Bazan NG (1989) Membrane docosahexaenoate is supplied to the developing brain and retina by the liver. Proc. Natl. Acad. Sci. U. S. A. 86(8):2903-7.

[13] Moore SA, Yoder E, Murphy S, Dutton GR, Spector AA (1991) Astrocytes, not neurons, produce docosahexaenoic acid (22:6 omega-3) and arachidonic acid (20:4 omega-6). J. Neurochem. 56(2):518-24.

[14] Sprecher H (2000) Metabolism of highly unsaturated n-3 and n-6 fatty acids. Biochim. Biophys. Acta. 1486(2-3):219-31.

[15] Nakamura MT, Nara TY (2004) Structure, function, and dietary regulation of delta6, delta5, and delta9 desaturases. Annu. Rev. Nutr. 24:345-76.

[16] Pereira SL, Leonard AE, Huang YS, Chuang LT, Mukerji P (2004) Identification of two novel microalgal enzymes involved in the conversion of the omega3-fatty acid, eicosapentaenoic acid, into docosahexaenoic acid. Biochem. J. 384(Pt 2):357-66.

[17] Moore SA, Yoder E, Spector AA (1990) Role of the blood-brain barrier in the formation of long-chain omega-3 and omega-6 fatty acids from essential fatty acid precursors. J. Neurochem. 55(2):391-402.

[18] Williard DE, Harmon SD, Kaduce TL, Preuss M, Moore SA, Robbins ME, Spector AA (2001) Docosahexaenoic acid synthesis from n-3 polyunsaturated fatty acids in differentiated rat brain astrocytes. J. Lipid Res. 42(9):1368-76.

[19] Rapoport SI, Chang MC, Spector AA (2001) Delivery and turnover of plasma-derived essential PUFAs in mammalian brain. J. Lipid Res. 42(5):678-85.

[20] Akbar M, Calderon F, Wen Z, Kim HY (2005) Docosahexaenoic acid: a positive modulator of Akt signaling in neuronal survival. Proc. Natl. Acad. Sci. U. S. A. 102(31):10858-63.

[21] Calderon F, Kim HY (2004) Docosahexaenoic acid promotes neurite growth in hippocampal neurons. J. Neurochem. 90(4):979-88.

[22] Hulbert AJ, Pamplona R, Buffenstein R, Buttemer WA (2007) Life and death: metabolic rate, membrane composition, and life span of animals. Physiol. Rev. 87(4):1175-213.

[23] Holman RT (1954) Autoxidation of fats and related substances, in: Holman RT, Lundberg WO, Malkin T. Progress in Chemistry of Fats and Other Lipids. London: Pergamon Press. pp. 2-51.

[24] Esterbauer H, Schaur RJ, Zollner H (1991) Chemistry and biochemistry of 4-hydroxynonenal, malonaldehyde and related aldehydes. Free Radic. Biol. Med. 11(1):81-128.

[25] Thorpe SR, Baynes JW (2003) Maillard reaction products in tissue proteins: new products and new perspectives. Amino Acids. 25(3-4):275-81.

[26] Pamplona R (2011) Advanced lipoxidation end-products. Chem. Biol. Interact. 192(1-2):14-20.

[27] Portero-Otin M, Pamplona R (2006) Is endogenous oxidative protein damage involved in the aging process? In: Pietzsch J, editors. Protein Oxidation and Disease. Kerala: Research Signpost. pp 91-142.

[28] Marnett LJ (2002) Oxy radicals, lipid peroxidation and DNA damage. Toxicol. 181-182:219-22.

[29] Portero-Otín M, Bellmunt MJ, Ruiz MC, Pamplona R (2000) Nonenzymatic modification of aminophospholipids by the Maillard reaction in vivo. In: Castell J, Garcia Regueiro JA. Research Advances in Lipids, vol.1. India: Global Research Network. pp. 33-41.

[30] Echtay KS, Esteves TC, Pakay JL, Jekabsons MB, Lambert AJ, Portero-Otín M, Pamplona R, Vidal-Puig AJ, Wang S, Roebuck SJ, Brand MD (2003) A signalling role for 4-hydroxy-2-nonenal in regulation of mitochondrial uncoupling. EMBO J. 22(16):4103-10.

[31] Wakabayashi N, Dinkova-Kostova AT, Holtzclaw WD, Kang MI, Kobayashi A, Yamamoto M, Kensler TW, Talalay P (2004) Protection against electrophile and oxidant stress by induction of the phase 2 response: fate of cysteines of the Keap1 sensor modified by inducers. Proc. Natl. Acad. Sci. U.S.A. 101(7):2040-5.

[32] Copple IM, Goldring CE, Kitteringham NR, Park BK (2008) The Nrf2-Keap1 defence pathway: role in protection against drug-induced toxicity. Toxicol. 246(1):24-33.

[33] Sanz A, Pamplona R, Barja G (2006) Is the mitochondrial free radical theory of aging intact? Antioxid. Redox Signal. 8(3-4):582-99.

[34] Saborido A, Naudí A, Portero-Otín M, Pamplona R, Megías A (2011) Stanozolol treatment decreases the mitochondrial ROS generation and oxidative stress induced by acute exercise in rat skeletal muscle. J. Appl. Physiol. 110(3):661-9.

[35] Caro P, Gomez J, Sanchez I, Naudí A, Ayala V, López-Torres M, Pamplona R, Barja G (2009) Forty percent methionine restriction decreases mitochondrial oxygen radical production and leak at complex I during forward electron flow and lowers oxidative damage to proteins and mitochondrial DNA in rat kidney and brain mitochondria. Rej. Res. 12(6):421-34.

[36] Gomez J, Caro P, Sanchez I, Naudi A, Jove M, Portero-Otin M, Lopez-Torres M, Pamplona R, Barja G (2009) Effect of methionine dietary supplementation on mitochondrial oxygen radical generation and oxidative DNA damage in rat liver and heart. J. Bioenerg. Biomembr. 41(3):309-21.

[37] Ferrer I (2009) Altered mitochondria, energy metabolism, voltage-dependent anion channel, and lipid rafts converge to exhaust neurons in Alzheimer's disease. J. Bioenerg. Biomembr. 41(5):425-31.

Chapter 11

Lipid Peroxidation and Polybrominated Diphenyl Ethers – A Toxicological Perspective

Mary C. Vagula and Elisa M. Konieczko

Additional information is available at the end of the chapter

1. Introduction

Understanding the structure and composition of plasma membrane is important as it guards the integrity and function of the cell as a whole. Lipid peroxidation is a process in which the lipids of the cell, particularly the membrane lipids, are degraded by oxidation resulting in the disruption of the entire cell. All cells are enclosed by a plasma membrane which serves as a boundary and it is the primary barrier between cell's internal and external environments. It is semi-permeable and plays a major role in regulating the transport of molecules into and out of the cells. This property of the plasma membrane is known as selective permeability. The fluid mosaic model describes the arrangement of the lipids and the proteins in the plasma membrane [1]. The lipids are smaller in size than the proteins. However, there are many more lipids than there are proteins. The lipids give the membranes their basic shape. Membrane lipids are amphipathic molecules, with hydrophilic and hydrophobic ends [2]. These lipids arrange in two layers (bilayer) such that the hydrophobic ends of one layer touches the hydrophobic ends of the other lipid layer. The hydrophilic ends of the phospholipids are in contact with the aqueous external and internal environments. Membrane lipids include phospholipids, cholesterol, sphingomyelin, triacylglycerides and glycolipids [3]. Several different types of phospholipids are found in cellular membranes: phosphatidylcholine (the most abundant), phosphatidylserine, phosphatidylinositol and phosphatidylethanolamime [4]. Each species of organisms has its own unique combination of these different phospholipids in its cellular membranes.

The proteins of the membrane float in the lipid bilayer. There are fewer proteins than lipids in the plasma membrane. However, they are much bigger than the lipids. The proteins can be associated with the lipids of the bilayer in one of the three ways [5]. Integral proteins are amphipathic molecules that span the entire width of lipid bilayer. Their hydrophilic and hydrophobic portions allow them to interact with both lipids and the aqueous environments on either side of the lipids. Membrane associated proteins are associated with either external or internal surfaces of the membrane. These proteins often contain a lipid anchor which helps to hold them in the membrane. Peripheral proteins are attached to the inner surface of the membrane through covalent bonds to another protein. For example, the proteins that link the plasma membrane to the cytoskeleton. Plasma membrane proteins have several functions [6]. They serve as transporters, receptors, enzymes and as anchors, tying the cell to the external environment or to other cells. Carbohydrates attached to some proteins serve as identification markers and help in the differentiation of self from non-self cells [7].

The composition and structure of the plasma membrane described above is affected by free radicals. So the next section deals with free radicals, on how they are generated in the cells and their effects on cellular macromolecules.

2. Free radicals and Reactive Oxygen Species (ROS)

A free radical may be defined as any atom or molecule that has one or more unpaired electrons in its outermost shell. Free radicals were discovered in biological material about 50 years ago [8] and since then, they have generated great curiosity for their involvement in disease and aging. Oxygen free radicals or reactive oxygen species (ROS) are defined as any free radical molecules containing oxygen. Some examples of the most common ROS include: hydrogen peroxide (H_2O_2), hydroxyl ($^\bullet OH$), and superoxide ($O_2^{\bullet -}$) anion. Under normal conditions, ROS are produced as necessary intermediates in white blood cells to defend against invading pathogens, and they also play a role in intracellular signaling. Owing to their unpaired electrons in their outermost shell, ROS are highly reactive and unstable and they exert damage on biological structures. Free radicals react quickly with other molecules and "steal" electrons to acquire stability. The molecule which lost its electron to a free radical will become a free radical itself, thus beginning a chain reaction. Free radicals modify the structure of other molecules and cause oxidative damage. Once the process is started, if not quenched in time, it can eventually result in cell damage. Additionally, excessive amounts of free radicals are generated when tissues are damaged, for example, in times of tissue hypoxia, exposure to smoke, ultraviolet radiation and pollutants.

2.1. Generation of reactive oxygen species

Most of the energy required to fuel metabolic functions of aerobic organisms is produced at the cellular level in the mitochondria via the electron transport chain. In addition to energy, reactive oxygen species (ROS) are also generated from electron transport chain as part of normal cellular metabolic reactions. In this transport chain, ROS are formed at the mitochondrial inner membrane. Under normal conditions, the oxygen molecule acts as a terminal electron acceptor in the electron transport chain and is reduced to form water. However, sometimes electrons leak prematurely and reduce oxygen to form ROS. The electron transfer between ubiquinone-cytochrome b is the most probable site of ROS formation [9]. Microsomes and membranes of the nucleus also involve electron transportation through the cytochromes P_{450} and B_5 which can produce free radicals [10]. The following few paragraphs discuss the generation of various ROS.

2.2. Superoxide anion ($O_2^{\bullet -}$)

When ground state oxygen accepts a single electron a superoxide anion ($O_2^{\bullet -}$) is formed.

$$O_2 + e^- \rightarrow O_2^{\bullet -}$$

Under normal physiological conditions cell organelles, namely, mitochondria, microsomes and peroxisomes, generate $O_2^{\bullet -}$. The half-life of superoxide anion is longer than that of another potent ROS, hydroxyl radical ($OH^{\bullet}$), and it can react with biological molecules for longer times [10]. Phagocytic cells produce superoxide during a phenomenon known as the "respiratory burst" which occurs when they encounter pathogens. Phagocytic cells known to produce superoxide anions are macrophages, monocytes, neutrophils and eosinophils [11]. Superoxide anions can trigger the formation of hydroxyl and peroxyl radicals, which in an acidic environment can form hydrogen peroxide (H_2O_2).

2.3. Hydrogen peroxide (H_2O_2)

The addition of a second electron to $O_2^{\bullet -}$ gives rise to the peroxide ion, O_2^{2-}. At physiological pH any peroxide ion formed is protonated to hydrogen peroxide (H_2O_2).

$$2O_2^{\bullet -} + 2H^+ \rightarrow H_2O_2 + O_2$$

Hydrogen peroxide can then diffuse far from the site of its production to other sites where its biological effects may be greater. The diffusion range is extended by H_2O_2 carriers formed spontaneously by hydrogen bonding with compounds such as amino and dicarboxylic acids, peptides, proteins, nucleic acid bases, and nucleosides. Equilibrium exists between an adduct-forming compound and H_2O_2. The hydrogen peroxide adducts (HPAs) retard the decomposition of H_2O_2 up to several hundredfold [12]. The overall charge on an HPA influences the cytotoxic and clastogenic effects of H_2O_2. The adducts, especially L- Histidine, play an important role in the stabilization and reduction in the reactivity of the hydrogen peroxide thereby preventing single strand breaks of the DNA in cell free DNA systems [13].

2.4. Hydroxyl radical ($^{\bullet}OH$)

The hydroxyl radical ($^{\bullet}OH$) can be formed by the homolytic fission of the O-O bond of the H_2O_2 molecule. Simple mixing of iron (II) salts with the H_2O_2 also forms the $^{\bullet}OH$ radical. This reaction was first reported by Fenton in 1894 and it is called Fenton's reaction. The hydroxyl radicals can also be produced by Haber-weiss reactions in the cells, in which H_2O_2 interacts with superoxide radical [10]. Additionally, copper salts can also react with H_2O_2 to generate $^{\bullet}OH$ radicals. All the three ways of $^{\bullet}OH$ radical generation is shown below:

By the decomposition of H_2O_2 by Fenton's reaction:

$$Fe^{+2} + H_2O_2 \rightarrow Fe^{+3} + {}^{\bullet}OH + OH^-$$

and H_2O_2 interaction with superoxide radical by the Haber-Weiss reaction:

$$O_2^{-} + H_2O_2 \rightarrow O_2 + H_2O + {^{\cdot}}OH$$

The hydroxyl radical can also be formed when copper salts react with H_2O_2 to make hydroxyl radicals as shown below:

$$Cu^{+} + H_2O_2 \rightarrow Cu^{2+} + {^{\cdot}}OH + OH^{-}$$

Although the hydroxyl radical (OH$^{\cdot}$) has the shortest half-life, it is one of the most reactive of all ROS [14, 15]. Due to this property it reacts with almost all macromolecules found in the cells such as proteins, nucleic acids (RNA and DNA), phospholipids, sugars and carbohydrates, and exerts deleterious effects on them.

2.5. Free radicals and the damage to cellular components

Excess generation of free radicals combined with the failure of cells to scavenge them through protective mechanisms (antioxidative agents and enzymes) results in oxidative damage. Oxidative damage is witnessed in the form of free radical mediated attack on macromolecules within their vicinity and causing disruption of these molecules. Severe oxidative stress can even result in apoptosis and cell death [16]. Free radical mediated damage to proteins, carbohydrates, nucleic acids and lipids is described below.

2.6. Free radical damage to proteins

The oxidization reaction of free radicals on amino acids causes changes in the physical properties of proteins that they compose. These physical changes are of three types: fragmentation, aggregation and susceptibility to enzyme digestion [17]. The fragmentation of albumin and collagen due to free radical mediated oxidization is a classic example of this phenomenon [18, 19]. Selective fragmentation at specific amino acids, especially at proline, histidine and arginine, is observed due to their close association with transition metals. The second type of physical change, protein aggregation, is believed to be caused predominantly by hydroxyl radicals due to their ability to form cross-links between the constitutive amino acids. Lastly, gross alteration in the conformation of protein structure may be the reason for the susceptibility of these altered proteins to enzyme digestion [10].

2.7. Free radical damage to carbohydrates

Hydroxyl radicals generated by Fenton reaction are reported to induce damage on simple carbohydrates [20]. It is also reported that oxidation of glucose may be a means of quenching hydroperoxide radicals while also generating a source of oxygen free radicals [21]. Another study indicated that rapid auto-oxidation of glucose results in the formation of dicarbonyl and H_2O_2 under physiological conditions [22]. Thus, oxidized glucose can react with proteins in a process called glucosylation or glucation leading to the disruption in the function of these macromolecules. Glucosylation of certain long-lived proteins by a non-enzymic reaction with free glucose may contribute to ageing.

2.8. Free radicals and the alteration of gene expression

Free radicals are extensively implicated for their role in DNA damage and alteration in the gene expression leading to various types of cancers [23]. It has been reported that about twenty types of changes are induced by ROS in the DNA. The damage in the DNA's deoxyguanosine residues is estimated to be in the range of 8 to 83 residues per million deoxyguanosine residues. This damage increases with age and seems to affect organs such as the liver, kidney and spleen while leaving brain tissue relatively undamaged [24]. Mitochondrial DNA experiences greater damage due to proximity to the ROS generated from the electron transport chain of mitochondrial membranes. Free radical induced damage to DNA includes a range of specifically oxidized purines and pyrimidines, alkali labile sites, single strand breaks and instability formed directly or by repair processes [25,26]. The pyrimidine residues, cytosine and thymine, of DNA are the most susceptible to the attack by hydroxyl radical, followed in decreasing levels of susceptibility by adenine, guanine and deoxyribose sugar [10]. Some of these modified bases have been found to possess mutagenic properties. Therefore, if not repaired they can lead to carcinogenesis.

2.9. Free radicals and lipid peroxidation

Free radicals steal electrons from the cell membrane's polyunsaturated fatty acids (PUFAs) and initiate an attack called lipid peroxidation. The peroxidation of lipids occurs in three steps, namely, initiation, propagation and termination. Free radicals target the carbon-carbon double bond of PUFAs, as this double-bond weakens the carbon-hydrogen bond, permitting easy dissociation of the hydrogen by a free radical. During the first step, a single electron is stripped by a free radical from hydrogen atoms associated with the carbon at the double-bond leaving the carbon with a single electron. Molecular arrangements occur in an attempt to stabilize, leading to the formation of conjugated dienes. In the propagation phase, the conjugated diene readily reacts with molecular oxygen and transitions to a hydroperoxide radical that begins stealing electrons from other lipid molecules, resulting in a lipid hydroperoxide [11]. The hydroperoxide molecules undergo decomposition with the aid of transition metals into alcoxyl and peroxyl radicals that, in turn, can initiate a chain reaction and further propagate lipid peroxidation and cause damage to the cell. The three phases or steps of lipid peroxidation are shown in the following equations:

Initiation reaction

$$Lipid - H + OH \rightarrow Lipid + H_2O$$

Hydroperoxide formation

$$Lipid + O_2 \rightarrow Lipid\text{-}O_2$$

Lipid hydroperoxide

$$Lipid\text{-}O_2 + Lipid - H \rightarrow Lipid + Lipid\text{-}O_2H$$

The important fatty acids that undergo peroxidation are: linoleic acid, arachidonic acid, and docosahexaenoic acids [14]. The hydroxyl radical damages cell membranes and other lipoproteins by lipid peroxidation. It is important to note that the low density lipoprotein based lipid peroxidation plays a significant role in atherosclerosis [27].

The generation of free radicals and their action on macromolecules thus leads to disruption of cellular functions. The following section discusses chemicals that are present in the body to protect against these damaging radicals.

3. Antioxidative protection against reactive oxygen species (Free radicals)

All aerobic cells have an armory of chemicals known as antioxidants which are capable of counteracting the damaging effects of oxidative free radicals and inhibiting the oxidation of other molecules in the body. Based on the mechanism of their action, these antioxidants are divided into two categories: chain breaking antioxidants and preventive antioxidants. Vitamin E, glutathione, vitamin C and beta-carotene are some examples of the chain breaking type as they donate their own electron to the existing free radical and prevent the continuation of oxidation chain reaction. These antioxidants are found in the plasma and interstitial fluid [28, 29]. The preventative category of antioxidants consists of enzymes that scavenge the free radicals before they initiate an oxidative attack.

Another way to categorize these antioxidants is based on their solubility in lipids or water. The antioxidants which are lipid soluble are located in the cellular membrane and lipoproteins, while the water soluble ones are located in the aqueous environments, such as the cytoplasm inside cells and the blood [30]. The following few sections discuss some of the important antioxidants of cells.

3.1. Glutathione

Glutathione (GSH) is one of the important cellular antioxidants. It is made up of three amino acids namely, cysteine, glutamic acid and glycine. All the cells of the body have the ability to synthesize glutathione, but it is found in high concentrations in the liver, lungs, and intestinal tract. The important functions of the glutathione include: anti-oxidation, detoxification, strengthening the immune system and signal transduction under stressful conditions. The diverse functions of GSH are due to the sulfhydryl group in cysteine, enabling it to chelate and detoxify harmful substances [31]. Additionally, glutathione also plays a role in processing medications and cancer-causing compounds (carcinogens), and in building DNA, proteins, and other important cellular components. Its usefulness as an antioxidative agent lies in its ability to donate electrons, and in that process it is oxidized to glutathione disulfide (GSSG) form. By donating an electron to a free radical, GSH can quench the unstable and highly reactive free radical and thus can prevent free radical mediated oxidative damage in the cells. GSSG is reduced to GSH by NADPH, an electron donor in this reaction. This reduction reaction is catalyzed by glutathione reductase. Thus the ratio between reduced and oxidized glutathione serves as an indicator of cellular toxicity mediated by oxidative damage. Large amounts of GSH are produced and stored in the liver, where it is used to detoxify harmful compounds. GSH levels are found to decline with age leaving the body susceptible to the damage by free radicals.

3.2. Vitamin C, vitamin E and beta carotene

Vitamin C is a water soluble antioxidant which plays an important role in protecting the body against oxidative damage. Chemically, it is a carbohydrate-like substance. Human beings, unlike other mammals, do not have the ability to synthesize Vitamin C. Therefore, it must be furnished through the diet. Vitamin C is easily oxidized and many functions in the body depend on this property. It stops the free radical chain reaction by donating an electron and prevents the reaction from progressing.

Carotenoids are a class of cellular antioxidants. Carotenoids are referred as provitamin A as they can potentially yield Vitamin A. Free radicals which can damage DNA and cause diseases like macular degeneration, coronary artery disease and cancer are neutralized by these carotenoids. Beta-carotene is perhaps the best known carotenoid. It was isolated in 1831 and its structure was determined in 1931.

Vitamin E is a fat soluble compound which also can impede oxidation to some degree [32]. Though there are eight forms in nature, only alpha-tocopherol form is functional in human beings. Vitamin E stops the production of ROS formation when fat undergoes oxidation. In addition to the above mentioned vitamins, essential trace elements, selenium, zinc, manganese, and copper, also play a critical role as antioxidants.

3.3. Antioxidant enzymes

Antioxidant enzymes inside the cells are the main defense against the free radicals. The most important antioxidant enzymes present inside the cells are superoxide dismutase (SOD), catalase, glutathione peroxidase (GPx), and glutathione S-transferase (GST). SOD catalyzes the dismutation of superoxide anion to O_2 and H_2O_2. This reaction is considered to be the body's primary defense against free radicals as it prevents further generation of free radicals [29]. SOD is found in high levels in the liver, adrenal glands, kidney and spleen [33]. GPx, catalyzes the reduction of hydroperoxides and serves to protect the cell from hostile free radicals. Catalase, another antioxidant enzyme, along with GPx, is essential for removing hydrogen peroxide formed during oxidation reactions, and is found in high levels in liver, kidney and red blood cells. Both catalase and GPx enzymes exhibit a great degree of cooperativity in their action. Glutathione S-Transferase (GST) is yet another antioxidant enzyme primarily responsible for cellular defense mechanism against ROS. GST metabolizes xenobiotics, whether they are of endogenous or exogenous origin [34]. GST along with glutathione and GPx neutralizes free radicals and lipid hydroperoxides especially at low levels of oxidative stress [35]. The undesirable end products of lipid peroxidation are thus detoxified by GST. All of these enzymes prove to be indispensable in the cellular antioxidant defense mechanism and they are all necessary for the survival of the cell, even in normal conditions. An appropriate equilibrium between antioxidant enzyme activities is vital to ensure the cell survival during increased oxidative stress.

However some exogenous factors, such as environmental contaminants, have the potential to disrupt this equilibrium and cause incredible damage to the cells. The following section describes a specific group of brominated flame retardants called polybrominated diphenyl ethers (PBDEs).

4. Polybrominated diphenyl ethers: A class of brominated flame retardants

Fire accidents in the United States have the potential to cause severe economic problems for the individuals and for society at large. Each year, several thousand Americans die, several thousand more are injured, and millions of dollars of loss are all due to fires. To recover from these losses, individuals, their insurance companies, and society as a whole spend significant resources on temporary housing, groceries and meals, medical costs, psychological support, and relocation costs, to name a few expenditures. These ancillary costs may be 10-times higher than the cost to rebuild structures actually gutted by fire. To give a rough idea, the annual losses from natural calamities such as floods, hurricanes, tornadoes, earthquakes, etc., all put together in the United States equal just a fraction of total loss due to fires [36]. To comply with fire safety regulations in the United States manufacturers must follow strict guidelines in order to reduce the flammability of their products. For years, manufacturers have added chemicals to plastics and fabrics in order to reduce combustibility and increase overall safety of the products. Different types of brominated flame retardants have the promise of slowing down fire and hence are being extensively used. Figure 1 shows the chemical structures of four typical bromine compounds, namely, polybrominated biphenyls (PBBs), polybrominated diphenyl ethers (PBDEs), hexabromocyclododecane (HBCDs) and tetrabromobisphenol A (TBBPAs) [37].

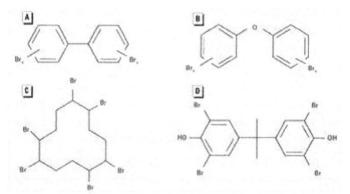

Figure 1. Chemical structures of BFRs. (A) PBBs, (B) PBDEs, (C) HBCD, and (D) TBBPA [37]

In the past few decades, use of PBDEs has become quite commonplace and these chemicals are added in large quantities to many commercial and household products, such as computers, TVs, foam mattresses, and carpets, to inhibit ignition and prevent fire accidents [37]. At high temperatures PBDEs liberate bromine atoms, which are effective at hindering the basic oxidizing reactions that drive combustion. Ever since PBDEs were introduced into the market, deaths due to fire accidents dropped significantly. Functionally, these flame retardants are similar to polychlorinated biphenyls (PCBs) that were banned from production in the U.S in the late 1970s due to potential adverse health effects and a persistent bio-accumulative nature [38]. Like PCBs, the PBDEs class contains 209 possible congeners. During the synthesis of PBDEs, a bromination reaction occurs on the biphenyl rings as one or more bromine atoms are attached to the biphenyl rings giving rise to one of the possible 209 possible congeners of PBDEs. The maximum number of bromine atoms that can attach to biphenyl rings is 10 and this brominated flame retardant is called decaBDE. Commercially these chemicals are available in three mixtures, pentaBDEs, octaBDEs and decaBDEs. Congeners with 1-5 bromine atoms are referred as lower BDEs and those with 6 - 10 are referred as higher BDEs.

4.1. Production of PBDEs

PBDEs were introduced into the market in the U.S. in the 1970s and peaked in the late 1990s. The production of various PBDEs in different regions of the world is shown in Table 1[37].

BDE	Americas	Europe	Asia	Rest of the World
Deca BDE	24,500	7,600	23,000	1,050
Octa BDE	1,500	610	1,500	180
Penta BDE	7,100	150	150	100
Total BDEs	33,100	8,360	24,650	1,330

Data from BSEF (2001)

Table 1. BFRs produced (metric tons) in various regions of the world [37]

Table 2 shows the percent composition of various commercial PBDE mixtures. Currently, deca BDE constitutes 97% of the total production of PBDEs worldwide and it is the only BDE which is in use in the US [39].

Product	Tetra	Penta	Hexa	Hepta	Octa	Nona	Deca
Penta	24-38	50-60	4-8				
Octa			10-12	44	31-35	10-11	<1
Deca						<3	<97

Table 2. Percent composition of commercial PBDE flame retardant mixtures [39].

Many research findings indicate that lower brominated congeners manifest greater affinity for lipids and accumulation. Hence they are more toxic and pose a greater health risk to humans and livestock than the higher brominated congeners [40].

4.2. Exposure to PBDEs

PBDEs are lipophilic and resistant to chemical and physical degradation. Due to their persistent nature, they have become widespread environmental contaminants and about 97% of the American adult population has detectable levels of PBDEs [41]. Over the last 30 years the levels of these compounds in humans have increased and their levels in the US population are 10-100 times higher than levels measured in the Europe and Asia. According to a recent report, their levels in humans may be leveling off or decreasing due to their ban in many states of the US and European countries[39]. They are reported to be present in sediments, soil, outdoor and indoor air, household dust, foods, birds, fish, and terrestrial organisms [42].They have been detected in human serum, adipose tissue, and breast milk, and long-term exposure to these contaminants may pose a human health risk, especially to the developing fetus.

Despite their ban in many countries, vast amounts of these compounds are found to persist in existing consumer products, potentially contributing to environmental and human burdens for years to come. When released into the environment PBDEs adsorb rapidly onto solid particles such as soil or sediments due to their low water solubility and vapor pressure. Therefore, sediments and sewage sludge serve as a sink for these pollutants. These contaminants are still bioavailable even after adsorbing to the sediment particles [43]. PBDEs exposure to human being occurs through inhalation of household and workplace dust and by eating PBDE-contaminated foods. Exposure to PBDEs is nearly impossible to avoid as they are found in air, water and food. The fetus is also exposed to these compounds in utero, and after birth the newborn continues to be exposed from breast milk, where the toxins are transferred from the mother to the baby. A 25-year Swedish study found that the concentration of PBDEs in breast milk doubled every five years during the 25-year period [44]. The PBDE levels in breast milk from North American women are much higher than in breast milk from Swedish women, indicating greater exposure to PBDEs in this continent.

4.3. Dietary intake of PBDEs

The dietary intake of PBDEs from food in humans between ages 0-60 years are presented in Table 3. The intake of PBDEs was the highest in the first year during nursing. Breast milk has the highest levels of PBDEs. The intake of PBDEs per kg body weight reduced gradually with age. Children have higher PBDE intake than adults due to higher food intake per kg body weight [45].

4.4. Adverse effects of PBDEs in laboratory animals

Few toxic studies on PBDEs are reported in laboratory animals [46,47]. Some studies report these compounds to be toxic, carcinogenic and endocrine disruptors [48]. There are some reports indicating that PBDEs can induce disruption in spontaneous behavior, reduced habituation, impaired memory and learning in rats exposed during critical developmental period [46, 47]. BDE-99 is a penta congener and it has been found to induce hyperactivity and impaired spermatogenesis in rats. Additionally, two recent in-vitro studies showed the induction of oxidative stress in rat neurons exposed to BDE-47 (another congener of penta-BDE congener) in primary cultured rat hippocampal neurons [49] and human cells [50] serving as note-worthy examples that call for more careful assessment of these toxic substances.

Age in Years	0-1	2-5	6-11	12-19	20-39	40-59	>60
Diary	0	427	293	372	250	203	175
Meat	0	1,839	1,170	1,721	1530	1,408	1,094
Fish	0	280	232	300	309	357	467
Eggs	0	69	38	55	48	53	58
Margarine	0	6	6	7	6	8	9
Butter	0	30	17	19	22	15	22
Human Milk	306,560	0	0	0	0	0	0
Total PBDE		2652	1755	2774	2164	2,084	1857

Table 3. Daily intake of PBDEs from food (pg/kg body weight) [45]

5. Lipid peroxidation during PBDE exposure

Despite many studies, exact mechanism of toxicity of PBDEs is not yet clear and there is no definite explanation as to how PBDEs trigger the above mentioned deleterious effects. There appeared a school of thought that PBDEs, by reason of their electron affinity favor free radical formation in various tissues [51]. In the face of increased free radical generation the antioxidant enzymes namely, SOD, GPx, Catalase, and GST, are expected to give protection to the cells against the oxidative damage. An appropriate equilibrium between the antioxidant enzyme activities and free radicals is vital to ensure the cell's survival during increased oxidative stress. In the process of investigating the toxic mechanism authors have surmised oxidative stress induction in mice exposed to BDE-209, one of the most abundantly used flame retardant worldwide. This is based on a recent study showing the induction of oxidative stress in rat neurons exposed to BDE-47, a penta congener [49]. To interpret the adverse effects, both an assessment of the magnitude of free radical production and a careful evaluation of the threshold of protection in terms of antioxidant enzyme activity is necessary. Antioxidant enzymes such as SOD, GPx, GST, and catalase could also be the key indicators that could disclose the sequence of biochemical dysfunctions and alterations in signal transduction brought about by BDE-209. In the following few paragraphs the research protocols that one of the authors (MCV) adopted to investigate the biological markers for oxidative stress following BDE-209 exposure are explicated. The study is performed on the liver and brain tissues of adult mice. The liver is a major bio-transforming organ and brain is the most sensitive target to lipid soluble toxicants. Studying these two organs would give a clear picture of how this particular BDE is handled by these two vital organs. Further, since the metabolism and excretion of many PBDEs or for that matter any xenobiotics are tissue, species, sex and age dependent [52] it is intended to understand the effects of BDE-209 on these two vital organs; thus the hepato- and neuro-toxic mechanisms could be simultaneously studied and correlated.

5.1. Materials and methods

5.1.1. Chemicals

BDE- 85- CS (certified standard) was procured from Cambridge Isotope laboratories, Inc., Andover, MA, USA. Assay kits for the determination of lipid hydroperoxides, SOD, GPx, catalase, and GST were purchased from Cayman Chemical Company, Ann Arbor, Michigan, USA. BCA™ protein assay kit was purchased from Pierce Biotechnology Inc., Rockford IL, USA, for the protein assay of the sample. All other chemicals needed for the experiments were obtained from Sigma Aldrich Company.

5.1.2. Animals

Adult male Swiss Webster mice were obtained from Taconic Inc (Hudson, New York) and used for oxidative stress studies. Mice weighing 33±3 g were maintained at $23^0C\pm2^0C$ with 12-hour light/12-hour dark photoperiods. Mice were provided food pellets and water ad libitum and were allowed to acclimatize for 15 days. The guidelines set forth by the Institutional Animal Care and Use Committee, Gannon University, Erie, Pennsylvania, were followed in the maintenance and use of mice. Mice were grouped into two groups randomly ($n = 9$). Group 1 was injected intraperitoneally with 0.25mg/kg body weight of BDE-209 for four consecutive days and group 2 served as control and received same volume of vehicle solution, corn oil. Oral route of administration was not considered in this study as the aim of this work is to study the biochemical alterations induced at this particular concentration and administering via oral route results in elimination of some fraction of this chemical through feces. The present dosage of BDE-209 was selected based on some reports on other penta-BDE congeners in rodents [42,52]. In addition, the dosage used in the present study also is commensurate with the average levels of PBDEs (except BDE-209) found in household dust [53]. On the fifth day mice were sacrificed and liver and brain tissues were separated, washed in phosphate-buffered saline (PBS), and weighed. For all the biochemical assays, a 5% tissue homogenate of liver or brain was prepared using appropriate homogenizing buffers. The protocols that came with the assay kits were used for all biochemical assays and they are described below. The protein content of all the enzyme samples is assayed using Pierce® BCA Protein Assay Kit. Bovine serum albumin was used as standard.

5.1.3. Lipid hydroperoxides

Lipid peroxidation is quantified by a new method which directly quantifies the lipid hydroperoxides utilizing the redox reactions with ferrous ions.The liver and brain tissues from control and experimental mice were homogenized in nanopure water and the lipid hydroperoxides from the homogenate were extracted into chloroform. According to the protocol accompanying the lipid hydroperoxide assay kit, supplied by the Cayman Chemical Company, this chloroform extract was then used for the determination of lipid hydroperoxides.

5.1.4. Superoxide Dismutase (SOD)

The liver and brain tissues from control and experimental mice were homogenized in 4-(2-hydroxyethyl)-1-piperazineethanesulfonic acid (HEPES) buffer containing 1 mmole/L ethylene glycol tetraacetic acid (EGTA), 210 mmole/L mannitol and 70 mmole/L sucrose and centrifuged. The supernatant was used for the assay of SOD activity and was determined from the quantity of tetrazolium salt formed. One unit of SOD is defined as the amount of enzyme needed to exhibit 50% dismutation of superoxide anion.

5.1.5. Glutathinoe Peroxidase (GPx)

GPx activity was measured indirectly by a coupled reaction with glutathione reductase. Oxidized glutathione, produced upon reduction of hydroperoxide by GPx is recycled to its reduced state by glutathione reductase and NADPH. The oxidation of NADPH to $NADP^+$ is accompanied by a decrease in absorbance at 340 nm. The rate of decrease in the A_{340} is directly proportional to the GPx activity in the sample. The tissues from control and exposed mice were homogenized in cold Tris- HCl with 5 mmole/L EDTA and 1 mmole/L dithiothreitol (DTT). The homogenate was centrifuged and the supernatant was used for the enzyme assay. The protocol described in GPx assay kit was followed.

5.1.6. Catalase

A tissue 5% homogenate of liver or brain was prepared in cold potassium phosphate buffer containing 1mmole/L EDTA. The supernatant that surfaced was used for catalase assay as described in the assay protocol provided by Cayman Chemical Company. The catalase assay kit used the peroxidatic function of CAT for determination of enzyme activity. The enzyme reacts with methanol in the presence of an optimal concentration of H_2O_2. The formaldehyde produced is measured spectrophotometrically ($\circledcirc$ = 540 nm) with 4-amino-3-hydrazino-5-mercapto-1,2,4-triazole (Purpald) as the chromogen. Purpald specifically forms a bicyclic heterocycle with aldehydes, which upon oxidation changes from colorless to a purple color. One unit is defined as the amount of enzyme that will cause the formation of 1.0 nmol of formaldehyde per minute at 250°C.

5.1.7. Glutathione S- transferase (GST)

Tissues from control and experimental mice were homogenized in cold potassium phosphate containing 2 mmole/L EDTA. Homogenates were centrifuged and the supernatants used for the assay of Glutathione S-transferase, GST. GST activity was measured from the quantity of 1-Chloro-2, 4-Dinitrobenzene (CDNB) conjugated to glutathione. The one unit of enzyme conjugates 1.0 nmole of CDNB with reduced glutathione/min/mg.

5.1.8. Statistical treatment of the data

Data presented in the Tables 4 and 5 are mean ± standard deviation (SD) of 9 animals per group. The data were subjected to analysis of variance (ANOVA) and when means were significantly different they were further subjected to Tukey test. The percentage variation from controls was calculated and presented in the following results section. A value of P < 0.05 was considered significant.

Oxidative Stress Marker	Control	BDE-209	% change
LPO	2.06 ± 0.63	2.94 ± 1.33	+42.72*
SOD	13.89 ± 1.77	19.25 ± 2.56	+38.58*
GPx	103.43 ± 14.77	85.92 ± 5.61	-16.93*
Catalase	58.78 ± 14.22	44.54 ± 5.9	-24.22▪
GST	0.22 ± 0.07	0.17 ± 0.05	-22.73▪

Table 4. The lipid hydroperoxides and antioxidant enzyme activities in the brain tissue. Values are mean ± Standard deviation (n = 9). *Significant at $p < 0.05$, ▪ not significant

Oxidative Stress Marker	Control	BDE-209	% change
LPO	22.52 ± 2.76	27.31 ± 1.98	+21.67*
SOD	33.37 ± 1.59	30.69 ± 1.77	-8.03▪
GPx	198.77 ± 26.87	143.78 ± 7.8	-28.01*
Catalase	134.75 ± 18.34	92.78 ± 10.54	-31.15*
GST	4.32 ± 0.9	2.97 ± 0.3	-31.3*

Table 5. The lipid hydroperoxides and antioxidant enzyme activities in the liver tissue. Values are mean ± Standard deviation (n = 9). *Significant at $p < 0.05$, ▪not significant

6. Results

The overall short term effect of BDE-209 over a period of four days is a remarkable disruption in the oxidant/antioxidant equilibrium. First, the levels of lipid hydroperoxides in the experimental tissues, liver and brain, have increased considerably. The percent increase in the levels of lipid hydroperoxides in liver was 27.67 % ($p < 0.05$) and in brain was 42.72% ($p < 0.05$) over controls. These results are shown in Table 4 and 5. The activities of SOD showed varied response in brain and liver. In liver, though there was a decrease in activity by 8.03 %, when compared to controls, it was not statistically significant. However in brain there was significant (($p < 0.05$) increase in activity (38.58%). Another anti-oxidant enzyme activity measured was the GPx. The activity of this enzyme has been reduced in both brain (16.93%) and liver tissues (28.01%). These percentage changes also were significant at $p < 0.05$ for both the tissues.

The enzyme catalase also suffered a similar reduction in activity over a four day exposure to BDE-209. Both in the brain and the liver tissues the catalase activity was lowered. In the brain there was a 24.22% of reduction and in liver it went down by 31.15%. Just like GPx and catalase, GST also can reveal oxidative damage. In the present work GST activities were decreased in exposed liver by 31.3% ($p < 0.05$) and in brain by about 22.73% (NS). All the above results clearly exhibited differential susceptibility of mice liver and brain tissues to BDE-209 – induced oxidative stress.

7. Discussion

The results of this study clearly indicate the susceptibility of mice brain and liver tissues to BDE-209 induced damage when exposed to 0.25 mg/kg body weight for four consecutive days. The oxidative stress markers namely, lipid hydroperoxides, were found to increase in both the exposed liver and brain tissues. The increase in the lipid hydroperoxide levels could be due to an immediate perturbation in metabolizing enzymes in liver [54].The degree to which this can happen is yet open for further research. Liver being an important metabolizing organ hydrolyzes PBDEs in its attempt to eliminate them from the body. In this process the metabolites formed might possess greater toxicity than the parent BDE [55], thus making the liver prone to excessive damage. In the present study, BDE-209 exposed brains also showed significant increase in lipid hydroperoxides. Similar observations of increased peroxidation in the brain regions of adult rats exposed to two different concentrations of BDE-99 [42] lend support to this observation. In another study an increased peroxidation was noted in rat hippocampal neurons during BDE-47 exposure [49]. Additionally, there are reports on BDE-99 exposure in rats causing an increased ROS production in the tissues [47,55]. The mechanism responsible for ROS generation during PBDEs is not clear; however, it is suggested that PBDEs are electron acceptors under standard conditions prevailing in biological systems [51]. Reports on PBDE exposure and their involvement in the release of [3H] arachidonic acid, protein kinase-C translocation, and disruption in calcium homeostasis [56,57] could further offer an explanation for the mechanism of ROS formation [49]. Thus the present observation of increased lipid hydroperoxide levels in the liver and brain tissues of mice substantiates higher lipid peroxidation process and oxidative stress following BDE-209 exposure.

The objective of the second part of this study was to examine how antioxidative enzymes responded to BDE-209 exposure in mice tissues. Although lipid hydroperoxides and lipid peroxidation have increased in the tissues, there was no corresponding increase in all the protective antioxidant enzymes studied, namely, SOD, GPx, catalase and GST in BDE-209 exposed mice tissues with the exception of brain SOD. The elevated lipid hydroperoxides under normal conditions should trigger an increase in antioxidant enzyme activity to suit the increased need to quench them. But in the present study there was a decrease in the activities of GPx, GST and catalse in the liver tissue. A decrease in catalase reflects an inability of the liver to remove hydrogen peroxide. As mentioned already, both catalase and GPx show high cooperativity in their action [58]. During the periods of increased hydroperoxide formation catalase activity depends on GPX activity to keep from being inactivated. On account of decreased GPx activity in BDE-209 exposed tissues of mice, the catalase might be inactivated by higher concentration of hydroperoxides. Glutathione-S-transferase is another antioxidant enzyme whose activity was reduced in liver tissue during BDE-209 exposure. BDE-99 exposure in rats showed a similar response of decreased antioxidant enzymes at 0.6 mg/kg bw [42]. In another report on BDE-99, rat liver and kidney showed an upsurge of lipid peroxidation and unstable antioxidant system [55], rendering support to our observations. The overall lowered activities of SOD, GPx and catalase in liver denote that liver tissue is not able to quench increased ROS and oxidative stress. These findings thus show that BDE- 209 brings about an oxidative stress in liver tissue not only through increased lipid hydroperoxide formation but also by reduced antioxidative enzyme protection.

The responses in the brain tissues were different with respect to SOD, catalase, and GST. Unlike in liver, SOD activity in brain was elevated indicating the presence of some degree of protection in the brain tissue against the oxidative damage. Though catalase and GST activities showed reduction (a trend as seen in the liver tissue) the values however, were not statistically significant. This variation in responses in brain compared to liver suggests that brain tissue was able to cope with increased lipid hydroperoxides to some degree, by way of increased SOD activity. The differences in responses in liver and brain tissues could be due to their intrinsic biochemical and functional differences. Brain with its high rate of O2 utilization, high levels of polyunsaturated lipids tends to respond differently [59]. A report on GST of rat brain regions after BDE-99 treatment showed similar response [47]. Further, the dosage used in this study could have resulted in this response in the brain tissue. A study [42] on the adult rat brains exposed to 2 different concentrations of BDE-99 showed contrasting results on TBARS (indicative of oxidative stress) levels. At lower dosage of exposure, the levels of TBARS were found to be reduced and at higher dosage they were increased. There are reports on rats showing decreased antioxidative enzymes in brain tissue during BDE-99 indicating that not all congeners have the same impact on the tissues. Another observation made by the authors (MCV) is BDE-209 is much less toxic compared to BDE-85 [60].

In summary, exposing mice to BDE-209 for four days has brought about an increased lipid hydroperoxide production as seen in the liver and brain tissues. The consequent peroxidation process has significantly impeded the activity of antioxidant enzymes, GPx, GST and catalase, in liver. Brain tissue on the other hand, showed elevated SOD, lowered GPx without significantly reducing catalase and GST activities. The results of this study indicate that BDE-209 congener of PBDEs has grave potential to disrupt the activity of antioxidant enzymes and induce oxidative stress. Additionally, liver seemed to be more susceptible to this toxicant than the brain. These differences in responses in these two tissues could be due to their inherent biochemical and functional differences.

8. Conclusion

Plasma membrane covers a cell and acts as a barrier between the internal and external environment thereof. It also regulates the movement of molecules across it. Chemically, the plasma membrane is a bi-layer composed of phospholipids and proteins. Membrane lipids form the bulk of the membrane itself and are more abundantly found than the proteins; however, they are smaller in size. Lipids serve as semipermeable barrier allowing only those materials that are lipid soluble and non-polar in nature to freely cross the plasma membrane. The membrane proteins give structural support and also have several functions including serving as transporters, receptors, enzymes and anchors. Carbohydrates which are attached to the membrane proteins serve as identification markers and help in the differentiation of self from non-self. Thus the plasma membrane is a very important structure of the cell and disruption in its structure leads to disruption in cellular function. The membrane lipids are prone to disruption and loss of selective permeability needed for normal metabolism especially through lipid peroxidation by free radicals.

Free radicals are atoms or molecules with one or more unpaired electrons in their outermost shell. Free radicals with oxygen are called reactive oxygen species (ROS). Examples of ROS include hydrogen peroxide, hydroxyl radicals, and superoxide anions. From time to time free radicals are generated as part of aerobic life during normal metabolism and in some instances these radicals are intentionally generated as part of normal signal transduction process. Superoxide anions are generated in cell organelles such as mitochondria and microsomes. These anions can react with biological molecules the most due to their longer half-life. Another most reactive free radical of the cell is hydroxyl radical. This radical is formed by the hemolytic fission of hydrogen peroxide. Free radicals, due to their high reactivity, attack other macromolecules within their vicinity and cause disruption of these molecules. When these free radicals attack proteins they change their biophysics thereby leading to functional disruption. They also earned notoriety in disrupting DNA and gene expression. It is reported that twenty types of alterations are introduced in the DNA by these ROS leading to variations in the gene expression. Last but not the least, ROS also can attack lipids especially the membrane lipids of the cell and cause extensive damage to the cellular function. Free radicals steal electrons from the polyunsaturated fatty acids (PUFAs) and other lipoproteins of the cells and initiate an attack called lipid peroxidation. The peroxidation of lipids occurs in three steps, namely, initiation, propagation and termination. In particular, lipid peroxidation characterized by low density lipoproteins plays an important role in atherosclerosis.

All cells have antioxidants in place to counteract the harmful effects of free radicals. Antioxidants are of two types, chain breaking antioxidants and preventive type. Vitamins such as E and C are examples for chain breaking type as they donate electrons to the existing free radicals and lessen their degree of reactivity to prevent further oxidative damage. The second category of antioxidants (the preventive type) consists of enzymes that scavenge the free radicals before they initiate oxidative attack. Glutathione is an important antioxidant of the cells whose function includes anti-oxidation, strengthening immune system, building DNA, and detoxification of medications and cancer causing compounds. The diverse functions of glutathione are mainly due to the sulfhydryl group in one of its amino acids known as cysteine. Vitamins C, E and provitamin A also play a role in protecting the body against the free radical induced damage. There are many enzymes inside the cells with antioxidant property. The most important antioxidant enzymes present inside the cells are superoxide dismutase (SOD), catalase, glutathione peroxidase, and glutathione S-transferase. SOD catalyzes the dismutation of superoxide anion to O_2 and H_2O_2. GPx catalyzes the reduction of hydroperoxides and serves to protect the cell from the hostile free radicals. Catalase, another antioxidant enzyme, along with GPx, is essential for removing the hydrogen peroxide formed during oxidation reactions. GST along with glutathione and GPx neutralizes free radicals and lipid hydroperoxides especially at low levels of oxidative stress.

Polybrominated diphenyl ethers (PBDEs) are organic brominated flame retardants that were introduced into the market in 1970s and are added in large quantities to many commercial and household products such as computers, TVs, foam mattresses, carpets, etc., to inhibit ignition and prevent fire accidents. Ever since they were introduced into the market, deaths due to fire accidents abated significantly. Commercially these chemicals are available in three mixtures, pentaBDEs, octaBDEs and decaBDEs. Currently, deca BDE constitutes 97% of the total production of PBDEs worldwide and it is the only BDE which is in use in the US. PBDEs are lipophilic and resistant to chemical and physical degradation. Due to their persistent nature, they have become widespread environmental contaminants. Avoiding exposure to PBDEs is nearly impossible as they are ubiquitously found in air, water and food. Few toxic studies on PBDEs are reported in laboratory animals. There are some reports indicating that PBDEs can induce disruption in spontaneous behavior, reduced habituation, impaired memory and learning in rats exposed during critical developmental period. When the toxic effect of BDE-209 was explored in the mice at 0.25 mg/kg body weight, it became clearly evident that this particular BDE triggers lipid hydroperoxide generation and increased lipid peroxidation process. Additionally, the antioxidant enzymes were diminished in liver, but in brain the levels of SOD increased without any change in catalase and GST. These results show that BDE-209 has the ability to elicit serious oxidative stress in mice with increased lipid peroxidation and decreased antioxidant protection. Research is underway to understand the effects of BDE-209 on the expression of genes and to find out if it are responsible for impairment of cell cycle regulation and causing apoptosis.

Author details

Mary C. Vagula* and Elisa M. Konieczko
Biology Department, Gannon University, Erie, PA, USA

Acknowledgement

The lead author (MCV) thanks Gannon University for granting the award of Faculty Research Grant. Authors also thank Prof. Gustafson and Dr. Masters for financial support and encouragement.

9. References

[1]   Singer S, Nicholson G (1972) The fluid mosaic model of the structure of cell membranes. Science. 175(4023): 720-731.

---

* Corresponding Author

[2]   van Meer, G (2006) Cellular lipidomics. EMBO Journal. 24:3159-3165.

[3]   Simons K, Sampaio J (2011) Membrane ogranization and lipid rafts. Cold Spring Harbor Perspectives in Biology. 3(10):a004697.

[4]   Henneberry A, Wright M, McMaster C (2002) The major sites of cellular phospholipid synthesis and molecular determinants of fatty acid and lipid group specificity. Molecular and Cellular Biology. 13:3148-3161.

[5]   Eisenberg D (1984) Three-dimensional structure of membrane and surface proteins. Annual Reviews in Biochemistry. 53:595-623.

[6]   Driscoll P, Vuidepot A (1999) Peripheral membrane proteins. FYVE sticky fingers. Current Biology. 9:R857-R860.

[7]   Gahmberg C, Tolvanen M (1996) Why mammalian cell surface proteins are glycoproteins. Trends in Biochemal Science 21:308-311.

[8]   Commoner B, Townsend J, Pake G (1954) Free radicals in biological materials. Nature 174: 689–691.

[9]   Tyler D (1975) Polarographic assay and intracellular distribution of superoxide dismutase in rat liver. Biochemistry Journal. 147:493-504.

[10] Ramos de Andrade Jr. D, Becco de Souza R, Alves dos Samtos S, Ramos de Andrade R (2005) Oxygen free radicals and pulmonary disease. Journal of Brazilian Pneumology. 31(1): 1-17.

[11] Gutteridge M, Halliwell B (1984) Oxygen toxicity, oxygen radicals, transition metals and disease. Biochemistry Journal. 219:1-14.

[12] Schibert J, Wilmer J (1991) Does hydrogen peroxide exist "free" in biological systems? Free radical biology. 11(6): 545-55.

[13] Oya-Ohta Y, Ochi T, Komoda Y, Yamamoto K (1995) The biological activity of hydrogen peroxide. VI. Mechanism of the enhancing effects of L-histidine: the role of the formation of a histidine-peroxide adduct and membrane transport. Mutation Research. 326(1):99-107

[14] Yu B (1994) Cellular defenses against damage from reactive oxidative species. Physiology Reviews. 74:139-62.

[15] Del Maestro R (1980) An approach to free radicals in medicine and biology. Acta Physiology Scandanavia. 492:153-68.

[16] Lennon S, Martin S, Cotter T (1991) Dose-dependent induction of apoptosis in human tumour cell lines by widely diverging stimuli. Cell Proliferation. 24 (2): 203–14.

[17] Griffiths H, Unsworth J, Blake D, Lunec J (1988) Free radicals in chemistry, pathology and medicine. London: Richelieu, pgs. 439-54.

[18] Marx G Chevion M (1986) Site-specific modification of albumin by free radicals. Reaction with copper (II) and ascorbate. Biochemistry Journal. 236:397-400.

[19] Wolff S, Dean S (1986) Fragmentation of protein by free radicals and its effect on their susceptibility to enzymic hydrolysis. Biochemistry Journal. 234:399-403.

[20] Morelli R, Russo-Volpe S, Bruno N, Lo Scalzo R (2003). Fenton-dependent damage to carbohydrates: free radical scavenging activity of some simple sugars. Journal of Agricultural and Food Chemistry 51(25): 7418-7425.

[21] Sagone Jr. A, Greenwald E, Kraut J, Bianchine J, Singh D (1983) Glucose: role as free radical scavenger in biological systems. Journal of Laboratory and Clinical Medicine. 101:97-104.

[22] Wolff S, Crabbe M, Thornalley P (1984) The autoxidation of glyceraldehydes and other simple monosaccharides. Exp basel. 40: 244-6.

[23] Ames B (1983) Dietary carcinogens and anticarcinogens. Oxygen radicals and degenerative diseases. Science. 221:1256-64.

[24] Fraga C, Shigenaga M, Park, J Degan, P, Ames B (1990) Oxidative damage to DNA during aging: 8-hidroixy-deoxyguanosine in rat organ DNA and urine. Proceedings of the National Academy of Sciences USA. 87:4533-7.

[25] Dizdaroglu M, Jaruga P, Birincioglu M, Rodriguez H (2002) Free radical-induced damage to DNA: mechanisms and measurement. Free Radicals in Biology and Medicine. 32:1102-15.

[26] Cooke M, Evans M, Dizdaroglu M, Lunec J (2003) Oxidative DNA damage: mechanisms, mutation, and disease. FASEB Journal 17:1195-214.

[27] Kerr M, Bender C, Monti E (1996) An introduction to oxygen free radicals. Heart and Lung: the Journal of Acute and Critical Care. 25(3):200-208.

[28] McDermott J (2000) Antioxidant nutrients: current dietary recommendations and research update. Journal of American Pharmacology Association. 11(1):28-42.

[29] Scheibmeir H, Christensen K, Whitaker S, Jegaethesan J, Clancy R, Pierce J (2005) A review of free radicals and antioxidants for critical care nurses. Intensive and Critical Care Nursing. 21:24-28.

[30] Clark S (2002) The biochemistry of the antioxidants revisited. Nutrition in Clinical Practice. 17:5-17.

[31] Jozefczak M, Remans T, Vangronsveld J, Cuypers A (2012) Glutathione is a key player in metal-induced oxidative stress defenses. International Journal of Molecular Science. 13(3):3145-75.

[32] Traber M, Vitamin E. In: Shils M, Shike M, Ross A, Caballero B, Cousins R (eds.) Modern Nutrition in Health and Disease, 10th ed. Baltimore, MD: Lippincott Williams Wilkins; 2006. P. 396-411.

[33] Halliwell B (1996) Antioxidants in human health and disease. Annual Review of Nutrition. 16:33-50.

[34] Sheweita S (1998) Heavy metal-induced changes in the glutathoine levels and glutathione reductase/glutathione-S-transferase activities in the liver of male mice. International Journal of Toxicology. 17:383-392.

[35] Yan H, Harding J (1997) Glycation induced inactivation and loss of antigenicity of catalase and superoxide dismutase. Biochemical Journal. 338:599-605.

[36] Fire in the United States, 2003-2007. Fifteenth edition, October 2009. FEMA. Executive summary. accessed at ://www.usfa.fema.gov/downloads/pdf/statistics/fa_325.pdf

[37] Birnbaum, L, Staskal, D (2004) Brominated flame retardants: Cause for concern? Environmental Health Perspectives. 112: 9–17.

[38] Hites R (2004) Polybrominated diphenyl ethers in the environment and in people: A meta-analysis of concentrations. Environmental Science Technology. 38: 945–956.

[39] Kimbrough K, Johnson W, Lauenstein G, Christensen J, Apeti D (2009) An Assessment of Polybrominated Diphenyl Ethers (PBDEs) in Sediments and Bivalves of the U.S. Coastal Zone. Silver Spring, MD. *NOAA Technical Memorandum* NOS NCCOS 94. 87.

[40] Gandhi N, Bhavsar S, Gewurtz S, Tomy G (2011) Can biotransformation of BDE-209 in lake trout cause bioaccumulation of more toxic, lower-brominated PBDEs (BDE-47, -99) over the long term? Environment International. 37(1):170-7.

[41] de Wit C (2002) An overview of brominated flame retardants in the environment. Chemosphere. 46 (5): 583–624.

[42] Belles M, Alonso V, Linares V (2010) Behavioral effects and oxidative status in brain regions of adult rats exposed to BDE-99. Toxicology Letters. 194(1-2): 1–7.

[43] Voorspoels S, Covaci A, Maervoet J, Schepens P (2004) PBDEs in marine and freshwater sediments from Belgium: levels, profiles and relations with biota. Journal of Environmental Monitoring. 6:914-918.

[44] Eriksson P, Jakobsson E, Fredriksson A (2004) Brominated Flame Retardants: A Novel Class of Developmental Neurotoxicants in our Environment? Environmental Health Perspectives 109(9): 903-908.

[45] Schecter A, Papke O, Harris T, Tung K, Musumba A, Olson J, Birnbaum L (2006) Polybrominated diphenyl ether (PBDE) levels in an expanded market basket survey of u.s. food and estimated pbde dietary intake by age and sex. Environmental Health Perspectives. 114(10) doi:10.1289/ehp.9121.

[46] Fischer C, Fredriksson A, Erilsson P (2008) Coexposure of neonatal mice to a flame retardant PBDE 99 (2,2′4,4′5-pentabromodiphenyl ether) and methyl mercury enhances developmental neurotoxic defects. Toxicology Science. 101(2):275-285.

[47] Cheng J, Gu J, Ma J, Chen X, Zhang M, Wang W (2009) Neurobehavioral effects, redox responses and tissue distribution in rat offspring developmental to BDE-99. Chemosphere 75: 963–968.

[48] McDonald T (2002) A perspective on the potential health risks of PBDEs. Chemosphere. 46: 745-755.

[49] He P, He W, Wang A, Xia T, Xu B, Zhang M, Chan X (2008) PBDE-47 induced oxidative stress, DNA damage and apoptosis in primary cultured rat hippocampal neurons. Neurotoxicology. 29(1):124-129.

[50] Kawashiro Y, Fukata H, Sato K, Aburatani H, Takigami H,Mori C (2009) Polybrominated diphenyl ethers cause oxidative stress in human umbilical vein endothelial cells. Human and Experimental Toxicology. 28(1):703-713.

[51] Zhao Y, Tao F, Zeng E (2008) Theoretical study on the chemical properties of polybrominated diphenyl ethers. Chemosphere. 70(5): 901–907.

[52] Chen L, Lebetkin E, Sanders J, Burka L (2006) Metabolism and disposition of 2,2′,4,4′,5 pentabromodiphenyl ether (BDE-99) following single or repeated administration to rats or mice. Xenobiotica. 36(6):515-534.

[53] Toms L, Hearn L, Kennedy K (2009) Concentrations of polybrominated diphenyl ethers in the matched samples of human milk, duct and indoor air. Environment International. 35(6):864-869.

[54] Wolkers H, Boily F, Fink-Gremmels J, Bavel B, Hammill M, Primicerio R (2009) Tissue specific contaminant accumulation and associated effects on hepatic serum analytes and cytochrome p450 enzyme activity in hooded seals (cystophora cristata) from the gulf of St. Lawrence. Archives of Environmental Contamination. 56: 360–370.

[55] Albina M, Alonso V, Linares V (2010) Effects of exposure to (BDE-99) on oxidative status of liver and kidney in adult rats. Toxicology. 271(1-2): 51–56.

[56] Kodavanti P, Derr-Yellin E (2002) Differential effects of polybrominated diphenyl ethers and polychlorinated biphenyl on [3 H] arachidonic acid release in rat cerebellar granule neurons. Toxicological Sciences. 68(2):451-457.

[57] Kodavanti P, Ward T 2005. Differential effects of commercial polybrominated diphenyl and polychlorinated biphenyl mixtures on intracellular signaling in rat brain in vitro. Toxicological Sciences. 85(2):952-962.

[58] Baud O, Greene A, Li J, Wang H, Volpe J, Rosenberg P (2004) Glutathione peroxidase–catalase cooperativity is required for resistance to hydrogen peroxide by mature rat oligodendrocytes. The Jouranl of Neuroscience. 24(7):1531-1540.

[59] Sayre, L, Perry G, Smith M (2008) Oxidative stress and neurotoxicity. Chemical Research in Toxicology. 21, 172–188.

[60] Vagula M, Kubeldis N, Nelatury C (2011) Effects of BDE- 85 on the Oxidative Status and Nerve Conduction in Rodents. International journal of toxicology. 30(4): 428-434.

Chapter 12

Use of CoA Biosynthesis Modulators and Selenoprotein Model Substance in Correction of Brain Ischemic and Reperfusion Injuries

Nina P. Kanunnikova, Natalya Z. Bashun and Andrey G. Moiseenok

Additional information is available at the end of the chapter

1. Introduction

Acute disruption in brain blood circulation is a widespread cause of death and the most frequent cause of health loss in most countries of the world. About 6 millions of people suffer from stroke every year and this number is constantly increasing. Stroke has a high mortality rate – up to 30% of patients die. Only about 20% of surviving patients manage to return to their previous occupation. Most of patients are unable to take care of themselves and need help of relatives or medical personnel [1, 2]. About 80-85% of all cases of stroke are ischemic strokes. Therefore protecting brain from ischemia-induced damage is in the focus of modern neuropathology and neurosurgery studies, especially due to the increase in the number of neurosurgical operations which might cause additional blood flow impairements.

The severity of injuries of physiological reactions and biochemical processes caused by blood flow impairements depends on the degree of blood flow disruption in brain (fig.1) [1, 3].

Blood brain flow, ml/100 g/min	Parameters
60-80	Standard
35-60	Decrease of protein synthesis, selective gene expression
20-35	Lactate acidosis, cytotoxic edema
10-20	Energic deficiency, glutamate excitoxicity
0-10	Anoxic depolarization, necrosis, apoptosis

Figure 1. Correspondence between brain tissue changes and blood flow disruption

Decline of partial pressure of oxygen, significant decrease of ATP and glucose levels, membrane depolarization, extremely high levels of extracellular glutamate and intracellular calcium ions - all these factors contribute to development of the aforementioned injuries in the nervous tissue [4, 5, 6]. For example, higher level of calcium ions leads to stimulation of phospholipases and proteases, and activation of glutamate NMDA-receptors which in turn increase activity of nNoS and eNOS isoforms. As the result, amplification of lipoperoxidation takes place.

Disruption of electron transfer and oxidative phosphorilation within mitochondria are the first manifestations of ischemia-induced damages in the brain [7, 8, 9, 10], and the basic object of the injuries are presumably mitochondrial membranes [11, 12, 13, 14]. On this basis, ability of the brain to restore its functions following ischemia and reperfusion depends mainly on three processes - depletion of energic resources, excessive accumulation of excitatory amino acids [15, 16, 17], and formation of reactive oxygen species caused by leakage of electrons from intermediate links of respiratory change [18, 19, 20].

2. Features of oxidative stress in brain tissue during ischemia-reperfusion

Brain ischemia causes formation of free radical forms of oxygen which induce damage of neuronal membranes and biomacromolecules, particularly nucleic acids and proteins. Brain tissue has heightened disposition to development of oxidative stress. Brain cell membranes have high concentrations of polyunsaturated fatty acids which are the main substrate of free radical reactions [21, 22, 23]. When a free radical appears in membrane chance for its interaction with fatty acid is increased as a number of unsaturated links is rised. Unsaturated fatty acids provide more fluidity for membranes, therefore their changes caused by more active lipoperoxidation lead to increase of their viscosity and injuries of their barrier functions. It is known that synapse plasmatic membranes contain higher level of polyunsaturated fatty acids than myelin membranes. Many functionally important neuronal proteins are membrane-bound and depend on lipid environment. Simultaneously, system of antioxidant protection in brain has obviously less capacity than in other tissues, and enzymatic components of the system in brain are more sensitive to oxidative action [24, 25].

The second danger of lipoperoxidative activation in brain lies in the fact that disruption of nerve membrane integrity leads to increasing release of "excitotoxic" transmitters, such as glutamate, aspartate, etc [15, 16, 26-28].High rate of biogenic amine metabolism in brain leads to formation of ROS [29]. For example, monoamine oxidase reaction is linked with H_2O_2 formation. This phenomenon may be an additional source for generation of active radical products which are able to initiate lipoperoxidation in the presence of metals with variable valence. Dopamine, its precursor L-DOPA, 5-hydroxytraptamine, and norepinephrine may generate $O^{2\bullet-}$ not only, but quinones/semiquinones, too, which may decrease GSH level and bound with protein SH-groups. Oxidation can be catalized by transitional metal ions. Maximal increase for free radical generation and following activation for lipoperoxidation takes place in postischemic time – during recovery of blood circulation in brain tissue [30-33]. Nevertheless, possibility for formation of free radicals at earlier stage of brain damage exists during ischemia, too [34]. Enhancement of redox state of mitochondrial respiratory chain in these conditions gives an opportunity for oxygen to interact with intermediate components of the respiratory chain, for example ubisemiquinone [7]. This process takes the path of one-electron reduction of molecular oxygen and leads to formation of superoxide-anion. Consequently, heightened formation of free radicals may take place in tissues with insufficient blood circulation and decreased partial oxygen pressure [35]. Studies in mice and rats with genetic deficiency of superoxide dismutase confirmed the important role of free radicals in neuronal death/survival during brain ischemia [36, 37].In the postischemic period (during recirculation) when oxygen actively absorbs by brain tissue, oxygen radicals generation is caused by activation of enzymatic processes, too (arachidonic acid cascade, xantine oxydase system, activation of NADPH-oxydase in polymorphonuclear leucocytes) [38-40].

The fact that ischemia itself is unable to increase level of lipoperoxidation intermediate products is not surprising because during hypoxia amount of molecular oxygen is insufficient for observable activation of lipoperoxidation in brain tissue. Nevertheless, in these conditions the amount of hydroxyperoxides is increased and lag period for lipoperoxidation activation becomes shorter, which serves as evidence for decrease of antioxidant protection and increased formation of superoxide oxygen anion in brain tissue [41]. Herewith level of endogenic antioxidants in brain may be unchanged [42, 43]. For example, α-tocoferol level in rat brain is unchanged following 80 min after occlusion of middle cerebral artery and subsequent reperfusion. Unchanged levels of antioxidants were observed following bilateral occlusion of arteria carotis and reperfusion in gerbils, too [44, 45].

Recently a concept on polyfunctional physiologic role of free radicals in organism and in brain especially, is declared [42, 43, 46-48]. On the one hand, they act on key cell enzymes and receptors inside cells and cause destructive processes in tissues. On the other hand, they play a role of second messengers and may help with cell adaptive reactions to changed environmental conditions. Therefore low efficiency of antioxidant therapy by substances which bound free radical excess during ischemic injury treatment is not surprising [49, 50, 51]. In addition, these drugs have low bioavailability and must be used for at least several weeks before any effect can be observed [52, 53, 54-56].

It is necessary to emphasize that the main defense from excessive amounts of free radicals formed within cells is the action of antioxidant protection enzymes, such as selenoproteins, but not the action of low molecular wieght antioxidants. Under normal conditions these enzymes are sufficient for maintenance of low safe levels of free radicals, but in reperfusional conditions their activity is insufficient for maintenance pro- and antioxidant balance. Earlier attempts to administer substances of superoxide dismutase and catalase enzymes to animals in experimental models were unsuccessful because they poorly penetrate blood-brain barrier and cell membranes [51, 53].

Further research of neuroprotection in this direction is not very promising because activation of lipoperoxidation in phospholipid structures of nerve cell membranes is eliminated by the system of superoxide dismutase – catalase to a small degree [53]. Detoxication in these structures is primarily carried by enzymes of glutathione cycle, selenium-cystein-comprising glutathione peroxidases [49, 54, 55, 57-59].

3. Role of glutathione in mechanisms of antioxidant protection in brain

Glutathione cycle is the most important antioxidant system in brain cells [59-66]. Glutathione protects cells against reactive oxygen, nitrogen and other species. As an antioxidant it is involved in the detoxication of malonic dialdehyde, 4-hydroxy-2-nonenal and other products of lipoperoxidation. The glutathione couple GSH/GSSG takes part in maintaining cellular redox status [67, 68]. Glutathione is presumably a key participant of the defense system in brain cell [69-71].

Increased level of oxidized form of glutathione and changes of glutathione system activity occur at early stages of oxidative stress and may be marks of the severity of oxidative stress [71-75]. Hydroxyl radical and nitric oxide or peroxynitrite may interact directly with GSH leading to GSSG formation. Hydrogen peroxide may be removed by catalase or by glutathione peroxidase [76].

GSH is present in cytoplasma, endoplasmatic reticulum, nucleus, and mitochondria. In most of the compartments GSH is found predominantly in highly reduced state (about 99% of the total level of glutathione). Glutathione peroxidase is localized mainly in cytosol, too, whereas catalase is found mostly in peroxisomes. The affinity of glutathione peroxidase for H_2O_2 is one to two orders of magnitude higher than that of catalase, and catalase is less active in brain than in other tissues.

GSH is synthesized from cysteine, glutamate and glycine. Neurones have lower GSH level and use a more limited list of substrates for GSH synthesis, but they use glutamine for GSH synthesis more effectively than astrocytes because they have glutaminase for formation of glutamate from glutamine. Neurones can not absorb cystine but they actively carry off cysteine [65], so availability of cysteine influences at the GSH level inside neurones. At the same time maximal rate for GSH synthesis within astrocytes is observed in the presence of glutamate, cystine and glycine.

GSSG restores by glutathione reductase into GSH in the presence of NADPH (salvage cycle), which originated mainly from pentose phosphate pathway of metabolism. About 3-5% of oxygen in brain is consumed this way.

In physiological conditions GSSG level both in neurones and astrocytes reach no more than 1% of total content of glutathione in tissue but during oxidative stress it may be about 40% of total value of the glutathione in the astrocyte culture. One astrocyte cell may effectively protect 20 neurons from peroxides but lack of glucose greatly decreases capacity of astrocytes to bind peroxides. It has been shown that pentose phosphate pathway in astrocyte culture is very sensitive to peroxide action.

Decrease of total glutathione content and decline of GSH/GSSG ratio are indicators of the severity of oxidative stress in ischemic brain tissue [67]. It is known that decrease of GSH level leads to aggravation of ischemia-induced injuries, while increase of its level leads to opposite result. Glutamate may facilitate decrease of GSH level because it inhibits use of cysteine which is required for glutathione synthesis by cells [67]. Genetic failure of a cell glutathione peroxidase makes rats more susceptible to neurotoxins and brain ischemia [68]. Excessive glutathione peroxidase expression in transgenic mice leads to prevention of irreversible hypoxia-induced changes. Decline of GSH concentration may weaken the stability of an organism to hypoxia both by inactivation of pentose phosphate way enzymes as by inhibition of thioenzymes of tissue respiration chain [76, 77]. These disruptions cause development of energy deficiency which is the main chain of biochemical mechanism of tissue hypoxia. In addition thiol-disulfide metabolism changes may form the basis for mechanisms of disconnection of the oxidation and phosphorilation processes [78-80]. As a result, use of oxygen in biological oxidation processes may be broken and become a base for pathogenetic component of intiation and generalization of oxidative stress.

4. Role of energy metabolism changes in mechanisms of brain tissue ischemia-induced injuries

Brain is very sensitive to disruptions of energy metabolism processes beacause brain tissue requires constant supply of energy substrates whereas sources for energy formation in brain are rather limited, turnover of metabolism is high, and metabolism is dependent on aerobic oxidation of glucose and constant supply of oxygen in a great extent [81, 82] Maintenance for electric neuron activity and rate of impulse passage depend directly on presence and availability of energy substrates, too [83, 84].

Brain tissue cannot metabolize fatty acids therefore the main source for energy formation in brain is glucose. Nevertheless, during focal brain ischemia, increase of glucose level does not help cells to prevent ischemic injuries and also promotes their structural and functional damage [85]. Mechanisms of these changes include shift of pH to an acid side inside cell, increase of permeability of blood-brain barrier, infiltration of brain parenchyma by neutrophiles, accumulation of extracellular glutamate, and unfavorable corticosterone action. Intensive metabolism of glucose in the penumbra region may promote increasing acidic reaction of the medium, promote attraction of neutrophiles in the region.

Limitation of metabolic consumption of glucose in ischemic brain tissue may have protective effect, especially in such conditions when its metabolism will be faster or other source of fuel will be used. Possibilities for replacement of glucose in brain tissue are rather limited [86-89]. Lactic acid may be an alternative source for energy formation in brain in certain conditions because the glucose is metabolized presumably in glial cells whereas in neurones energy metabolism is based presumably on lactate oxidation [90, 91]. There is some evidence that the process is particularly important for maintenance of vital functions during postischemic time. For example, decrease of lactate transfer through plasmatic mebrane in brain following preliminary whole ischemia causes neurone injuries. From clinic practice it is known that consumption of lactate or pyruvate during brain ischemia show neuroprotective actions of the substances [90, 91]. Presence of adequate concentrations of pyruvate facilitate for maintenance of stable level of membrane potentials and proton gradient on vesicular membranes [87, 92].

Brain ischemia is different from other types of ischemia because oxygen deficiency causes significant changes in the oxidating process of energy substrates which are present in brain in suffucient quantities [20, 42]. Anaerobic glycolisis as alternative way for energy supply is not substantial for supporting ATP stock in nerve tissue during compensated and decompensated stages of hypoxia [11].

Aerobic energy formation is the basic process for nerve tissue, but starts to fail before oxygen concentration falls below critical level, because hypoxia influences kinetic properties of respiratory chain enzymes. During early stages of ischemia energy functions of mitochondria already start changing: conjugacy of oxidative phosphorilation process and regulatory control by ATP becomes weaker, rate of inphosphorilated respiration increases. Shift of ratio NAD/NADH occurs to the side of NADH, as a result final stages of the Krebs' cycle are inhibited, and activation of succinate oxidase stage takes place. This way allows to maintain oxidative phosphorilation and respectively macroergic substance production at sufficient level for some time. "Oxygen hunger" already at early stages of hypoxia leads to beginning of relative "substrate hunger" - energy substrates are not being oxydated while they are still available. This is a characteristic property for ischemia [93].

There are only 2 ways of restoration of brain metabolism after stroke and hypoxia – restoration of NAD-dependent part of the Krebs' cycle and stimulation of alternative path of metabolism, succinate oxidation.

Succinic acid is an intermediate of the cytric cycle which supports formation of macroergic phosphates and reductive equivalents in the conditions with physical loadings and stress [94-97]. Oxidation of succinic acid is the most potent energy process inside mitochondria, and during stress this process becomes even more important due to succinate dehydrogenase activation. Depletion of endogenic succinic acid may be a reason explaining inability of tissue to maintain reaction of activation of energy processes for a long time [98]. If NADH and CO_2 are present in excess, conversion of reactions of second part of the cytrate cycle in which NADH is consumed takes place - from oxaloacetate to succinate, and that fact supports reactions of the first part of the cytrate cycle reactions which require oxidized NAD and promote for additional accumulation of succinate.

Another important result of the bioenergic hypoxia is damage of ion pump action and ion imbalance in the form of excessive accumulation of intracellular calcium, sodium, chlorine [99]. The intracellular calcium excess leads to activation of phospholipase A, damage of cell membranes and release of arachidonic acid take place. As a result, lipoperoxidation activates and causes following cell membrane damages, neuron depolarization and release of excitotoxic amino acids, especially glutamate, in extracellular space [99].

Thereby adaptive effects of succinate derivatives are related to their property to induce compensatory metabolic flows in mitochondrial respiratory chain ("succinate oxidase" way) in extreme conditions, to provide replenishment for cytoplasmatic pool by reduced forms of NAD and NADP, to accelerate ATP formation, change over energy formation from NAD-dependent to FAD-dependent way, eliminate an excess of acetyl-CoA, support activity of the Krebs' cycle in hypoxic conditions, stabilize membrane potential of mitochondria and cell membranes. Advantage of succinate oxidase way versus NAD-dependent substrates in competition for respiratory chain is amplified in the hypoxia conditions because flavines (flavoproteins) continue in oxygenated form longer than pyridine nucleotides.

Disruption of energy metabolism can be mainly observed at the stage of succinate formation. That may be caused by oxidative stress-induced changes of stable state of plasmatic and mitochondrial membranes and changes in activity of membrane-bound enzymes of the Krebs cycle and GABA bypass [100, 101]. Significant activation of the GABA bypass enzymes takes place during ischemia-reperfusion which not only causes raise of succinate formation, but also leads to increasing formation of gamma-hydroxybutyric acid through reductase reaction. GHBA has protective effect against changes of energy formation processes in brain tissue during hypoxia [102]. As activity of glutamate dehydrogenase in brain is rather low compared with other tissues, GABA bypass plays a key role in compensatory maintenance of succinate level sufficient for adequate metabolism in "succinate oxidase" way in different extremal situations, for example in brain ischemia-reperfusion conditions.

Succinic acid derivatives are effective modern antioxidants in the brain because succinate regulates activity of SDH in the Krebs' cycle and restores activity of respiratory mitochondrial chain not only, but increases microcirculation in tissues.

5. Role of CoA in mechanisms of neuroprotection in brain ischemia-reperfusion conditions

Beneficial effects of precursors of CoA biosynthesis, such as pantothenic acid and its derivatives, include protection from lipoperoxidation and supporting membrane structure, and these effects have been observed in radiation injury, miocard ischemia, diabetes mellitus, CCl_4 -intoxication, heavy hypothermia, etc [103-106]. Protective action of pantothenate derivatives have been reported in situations accompanied by oxidative stress, for example, in experimental ischemia-reperfusion of myocardium. It has been shown that antioxidative and membrane-protective effects of the pantothenate derivatives are accompanied by an activation of biosynthesis CoA system and increasing of intracellular level of a free CoA [108, 109]. Presumably, the mechanism of cell protection is CoA-dependent or realized through CoA-(acyl-CoA)-dependent biochemical reactions, including rise of intracellular glutathione level and maintenance of its redox status.

It is believed that the physiological function of CoA system is participation in formation of redox potential of glutathione and proteins, redox signaling and maintenance of biological membrane stability, especially in brain tissue [104, 105, 106].

The CoA biosynthesis system is a group of very stable continuously active self-regulated processes focused on maintaining stability of intramitochondrial CoA-SH (up 70-80% of the total cell value). This function maintains constant flow of oxidative substrates and their effective using for ATP formation in the citric acid cycle [109, 110, 111].

The lesser CoA pool in cytosol where acyl-CoA is used in biosynthetic processes (biosynthesis of phospholipids, fatty acids) is studied to a lesser extent. "Turnover" pool of the coenzyme takes part in reactions of carnitine-dependent transfer of fatty acid residues and acetate through mitochondrial membranes [105]. The main events for interrelations between specifically bound cytosolic CoA-S-S-protein, CoA-S-S-glutathione, free and proteidized glutathione take place within cytosolic compartment (including endoplasmatic reticulum) presumably due to thiol-disulfide-exchange reactions which provide stability during limited variations of redox potential and support a realization of redox sygnaling. Based on this hypothesis, the capability of CoA biosynthesis precursors in low concentrations (0.1-1 мМ) or in vivo experiments prevents lipoperoxidation activation, damages of membrane integrity initiated by different physical or chemical factors. The obligatory condition of the above-mentioned defensive effect is biotransformation of pantothenate derivatives into CoA and significant increase of intracellular GSH level. The process is highly specific because homopantothenic acid which is similar to pantothenic one in terms physical and chemical properties can not transform into CoA, does not increase intracellular glutathione level and does not protect plasmatic membrane stability in cell culture [112].

Additional effects of the CoA precursors in defense of lipoperoxidative activation have also been observed. These include rapid initiation of lipid biosynthesis from labeled precursors, positive influence on mitochondrial energy parameters, as well as protection against apoptosis activation caused by free radical oxidation products [113, 114]. Redox sygnaling process controls the initiation and direction of these processes. The redox potential is determined by the glutathione system predominantly [105, 115]. This data may confirm that the CoA biosynthesis system is the most important factor of intracellular stability of GSH level [103, 105].

Maintenance of sufficiently high CoA biosynthesis activity has an important role in brain because acetyl-CoA is used not only as the main way for glucose intake into the Krebs cycle but is also a substrate for acetylcholine synthesis. The relationship between ability for CoA biosynthesis and activity of acetylcholine metabolism within cholinergic neurones may be an important factor in modulation of their sensitivity to damaging influences [103].

Among the necessary conditions for successful biosynthesis of acetyl-CoA, are presence of CoA precursors inside mitochondria, and also the presence of carnitine which transfers acetyl radicals into mitochondria. Under oxidative stress conditions when a lot of lipoperoxide products are released from membranes as a result of phospholipase activation, the potential for CoA sequestration increases, which includes appearance of hard to metabolize acyl-CoA derivatives. Under these conditions the role of carnitine increases. CoA and L-carnitine are among the key factors of intramitochondrial metabolism of fatty and organic acids, and relationship between their levels represents an essential mechanism for cytosol-mitochondrial process of acyl residue activation and transfer [110, 116]. Based on the main localization of a total CoA within mitochondrial matrix, while carnitine is located in cytosol, the molar ratio of CoA/carnitine may have significant functional role for decrease of long-chain acyl-CoA in cytosol and their accelerated utilization in a β-oxidation process.

Generation and use of succinyl-CoA in mitochondria have a special role for mitochondrial oxidation regulation during oxidative stress caused by ischemia-reperfusion [107, 108]. Chances for alternative succinyl-CoA biosynthesis increase significantly when CoA biosynthesis processes are activated in cytosol. In view of this, data on the effects of carnitine on the activity of the key enzyme of CoA biosynthesis, namely patothenate kinase, has high significance. It has been shown that L-carnitine cancels out inhibitory action of physiological concentrations of dephospho-CoA, CoA-SH, and acetyl-CoA on pantothenate kinase. This enables directed regulation of CoA-dependent metabolic processes following simultaneous injection of carnitine substances and pantothenate derivatives - precursors of CoA biosynthesis - namely, panthenol [117, 118].

Study of changes of CoA level during ischemia or ischemia-reperfusion showed markedly stable ratio and levels of free CoA, short-chain acyl-CoA, and on the whole the acid-soluble fraction of CoA, in hemispheres during ischemic damage [117]. Following 2-3 h of brain ischemia, the free CoA level declines. This diminishment with simultaneous decrease of the acid-soluble CoA fraction achieves maximal reduction within 24 h under continued conditions of reperfusion (reoxygenation). These results confirm significance of the CoA system in pathogenic mechanisms of reoxygenation-reperfusion syndrome development.

CoA is one of the fundamental metabolism factors, and its biosynthesis and catabolism are subject to rigid control on the cell level. Therefore, as a rule, changes of particular CoA forms may happen only under extreme conditions and after prolonged period of time, sufficient to cause imbalance in metabolism regulation systems in the cell. Such imbalance starts to influence the CoA system during ischemia no earlier than one hour after occlusion of arteria carotis.

Data on the key role of the CoA biosynthesis system in maintaining redox potential of the glutathione system, neuronal membrane stability and defense of nitroperoxide acyl-CoA gives rationale to the use of CoA precursors in treatment of ischemia and ischemia-reperfusion-induced damages in the brain tissue.

6. Role of selenium in mechanisms of antioxidant protection for brain

Selenium is an essential microelement in different brain functions [119-124]. Neuroprotective potential of selenium is realized through the expression of selenoproteins: glutathione peroxidase, thioredoxine reductases, methionine sulfoxide reductases, selenoproteins P and R, which participate in regulation of the oxidation-reduction state of the neurones and glial cells under both physiologic conditions, and during oxidation stress [125, 126, 128]. Selenium regulates antioxidative processes in the CNS, protects brain tissue from neurodegenerative injuries during Alzheimer and Parkinson diseases, prione diseases, has antiischemic and angiogenic actions, etc. Insufficient level of selenium intensifies damages of neuron functions and structure caused by different endogenic and exogenic affections and leads to some neurodegenerative pathologies [122, 129-132].

The biological role of selenium is explained by the selenium presence in active sites of selenium-related enzymes [121, 133], which protect brain tissue during oxidative stress. Expression extremely diverse Se-containig proteins is observed in the brain. Selenoprotein P is required for transfer of selenium into the brain, and the brain selenium level is strictly dependent from an expression of selenoprotein [129]. Activity of Se-dependent enzymes in the brain is maintained at rather stable levels even during profound selenium deficiency, owing to the presence of unique Se-transport system in the brain (proteins containing selenium-cysteine, Se-transported protein of a Golgi apparatus). This system achieves its maximum value in hypothalamus.

Injections of selenium-containing compounds lead to an increase of activities of glutathione peroxidase and thioredoxin reductase, decrease of lipoperoxidation processes, cell defense from apoptosis [122, 126, 127]. Selenium ions activate oxidative-reductive enzymes of mitochondria and microsomes, take part in ATP synthesis, in electron transfer from hemoglobin to oxygen, maintain cysteine turnover, enhance α-tocoferol action.

7. Metabolic approaches to correction for brain ischemia-reperfusion-caused injuries

Steady advances in the neurosciences have elicidated the pathophysiological mechanisms of brain ischemia and have suggested many therapeutic approaches to achieve neuroprotection in the acutely ischemic brain that are directed at specific injury mechanisms [134-136]. Nevertheless, methods of protection of ischemia and reperfusion-induced damages are still lacking [51, 137, 138]. Search for new ways of neurodefense during brain ischemia-reperfusion is necessary due to the absence of sufficient protective activity in the most substances with specific focus in clinical conditions: controlling excitotoxic effects of neurotransmitter amino acids (modulators of glutamic acid receptor activity and Ca-channels), regulating redox status of cells, as well as presence of high toxicity in the most anti-ischemic medicines [51, 139-141]. In the past two decades, numerous attempts were made to use different substances with the effect on Ca level in a cell and glutamate extracellular level, aiming to apply these as drugs for ischemia-induced injuries treatment, but they have not been successful in men [51, 139-141]. For example, in experimental models, blockators of NMDA- and AMPA-receptors of glutamate exhibit high protective action, but they have strong side-effects and weak protective effects in humans, especially blockators of NMDA-receptors. The role of glutamate in neurotoxic phenomena during ischemia is known to be significant, but usage of glutamate receptor antagonists is rather problematic [51]. There are ongoing studies of Mg substances which block NMDA-receptors, as well as with blockators of AMPA-receptors. There have been recent proposals to combine usage of several drugs with different mechanisms of action. All of the above drugs have a common property – rather high toxicity. Therefore, usually a certain combination of drugs is applied in order to minimize their toxicity and maximize effectiveness [140].

Substances for metabolic therapy may be particularly useful during treatment of brain blood circulation injuries in the case of their simultaneous application with specific medicines because they have no toxic effects and may be used safely for prolonged period [51, 142, 143]. Apart from these drugs, compounds for so-called restoration therapy may be used. Their effects include restoration of metabolism and blood flow in damaged region. Application of the metabolic substances that help to maintain energy metabolism and redox status of glutathione system may be useful for remedying damages to the brain after ischemia-reperfusion [51, 143]. Previously we have shown high efficacy of pantothenic acid derivatives – CoA precursors, as a means of protecting cell membranes from different types of oxidative stress [117, 118, 144]. D-panthenol presents an important substance in this respect because it penetrates into the brain through blood-brain barrier easily and is converted into pantothenic acid, 4-phosphopantothenic acid, CoA, and after that into acyl-CoA (acetyl-, malonyl-, succinyl-CoA), which have high metabolic activity. These effects create the preconditions for stabilization of CoA-dependent processes of membrane phospholipid biosynthesis, neurotransmitter biosynthesis, regulation of energic processes, etc [118].

The efficacy of panthenol as a neuroprotector within a stroke model in rats has been demonstrated [117, 118, 144]. Panthenol not only decreased the volume of infarction, but also diminished neurological deficiency in animals [103]. Fairly high protector activity of D-panthenol was observed in respect to changes of energic metabolism and glutathione system activity during brain ischemia-reperfusion. Protective effects of pantothenic derivatives is not related to their action as free radical scavengers, however. They act primarily as CoA precursors, whereas CoA accelerates various metabolic pathways, such as biosynthesis of glutathione, which constitutes one of the main systems of cell protection against oxidative stress.

Succinic acid is essential for keeping energy formation processes stable in the brain under extreme conditions [94-97]. Consequently, injections of panthenol and succinate following brain ischemia-reperfusion stabilize levels of lipoperoxidation in blood and in brain hemispheres, stabilize levels of protein SH-groups in blood, lead to significant decrease of the GSSG level and normalization of glutathione enzyme activities, as well as glutamate and glutamine metabolism in the brain to control values [118]. D-panthenol and succinate ammonium injection served to partially remedy the injuries and restore these parameters to their normal levels, especially if administered together. These effects are likely linked to activation of succinyl-CoA biosynthesis.

Attempts were made to use selenium-containing compounds for prevention of ischemia-indused injuries, such as ebselen (2-phenyl-1,2-benzisoselenozol-3), which imitated glutathione peroxidase activity [146-148]. However, under clinical conditions the ebselen was not effective. Di-(3-methylpyrazolil-4)-selenide (selecor) imitates effects of selenoproteins, has low toxicity, and satisfactory bioavailability. Additional injections of selecor increase effects of the panthenol and succinate, especially on the lipoperoxidation parameters and activities of glutathione system and selenium-bound enzymes, on ischemia-reperfusion- induced injuries [149].

Effects of D-panthenol and succinate on decrease of lipoperoxidation activities contribute to the overall protective effects of the composition. However, it is evident that metabolic actions of the substances are related to their capacity for regulation of energy metabolism and mitochondrial respiration activity, restoration of the CoA-SH level and cell redox status, membrane-protective activity of the panthenol [118]. Addition of di-(3-methylpyrazolil-4)-selenide (selecor) to D-panthenol and succinate does has limited effect on protective antioxidant properties of the composition. It is likely that this provides additional evidence for significance of specifically metabolic effects of the composition. Increase of selenium level in blood plasma, which may contribute in maintaining of antioxidant activity of glutathione system, takes place in the absence of selenoprotein substrates, after injections of panthenol and succinate. Nevertheless, addition of a selenium source to panthenol and succinate strengthened protective potential of the substances with respect to changes for enzyme activities of glutamate and glutamine which play an important role in maintaining of energy supply and detoxication in ischemic brain tissue and confirms the antiischemic effect of the substances. Effects of di-(3-methylpyrazolil-4)-selenide may be explained less by selenium supply as a selenoprotein component rather than by its modeling of selenoprotein activity, as is known to be the case with ebselen [146-148].

Therefore it is expected that the tested substances, such as panthenol, succinic acid, selecor, and potentially other metabolic therapy drugs may have high efficacy as neuroprotectors in brain ischemia and reperfusion-induced damages.

Author details

Nina P. Kanunnikova and Natalya Z. Bashun
Department of Zoology and Physiology of Men and Animals,
Yanka Kupala's Grodno State University, Grodno, Republic of Belarus

Andrey G. Moiseenok
National Center for Foodstuffs, National Academy of Sciences of Belarus, Minsk, Republic of Belarus

8. References

[1] Gusev EI, Skvortsova VI (2001) Brain ischemia. Moscow. Medicina. 328 p (rus).

[2] Breton RR, Rodriguez JCG (2012) Excitoxicity and oxidative stress in acute ischemic stroke. Acute ischemic stroke. Ed.JCG Rodriguez: 29-58.

[3] Heiss WD (1983) Flow threshold of functional and morphological damage of brain tissue. Stroke. 14: 329-331.

[4] Choi DW (1995) Calcium: still center stage in hypoxic-ischemic neuronal death. Trends Neurosci. 18: 58-60.

[5] Rayevsky KS, Bashkatova VG (1996) Oxidative stress, apoptosis and brain damage. Neirokhimia 13(1): 61-64 (rus).

[6] Rothman SM, Olney JW (1987) Excitotoxicity and the NMDA receptor // Trends Neurosci. 10: 299-302.

[7] Doubinina EE (2003) Role of oxidative stress in pathological states of nervous system. Uspekhi Funkc.Neirokhimii.Ed. Dambinova SA, Arutiunian AV. StPetersburg.: 285–301 (rus).

[8] Lazarewicz J, Strosznajder J, Gromek A (1972) Effects of ischemia and exogenic fatty acids on the energy metabolism in brain mitochondria. Bull.Acad.Pol.Sci.Biol. 20: 599-606.

[9] Strosznajder J (2005) Czynniki biochemiczne inicjuace niedokrwienie. Mozg a niedokrwienie. Pod red. J. Strosznajder, Z. Czernicki. Wydawnictwo PLATAN. Krakow: 7-15.

[10] Zavalishin IA, Zakharova MN (1996) Oxidative stress is a general mechanism of damages in nerve system diseases. Zh. Neurol. Psychiatr. 96(2): 111–114 (rus).

[11] Lin MT, Beal MF (2006) Mitochondrial disfunction and oxidative stress in neurodegenerative diseases. Nature. 443: 787-795.

[12] Andreyev AYu, Kushnareva YuE, Starkov AA (2005) Mitochondrial metabolism of reactive oxygen species. Biochemistry. 70 (2): 200–214.

[13] Beal MF (1996) Mitochondria, free radicals, and Neurodegeneration. Curr. Opin. Neurobiol. 6: 661–666.

[14] Shinder AF, Olson EC, Spitzer NC, Montal M (1996) Mitochondrial disfunction is a primary event in glutamate neurotoxicity. J.Neurosci. 16: 6125-6133.

[15] Sopala M, Danysz W (2005) Rola ukladu glutaminianergicznego w ischemicznym uszkodzeniu mozgu // Mozg a niedokrwienie / Pod red. J. Strosznajder, Z. Czernicki // Wydawnictwo PLATAN. Krakow: 47-65.

[16] Benveniste H et al (1984) Elevation of extracellular concentrations of glutamate in rat hippocampus during transient cerebral ischemia monitored by intracerebral microdialysis. J.Neurochem. 43: 1369-1374.

[17] Choi DW (1988) Glutamate neurotoxicity and diseases of the nervous system. Neuron 1: 623-634.

[18] Pelligrini-Giampietro DE, Cherici G, Alesiani M, Carla V, Moroni F. Excitatory amino acid release and free radical formation may cooperate in the genesis of ischemia-induced neuronal damage. J.Neurosci. 10: 1035-1041.

[19] Boldyrev AA (2003) Role of active species of oxygen in neuron vital functions. Uspekhi physiol.nauk. 34(3): 21–34 (rus).

[20] Boldyrev AA, Kukley ML (1996) Free radicals in normal and ischemic brain. Neirokhimia. 13(4): 271–278 (rus.).

[21] Kinuta Y, Kikuchi H, Ishikawa M, Kimura M, Itokawa Y (1989) Lipid peroxidation in focal cerebral ischemia. J.Neurosurg. 71: 421-429.

[22] Boldyrev AA (1998) Mechanisms of brain defense from oxidative stress. Priroda. 3: 26-34 (rus).

[23] Hayashi M. (2009) Oxidative stress in developmental of brain disorders. Neuropathology. 29(1): 1–8.

[24] Yamamoto M [et al.] (1983) A possible role of lipid peroxidation in cellular damages caused by cerebral ischemia and the protective effect of alpha-tocopherol administration. Stroke. 14: 977-982.

[25] White BC et al (2000) Brain ischemia and reperfusion: molecular mechanisms of neuronal injury. J.Neurol.Sci.179(S1-2): 1-33.

[26] Siesjo BK, Zhao Q, Pahlmark K, Siesjo P, Katsura K, Folbergrova J (1995) Glutamate, calcium and free radicals as mediators of ischemic brain damage. Ann.Thorac.Surg. 59: 1316-1320.

[27] Dugan LL, Choi DW (1994) Exitotoxicity, free radicals and cell membrane changes. Ann. Neurol. 35: 17-21.

[28] Oh SM, Betz L (1991) Interaction between free radicals and excitatory amino acids in the formation of ischemic brain edema in rats. Stroke 22: 915-921.

[29] Zeevalk GD, Bernard LP, Nicklas WJ (1998) Role of Oxidative Stress and the Glutathione Systems in Loss of dopamine Neurons Due to Impairment of Energy Metabolism. J. Neurochem. 70(4): 1421–1429.

[30] Fiodorova TN, Boldyrev AA, Gannushkina IV (1999) Lipoperoxidation in experimental brain ischemia. Biokhimia. 64(1): 94–98 (rus).

[31] Kukley ML, Stvolinsky SL, Shavratsky VCh (1995) Lipoperoxidation in rat brain during ischemia. Neirokhimia. 12(2): 28–35 (rus).

[32] Rafols JA, Krause GS (2000) Brain ischemia and reperfusion: molecular mechanisms of neuronal injury. J.Neurol.Sci.179(S1-2): 1-33.

[33] Sakamoto A, Ohnishi ST, Ohnishi T, Ogawa R (1991) Relationship between free radical production and lipid peroxidation during ischemia-reperfusion injury in the rat brain. Brain Res. 554: 186-192.

[34] Kitagawa RM [et al.] (1990) Free radical generation during brief period of cerebral ischemia may trigger delayed neuronal death. Neuroscience. 35: 551–558.

[35] Fridovich I (1995) Superoxide radical and superoxide dismutases. Annu.Rev. Biochem. 64: 97-112.

[36] Chan PH (2004) Oxidant stress as a molecular switch in neuronal death/ survival after stroke. J.Neurochem. 88(Suppl): S6-4.

[37] Murakami K, Kondo T, Kawase M, Sato S, Chen SF, Chan PH (1998) Mitochondrial susceptibility to oxidative stress exacerbated cerebral infarction that follows permanent focal cerebral ischemia in mutant mice with manganese superoxide dismutase deficiency. J Neurosci. 18: 205-213.

[38] Clemens JA (2000) Cerebral ischemia: gene activation, neuronal injury, and the protective role of antioxidants. Free Rad.Biol.Med. 28: 1526-1531.

[39] Nita D Al et al. (2001) Oxidative damage following cerebral ischemia depends on reperfusion – a biochemical study in rat. Journal Cell. Mol. Med. 5(2): 163–170.

[40] Skvortsova VI [et al.] (2007) Oxidative stress and oxygen status in ischemic stroke. Zh. Nevrol. Psikhiatr. Im. S.S. Korsakova. 107(1): 30–36 (rus).

[41] Turrens JF (1997) Superoxide production by the mitochondrial respiratory chain. Biosci.Rep. 17: 3-8.

[42] Boldyrev AA (1995) Dual role of free radical species of oxygen in ischemic brain. Neirokhimia. 12(3): 3–13 (rus).

[43] Gulyaeva NV, Yerin AN (1995) Role of free radical processes in development of neurodegenerative diseases. Neirokhimia. 13(2): 3–15 (rus).

[44] Hall ED, Braughler JM, Pazak KE (1993) Hydroxyl radical production and lipid peroxidation parallels selective post-ischemic vulnerability in gerbil brain. J. Neurosci. Res. 45(1): 107–112.

[45] Oliver CN [et al.] (1990) Oxidative damage to brain proteins loss of glutamine synthesis activity and production of free radicals during ischemia–reperfusion induced injury to gerbil brain. Proc. Natl. Asad. Sci. USA. 87: 5144–5147.

[46] Halliwell B (1992) Reactive oxygen species and the central nervous system. Free Radical in the brain. Aging. Neurological and Mental Disorders/ Eds Packer L, Christen Y. Springer – Verlag. Berlin, N.–Y., London: 21–41.

[47] Halliwell B (2006) Oxidative stress and neurodegeneration: where are we now? J. Neurochem. 97: 1634–1658.

[48] Halliwell B (2001) Role of free radicals in the neurodegenerative diseases: therapeutic implications for antioxidant treatment. Drugs Aging. 18(3): 685–716.

[49] Gupta R, Singh M, Sharma A (2003) Neuroprotective effect of antioxidants on ischaemia and reperfusion-induced cerebral injury. Pharmacol. Res. 48(Is.2): 209–215.

[50] Halliwell B. (1996) Antioxidants in human health and disease. Ann. Rev. Nutr. 16: 33–50.

[51] Grieb P (2005) Szanse na skuteczne neuroprotekcyjne leczenie udaru niedokrwiennego mozgu. Mozg a niedokrwienie. Pod red. J. Strosznajder, Z. Czernicki. Wyd. PLATAN. Krakow: 237-260.

[52] Gruener N [et al.] (1994) Increase in superoxide dismutase after cerebrovascular accident. Life Sci. 54(11): 711–713.

[53] Kumari NK, Panigrahi M, Prakash Babu P (2007) Changes in endogenous antioxidant enzymes during cerebral ischemia and reperfusion. Neurol. Res. 29(8) 877–883.

[54] Suslina ZA et al (2000) Antioxidant therapy in ischemic insult. Zh. Neurol. Psychiatr. 100(10): 34-38 (rus).

[55] Shapoval GS, Gromovaya VF (2003) Mechanisms of antioxidant protection during action of reactive oxygen species. Ukr.Biokhim.Zh. 75(2): 5–13 (rus).

[56] Valko M, Leibfritz D, Moncol J, Cronin MTD, Mazur M, Telser J (2007) Free radicals and antioxidants in normal physiological functions and human disease. Int. J. Biochem.Cell Biol. 39: 44-84.

[57] Kalinina EV et al. (2010) Modern conceptions on antioxidant role of glutathione and glutathione-dependent enzymes. Vestnik RAMN. 3: 46–54 (rus).

[58] Sies H (1997) Oxidative stress: oxidants and antioxidants. Exp. Physiology. 82: 291-295.

[59] Sies H (1999) Glutathione and its role in cellular functions. Free Radical Biology and Medicine. 27(Iss.9–10): 916–921.

[60] Schulz JB Lindenau J, Seyfried J, Dichgans J (2000) Glutathione, oxidative stress, and neurodegeneration. Eur. J. Biochem. 267(16): 4904–4911.

[61] Lorenc-Koci E. (2003) Neuroprotekcyjne wlasciwosci glutationu I ich znaczenie w schorzeniach neurodegeneracyjnych. Neuroprotekcji. XX Zimowa Szkola Instytutu Farmakologii PAN. Pod red. M.Smialowskiej. – Mogilany: 169-185.

[62] Cooper AJL, Kristal BS (1997) Multiple roles of glutathione in the central nervous system. J.Biol.Chem. 378: 793-802.

[63] Cruz R, Almaguer Melian W, Bergado JA Rosado J (2003) Glutathione in cognitive function and neurodegeneration. Rev. Neurol. 36(9): 877–886.

[64] Dringen R (2000) Metabolism and functions of glutathione in brain. Prog. Neurobiol. 62: 649-671.

[65] Dringen R, Hirrlinger J (2003) Glutathione pathways in the brain Biol. Chem. 384(4): 505–516.

[66] Lushchak VI (2012) Glutathione homeostasis and functions: potential targets for medical interventions. J Amino Acids. 2012:
http://www.hindawi.com/journals/jaa/2012/736837/

[67] Jones DP (2002) Redox potential of GSH/GSSG couple: assay and biological significance. Methods in enzymology. 348: 93-112.

[68] Makarov P, Kropf S, Wiswedel I, Augustin W, Schild L (2006) Consumption of redox energy by glutathione metabolism contributes to hypoxia/reoxygenation-induced injury in astrocytes. Mol. Cell Biochem. 286 (1-2): 95-101.

[69] Bains JS, Shaw CA (1997) Neurodegenerative disorders in humans: The role of glutathione in oxidative stress-mediated neuronal death. Brain Res. Rev. 25: 335–358.

[70] Townsend DM, Tew KD, Tapiero H (2003) The importance of glutathione in human disease. Biomed.Pharmacotherapy. 57(3): 145-155.

[71] Mari M, Morales A, Colell A, Garcia-Ruiz C, Fernandez-Checa JC (2009) Mitochondrial glutathione, a key survival antioxidant. Antioxidants and Redox Signalling. 11(11):2685-2700.

[72] Beer SM, Taylor ER, Brown SE et al. (2004) Glutaredoxin 2 catalyzes the reversible oxidation and glutathionylation of mitochondrial membrane thiol proteins: implications

Use of CoA Biosynthesis Modulators and Selenoprotein Model Substance in Correction of Brain
Ischemic and Reperfusion Injuries

273

for mitochondrial redox regulation and antioxidant defense. J Biol Chem. 279(46): 47939-47951.

[73] Kumar C, Igbaria A, D'Autreaux B et al. (2011) Glutathione revisited: a vital function in iron metabolism and ancillary role in thiol-redox control. EMBO J. 30(10): 2044-2056.

[74] Kramer K et al. (1992) Glutathione mobilization during cerebral ischemia and reperfusion in the rat. Gen. Pharmacol. 23(1): 105–108.

[75] Bessonova LO, Verlan NV, Kolesnichenko LS ЛC (2008) Role of glutathione system in antioxidant protection during complex pathology of hypoxic genesis. Sibirskij Med.Zh. 6: 19–21 (rus).

[76] Mizui T, Kinouchi H, Chan PH (1992) Depletion of brain glutathione by buthionine sulfoximine enhances cerebral ischemic injury in rats. Am. J. Physiol. 262(2): 313–317.

[77] Islecel S [et al.] (1999)Alterations in superoxide dismutase, glutathione peroxidase and catalase activities in experimental cerebral ischemia-reperfusion. Res. Exp. Med. 199: 167–176.

[78] Anderson MF, Sims NR (2002) The effects of focal ischemia and reperfusion on the glutathione content of mitochondria from rat brain subregions. J. Neurochem. 81: 541–549.

[79] Holmgren A (2000) Antioxidant function of thioredoxin and glutaredoxin systems. Antioxidant Redox Signal.2: 811-820.

[80] Shivakumar BR, Kolluri SV, Ravindranath V (1995) Glutathione and protein thiol homeostasis in brain during reperfusion after cerebral ischemia. J. Pharmacol. Exp. Ther. 274(3): 1167–1173.

[81] Shivakumar BR, Kolluri SR, Ravindranath V (1992) Glutathione homeostasis in brain during reperfusion following bilateral carotid artery occlusion in the rat. Moll. Cell. Biochem. 111(1–2): 125–129.

[82] Suh SW, Shin BS, Ma H, Van Hoecke M, Brennan AM, Yenari MA, Swanson RA (2008) Glucose and NADPH oxidase drive neuronal superoxide formation in stroke. Ann.Neurol. 64: 654-663.

[83] Zauner A, Daugherty WP, Bullock MR, Warner DS (2002) Brain oxygenation and energy metabolism. Neurosurgery 51: 289-301.

[84] Erecinska M, Nelson D, Chance B. (1991) Depolarization induced changes in cellular energy production // Proc. Natl Acad.Sci.USA. 88: 7600–7604.

[85] Kozuka M (1995) Changes in brain energy metabolism, neurotransmitters, and choline during and after incomplete cerebral ischemia in spontaneously hypertensive rats. Neurochem. Res. 20(1): P. 23–30.

[86] Martin RL, Lloyd HG, Cowan AI (1994) The early events of oxygen and glucose deprivation: setting the scene for neuronal death? Trends Neurosci 17: 251-257.

[87] Elman I. et al. (1999) The effects of pharmacological doses of 2-deoxyglucose on cerebral blood flow in healthy volunteers. Brain Res. 815: 243-249.

[88] Gonzalez-Falcon A. et al. (2003) Effects of pyruvate administration on infarct volume and neurological deficits following permanent focal cerebral ischemia in rats // Brain Res. 990: 1-7.

[89] Suzuki M et al.(2001) Effect of of beta-hydroxybutyrate, a cerebral function improving agent, on cerebral hypoxia, anoxia and ischemia in mice and rats. Japan J.Pharmacol. 87: 143-150.

[90] Wey J, Cohen DM, Quast MJ (2003) Effects of 2-deoxy-d-glucose on focal cerebral ischemia in hyperglicemic rats // J.Cereb.Blood Flow Metab. 23: 556-564.

[91] Bouzier-Sore AK et al (2003) Lactate is a preferential oxidative energy substrate over glucose for neurons in culture. J.Cereb.Blood Flow Metab. 23: 1298-1306.

[92] Chih CP, Lipton P, Roberts EL (2001) Do active cerebral neurons really use lactate rather than glucose? Trends Neurosci. 24: 573-578.

[93] Hsueh-Meei Huang [et al.] (2003) Inhibition of the α-ketoglutarate dehydrogenase complex alters mitochondrial function and cellular calcium regulation. Biochim. Biophys.Acta.1637: 119–126.

[94] Niizuma K, Endo H, Chan PH (2009) Oxidative stress and mitochondrial disfunction as determinants of ischemic neuronal death and survival. J.Neurosci. 109 (Suppl.1): 133-138.

[95] Kondrashova MN (1991) Cooperation for processes of transaminations and oxidation of carbonic acids in different cell functional states. Biokhimia. 56(3): 388-405 (rus).

[96] Kondrashova MN, Grigorenko EV, Babski AM, Khazanov VA (1987) Homeostasis of physiologic functions at the level of mitochondria. In: Mol.mechanisms of cell homeostasis. Novosibirsk: 40–66 (rus).

[97] Khazanov VA (2004) Past, present, and future of bioenergic pharmacology. In: Regulators of energic metabolism. Clinical and pharmacologic aspects. Tomsk: 3-7 (rus).

[98] Khazanov VA (1997) Oxidation of succinic acid in brain mitochondria. In: Succinic acid in medicine, food industry, agriculture. Ed. Kondrashova MN. Pushchino: 74–78 (rus).

[99] Mayevsky EI, Rosenfeld AS, Vazatashvili MV, Tchilaya SM (1997) Possibility for oxidation of exogenic succinic acid in vivo. In: Succinic acid in medicine, food industry, agriculture. Pushchino: 52–57 (rus).

[100] Lazarewicz J, Salinska E (2005) Udzial ionow wapnia w pathologii niedokrwiennej mozgu. Mozg a niedokrwienie. Pod red. J. Strosznajder, Z. Czernicki // Wydawnictwo PLATAN. Krakow: P.16-46.

[101] Villa RF, Gorini A, Hoyer S (2009) Effect of aging and ischemia on enzymatic activities linked to Krebs' cycle, electron transfer chain, glutamate and amino acids metabolism of free and intrasynaptic mitochondria of cerebral cortex. Neurochem.Res. 28: P.347–351.

[102] Kanunnikova NP, Omeyanchik SN, Bashun NZ, Shalavina EG, Doroshenko EM, Zolotukhin MM, Karayedova LM, Artiomova OV, Moiseenok AG (2003) Metabolism of

GABA abd glutamate in the rat brain in experimatnal ischemia and reperfusion. Neirokhimia. 20(3): 196-200 (rus).

[103] Mutuskina EA, Zarzhetski YuV, Trubina IE, Avrushchenko MA, Volkov LV, Onufriev MN, Lazareva NA, Stepanichev My, Gulyaeva NV, Gurvich AM (1997) Influence of ammonium succinate on functional, biochemical, and morphologic parameters of the central nervous system recovery in rats after 10 min break of blood flow. In: Succinic acid in medicine, food industry, agriculture. Pushchino: 145-150 (rus).

[104] Moiseenok AG, Komar VI, Khomich TI, Kanunnikova NP, Slyshenkov VS (2000) Pantothenic acid in maintaining thiol and immune homeostasis. Biofactors. 1: 53-55.

[105] 104] Brass EP (1994) Overview of Coenzyme A metabolism and its role in cellular toxicity. Chem. Biol. Interact. 90(3): 203-214.

[106] Moiseenok AG et al. (2004) CoA biosynthesis is an universal mechanism for conjugation of exogeneity and multiplicity of pantothenic acid functions. Ukr. Biokhim.Zh. 76(4): 68–81 (rus).

[107] Moiseenok AG (2003) Pantothenic acid: from universal distribution to universal functions. In: Biochemistry, pharmacology, and clinical use of pantothenic acid derivatives. Grodno: 107-114 (rus).

[108] Omelyanchik SN, Kanunnikova NP, Bashun NZ (2007) Changes of CoA pool structure in experimental brain ischemia and reperfusion and their correction by CoA biosynthesis precursors. Vestnik GrDU. 2(2): 89-93 (rus).

[109] Rabin O [et al.] (1997) Changes in cerebral acyl–CoA concentrations following ischemia–reperfusion in awake gerbils. J.Neurochem. 68: 2111–2118.

[110] Leonardi R, Zhang Yong-Mei, Rock ChO., Jackowski S (2005) Coenzyme A: Back in action. Progress in Lipid Res. 44: 125-153.

[111] Moiseenok AG, Rolevich IV, Kanunnikova NP, Khomich TI, Slyshenkov VS, Omelyanchik SN (1998) Simultaneous use of D-pantothenic acid substances and carnitine: foundation, and perspectives of use in prophylactic and therapeutic treatment. In: Panthenol and other derivatives of pantothenic acid: biochemistry, pharmacology, and medical use. Grodno: 130-137 (rus.).

[112] Grishina EV et al. (2007) Inhibition of mitochondrial energetics by CoA derivatives. Toxic effects of saturated fatty acids. In: Reception and intracellular signalling. Pushchino: 215-218 (rus).

[113] Slyshenkov VS, Rakowska M, Moiseenok AG, Wojtczak L. (1995) Pantothenic acid and its derivatives protect Ehrlich ascites tumor cells against lipid peroxidation. Free Rad. Biol. Med.19(66): 767-772.

[114] Slyshenkov VS, Piwocka K, Sikora E, Wojtczak L (2001) Pantothenic acid protects jurkat cells against ultraviolet light-induced apoptosis. Free Rad. Biol. Med. 30(11): 1303-1310.

[115] Slyshenkov VS, Dymkowska D, Wojtczak L (2004) Pantothenic acid and pantothenol increase biosynthesis of glutathione by boosting cell energetics. FEBS Letters 569: 169-172.

[116] Etensel B. et al. (2007) Dexpanthenol attenuates lipid peroxidation and testicular damage at experimental ischemia and reperfusion injury. Pediatr. Surg. 23(2): 177-181.

[117] Kanunnikova NP, Bashun NZ, Raduta EF, Slyshenkov VS, Omelyanchik SN (2007) Influence of panthenol and carnitine on parameters of lipoperoxidation and glutathione in experimental brain ischemia-reperfusion. In: Ecolog.Antropologia. Minsk: 154-156 (rus).

[118] Omelyanchik SN, Kanunnikova NP, Slyshenkov VS, Bashun NZ (2003) Role of panthenol in correction of some damages caused by experimental brain ischemia-reperfusion. In: Biochemistry, pharmacology, and clinical use of pantothenic acid derivatives. Grodno: 144-150 (rus).

[119] Bashun NZ, Raduta HF, Balash ZhI, Kirvel PCh, Sushko LI, Kanunnikova NP, Moiseenok AG (2007) Correction of postischemic disturbances in the hemispheres of the brain using precursors of succinyl-CoA biosynthesis. Neurochemical Journal. 1(3): 249-252.

[120] Schweizer U, Brauer A, Kohrle J, Nitsch R, Savaskan NE (2004) Selenium and brain function: a poorly recognized liaison. Brain Res. Rev. 45: 164–178.

[121] Brauer AU, Savaskan NE (2004) Molecular actions of selenium in the brain: neuroprotective mechanisms of an essential trace element. Rev.Neurosci. 15: 1-19.

[122] Gromova OA, Rebrov VG, Solupayeva LV (2004) Biological role of selenium. In: Selenium substances and health. Ed. Sanockij IV. Moscow: 12–42 (rus).

[123] Arteel GE, Sies H (2001) The biochemistry of selenium and the glutathione system. Environmental Toxicol.Pharmacol. 10: 153–158.

[124] Whanger PD (2001) Selenium and the brain: a review. Nutr. Neurosci. 4(2): 81–97.

[125] Baraboj VA (2004) Selen: biological role and antioxidant activity. Ukr.Biokhim.Zh. 76(1): 23-31(rus).

[126] Tapiero H, Townsend DM, Tew KD (2003) The antioxidant role of selenium and seleno–compounds. Biomedicine and Pharmacoterapy. 57: 134–144.

[127] Chen J, Berry MJ (2003) Selenium and selenoproteins in the brain and brain diseases. J. Neurochem. 86(1): 1–12.

[128] Steinbrenner H [et al.] (2006) Involvement of selenoprotein P in protection of human astrocytes from oxidative damage. Free Radical Biol. Med. 40(9): 1513–1523.

[129] Ozbal S [et al.] (2008) The effects of selenium against cerebral ischemia-reperfusion injury in rats. Neurosci. Lett. 438(Is.3): 265–269.

[130] Holmgren A, Lu J (2010) Thioredoxin and thioredoxin reductase: current research with special reference to human disease. Biochem. Biophys. Res.Communs. 396(1): 120-124.

[131] Ansari M.A. [et al.] (2004) Selenium protects cerebral ischemia in rat brain mitochondria. Biol. Trace Elem. Res. 101(1): 73–86.

[132] Yousuf S [et al.] (2007) Selenium plays a modulatory role against cerebral ischemia-induced neuronal damage in rat hippocampus. Brain Res. 114(25): 218–225.

[133] Huang K., Lauridsen E, Clausen J (1994) The uptake of Na-selenite in rat brain. Localization of new glutathione peroxidases in the rat brain. Biol. Trace Elem.Res. 46(Is.1–2): 91–102.

[134] Moiseenok AG (2004) Efficiency of sodium selenite and dimethyldiprosolilselenid (selecor) in protection of development of oxidative stress during endogenic intoxication. In: Selenium substances and health. Ed. Sanotsky. Moscow: 79–88 (rus).

[135] Moustafa RR, Baron J-C (2008) Pathophysiology of ischaemic stroke: insights from imaging, and implications for therapy and drug discovery. Br J Pharmacol 153: S44-S54.

[136] Adamek D (2003) Neuroprotekcja w urazach mozgu I rdzenia. Neuroprotekcija. XX Zimova Szkola Farmakol. PAN. Pod red. Smialowskiej M. Mogilany: 107-125.

[137] Zaleska MM, Mercado MLT, Chavez J, Feuerstein GZ, Pangalos MN, Wood A (2010) The development of stroke therapeutics: promising mechanisms and translational challenges. Neuropharmacol. 56: 329-341.

[138] Ginsberg MD (1995) Neuroprotection in brain ischemia: an update (Part II). The Neuroscientist. 1(3): 164-175.

[139] Cheng YD, Al-Khoury L, Zivin JA (2004) Neuroprotection for ischemic stroke: two decades of success and failure. NeuroRes1:36-45.

[140] Zhang ZG, Chopp M (2009) Neurorestorative therapies for stroke: uderlying mechanisms and translation to the clinic. Lancet Neurol 8: 491-500.

[141] Mehta SL, Manhas N, Raghubir R (2007) Molecular targets in cerebral ischemia for developing novel therapeutics. Brain Res. Rev 54: 34-66.

[142] Bacigaluppi M, Hermann DM (2008) New targets of neuroprotection in ischemic stroke. Scientific World Journal. 13(8): 698–712.

[143] 142] Szczudlik A (2003) Neuroprotekcja jako kierunek leczenia niedokrwienia mozgu. Neuroprotekcji. XX Zimowa Szkola Instytutu Farmakologii PAN. Pod red. M.Smialowskiej. - Mogilany: 99-105.

[144] Kanunnikova NP (2009) Role of metabolic therapy in complex treatment of ischemic and reperfusion injuries in brain. Zh. Grodn.Gos.Med.Univ. 4: 16-19 (rus).

[145] Slyshenkov VS, Shevalye AA, Moiseenok AG (2007) Pantothenate Prevents Disturbances in the Synaptosomal Glutathione System and Functional State of Synaptosomal Membrane under Oxidative Stress Conditions. Neurochem. J. 1(3): 235–239.

[146] Onufriev MV, Stepanichev MYu, Lazareva NV, Katkovskaya IN, Tishkina AO, Moiseenok AG, Gulyaeva NV (2010) Panthenol as neuroprotectant: study in a rat model of middle cerebral artery occlusion. Neurochem.J. 4(2): 148-152.

[147] Imai H, Graham DJ, Masayasu H, Macrae IM (2003) Antioxidant ebselen reduces oxidative damage in focal cerebral ischemia. Free Rad.Biol.Med. 34: 56-63.

[148] Parnham M, Sies H (2000) Ebselen: prospective therapy for cerebral ischaemia. Expert Opin.Invest.Drugs. 9: 607-619.

[149] Yamagata K [et al.] (2008) Protective effect of ebselen, a seleno-organic antioxidant on neurodegeneration induced by hypoxia and reperfusion in stroke-prone spontaneously hypertensive rat. Neuroscience. 153(2): 428–435.

[150] Kanunnikova NP, Raduta EF, Yelchaninova VA, Katkovskaya IN, Kovalenchik IL, Lukiyenko EP, Pehovskaya TA, Bashun NZ, Moiseenok AG (2010) Influence of panthenol, succinate ans selecor on changes of activities of lipoperoxidation and glutathione system in rat brain and blood after brain ischemia and reperfusion. Novosti med.-biol.nauk. 2(4): 156-162 (rus).

Permissions

The contributors of this book come from diverse backgrounds, making this book a truly international effort. This book will bring forth new frontiers with its revolutionizing research information and detailed analysis of the nascent developments around the world.

We would like to thank Angel Catala, for lending his expertise to make the book truly unique. He has played a crucial role in the development of this book. Without his invaluable contribution this book wouldn't have been possible. He has made vital efforts to compile up to date information on the varied aspects of this subject to make this book a valuable addition to the collection of many professionals and students.

This book was conceptualized with the vision of imparting up-to-date information and advanced data in this field. To ensure the same, a matchless editorial board was set up. Every individual on the board went through rigorous rounds of assessment to prove their worth. After which they invested a large part of their time researching and compiling the most relevant data for our readers. Conferences and sessions were held from time to time between the editorial board and the contributing authors to present the data in the most comprehensible form. The editorial team has worked tirelessly to provide valuable and valid information to help people across the globe.

Every chapter published in this book has been scrutinized by our experts. Their significance has been extensively debated. The topics covered herein carry significant findings which will fuel the growth of the discipline. They may even be implemented as practical applications or may be referred to as a beginning point for another development. Chapters in this book were first published by InTech; hereby published with permission under the Creative Commons Attribution License or equivalent.

The editorial board has been involved in producing this book since its inception. They have spent rigorous hours researching and exploring the diverse topics which have resulted in the successful publishing of this book. They have passed on their knowledge of decades through this book. To expedite this challenging task, the publisher supported the team at every step. A small team of assistant editors was also appointed to further simplify the editing procedure and attain best results for the readers.

Our editorial team has been hand-picked from every corner of the world. Their multi-ethnicity adds dynamic inputs to the discussions which result in innovative outcomes. These outcomes are then further discussed with the researchers and contributors who give their valuable feedback and opinion regarding the same. The feedback is then collaborated with the researches and they are edited in a comprehensive manner to aid the understanding of the subject.

Apart from the editorial board, the designing team has also invested a significant amount of their time in understanding the subject and creating the most relevant covers. They scrutinized every image to scout for the most suitable representation of the subject and create an appropriate cover for the book.

The publishing team has been involved in this book since its early stages. They were actively engaged in every process, be it collecting the data, connecting with the contributors or procuring relevant information. The team has been an ardent support to the editorial, designing and production team. Their endless efforts to recruit the best for this project, has resulted in the accomplishment of this book. They are a veteran in the field of academics and their pool of knowledge is as vast as their experience in printing. Their expertise and guidance has proved useful at every step. Their uncompromising quality standards have made this book an exceptional effort. Their encouragement from time to time has been an inspiration for everyone.

The publisher and the editorial board hope that this book will prove to be a valuable piece of knowledge for researchers, students, practitioners and scholars across the globe.

List of Contributors

Juliann G. Kiang, Risaku Fukumoto and Nikolai V. Gorbunov
Radiation Combined Injury Program, Armed Forces Radiobiology Research Institute, Uniformed Services University of the Health Sciences, Bethesda, Maryland, USA

Yaşar Gül Özkaya
School of Physical Education and Sports, Akdeniz University, Antalya, Turkey

María Cristina Carrillo, María de Luján Alvarez, Juan Pablo Parody, Ariel Darío Quiroga and María Paula Ceballos
Institute of Experimental Physiology (IFISE-CONICET), Faculty of Biochemistry and Pharmacological Sciences, National University of Rosario, Rosario, Argentina

Maria Armida Rossi
Department of Experimental Medicine and Oncology, University of Turin, Turin, Italy

Teresa Sousa, Joana Afonso and António Albino-Teixeira
Department of Pharmacology and Therapeutics, Faculty of Medicine, University of Porto, Portugal

Félix Carvalho
REQUIMTE, Laboratory of Toxicology, Department of Biological Sciences, Faculty of Pharmacy, University of Porto, Portugal

Nicolas J. Pillon and Christophe O. Soulage
Université de Lyon, INSA de Lyon, CarMeN, INSERM U1060, Univ Lyon-1, F-69621, Villeurbanne, France

Juliana C. Fantinelli, Ignacio A. Pérez Núñez, Luisa F. González Arbeláez and Susana M. Mosca
Centro de Investigaciones Cardiovasculares, Facultad de Ciencias Médicas, Universidad Nacional de La Plata, La Plata, Buenos Aires, Argentina

Ichiro Shimizu, Noriko Shimamoto, Katsumi Saiki, Mai Furujo and Keiko Osawa
Showa Clinic, Shin Yokohama, Kohoku-ku, Yokohama, Kanagawa, Japan

Bojana Kisic and Dijana Miric
Institute of Biochemistry, Serbia

Lepsa Zoric
Clinic for Eye Diseases, Serbia

Aleksandra Ilic
Institute of Preventive Medicine, Faculty of Medicine, Kosovska Mitrovica, Serbia

Alba Naudí, Mariona Jové, Victòria Ayala, Omar Ramírez, Rosanna Cabré, Joan Prat, Manuel Portero-Otin, Isidre Ferrer and Reinald Pamplona
Department of Experimental Medicine, University of Lleida-Biomedical Research Institute of Lleida, Lleida, Spain

Mary C. Vagula and Elisa M. Konieczko
Biology Department, Gannon University, Erie, PA, USA

Nina P. Kanunnikova and Natalya Z. Bashun
Department of Zoology and Physiology of Men and Animals, Yanka Kupala's Grodno State University, Grodno, Republic of Belarus

Andrey G. Moiseenok
National Center for Foodstuffs, National Academy of Sciences of Belarus, Minsk, Republic of Belarus

9 781632 394507